HURRICANE!

Coping with Disaster

Progress and Challenges since Galveston, 1900

Robert Simpson, Editor

Richard Anthes and Michael Garstang, Co-Editors

Joanne Simpson, Assistant Editor

American Geophysical Union
Washington, DC 2003

Library of Congress Cataloging-in-Publication Data

Hurricane!: coping with disaster: progress and challenges since Galveston, 1900/ Robert Simpson, editor, Richard Anthes and Michael Garstang, co-editors, Joanne Simpson, assistant editor.

p.cm.

Includes bibliographical references and index.

ISBN 0-87590-297-9

1. Hurricanes--United States--Congresses. 2. Hurricane protection--United States--Congresses. 3. Natural disaster warning systems--Congresses. 4. Weather forecasting--United States--Congresses. I. Simpson, Robert H. II. Anthes, Richard A. III. Garstang, Michael.

QC945.H864 2002
363.34'9227'973--dc21

2002033280

ISBN 0-87590-297-9

Printed in the United States of America.

Contents

PART III. Potential Impact of Science and Technology, Present and Future

Foreword

On average, 80 tropical cyclones occur around the globe every year. While most spend their entire existence over the open ocean, a few make landfall, often causing considerable damage and loss of life. In 1998, Hurricane Mitch caused the loss of over 10,000 lives in Central America. While the United States is better prepared to cope with such natural disasters, they continue to pose one of the most serious threats to our society, particularly as population growth escalates along U. S. coastal areas.

This book discusses the astonishing achievements made during the 20th Century in improving our understanding and ability to monitor and predict these storms. The Editors of the Volume have pioneered many of these accomplishments, and are the most appropriate individuals to assemble the collection of chapters that appear here. Rick Anthes is universally recognized for his seminal tropical cyclone and regional numerical modeling developments, while Mike Garstang is internationally known for his pioneering observational studies of tropical weather systems. During his tenure as Director of the National Hurricane Center, Bob Simpson was responsible for implementing the latest models and observational concepts for tracking and predicting hurricanes and tropical storms. His adoption of new understanding and new technologies in this role has contributed significantly to the reduction in loss of life in the Western Hemisphere during the late 20th Century. The chapter authors selected by the Editors reflect continual advancement in our ability to cope with hurricanes. Most of the authors are world-renowned authorities on hurricanes. Bill Gray's innovative seasonal hurricane forecasts receive wide media distribution each year, while his observational analyses of tropical cyclones have provided a fundamental scientific basis for progress in understanding during the late 20th Century. As he reports in his chapter, worldwide tropical cyclones caused three times as much damage as earthquakes, and resulted in about one and a half times more deaths than earthquakes, in the years 1953-1992. The loss of life would be even greater, however, if it were not for the operational tropical cyclone forecast centers around the world. Bob Sheets and Neil Frank served as Directors of the National Hurricane Center and continue to apply their experience and dedication to this life-saving profession.

Mark DeMaria, Hugh Willoughby, Kerry Emanuel, and Jim Gross are extending hurricane modeling of track and intensity change. Chris Velden, Greg Tyrrell, Greg Holland, Dave Emmitt, Elizabeth Ritchie and their colleagues are advancing the ability to monitor hurricanes with a variety of on-site and remote sensing technologies. These tools are essential for improvements in monitoring these storms, and as initial value data for numerical models.

The transfer of improved scientific understanding as well as technological advancements for the benefit of society is exemplified by the studies of Mary Frances Myers, Gilbert White, Herbert Saffir, and Roger Pielke, Jr. We need this capability for guidance of the public, as well as emergency managers and engineers. Herbert Saffir, for example, worked with Bob Simpson to create the widely respected Saffir/Simpson Scale that describes hurricane damage potential. This scale provides a numerical damage potential associated with tropical cyclones, which can be used in both warnings and in the design of structures in coastal regions exposed to these storms. Joanne Simpson, who is one of the outstanding scientists of the 20th Century, contributes her extensive knowledge of hurricanes to this Volume. Her work encompasses a broad spectrum, including both observational and modeling studies. Ed Rappaport uses extensive knowledge of the history of hurricane research and applications to educate the reader on the role of this work in the broad societal context of the last 100 years. During this period, society experienced the advantages of such new technologies as reconnaissance aircraft, radar, and satellites - tools adapted within the atmospheric science community to vastly improve the monitoring and prediction of hurricanes.

The editorship and authorship of this Volume guarantee that this publication will be a classic throughout the 21st Century.

Dr. Roger A. Pielke, Sr.
Professor, Department of Atmospheric Sciences
Colorado State University
Fort Collins, Colorado

Preface

On September 8, 1900 an unheralded hurricane virtually destroyed the city of Galveston, Texas, with a loss of approximately 8000 lives — the deadliest natural disaster in this nation's history. Thomas Alva Edison, who had rushed to the scene with his famous hand cranked movie camera, captured on film graphic evidence of the tragedy: the incredible damage wrought by the hurricane and rescue activities then underway, including footage of bodies being removed from the wreckage. These are poignant illustrations of what can happen when inadequate scientific understanding, insufficient supporting technology, or poor communications combine to vitiate local abilities to warn and prepare residents to cope with a hurricane threat.

The Galveston disaster is an appropriate historical anchor for evaluating a century of progress in the United States to help those in harm's way not only survive but cope effectively with the threat of hurricanes. Coping, however, is a multifaceted problem. Scientific understanding is nearly always harnessed to and constrained by the availability of technologies needed to acquire such understanding. As we shall see, for sustained and effective growth, a partnership between science and technology — in which one feeds upon the success of the other – is a must.

Ironically, however, when the cutting edge of new science is put to work, the first of several "catch 22s" appears. In meteorology, this happens when it becomes apparent that improved scientific understanding does not always foreshadow commensurate improvements in prediction and warning. For the public, this is often reflected in their perception of the relatively slow progress made in reducing the ratio of coastal areas warned to that of coastal areas where dangerous winds and tides occurred. This is a critical factor for the forecaster when a hurricane approaches landfall at a small angle to the coast. An error of less than five degrees in direction of movement, over the 36 hour period needed to relocate coastal residents, can shift the point of greatest vulnerability hundreds of miles leaving large stretches of coastline warned without subsequent evidence of the need.

The cascade of "catch 22s" extends through a plethora of problems that begin with the natural reluctance of coastal residents to leave home if they perceive a reasonable chance to survive without doing so. Societal problems extend to questions of warning credibility after a few cases of perceived over-warning. Finally, the uncertainty of transportation to reach safe shelters within the time available complicates decisions to seek safer shelter.

With these obstacles impeding progress, it is indeed remarkable that during the twentieth century it has been possible to provide as many bulwarks that we have to prevent the recurrence of a disaster comparable to that at Galveston in 1900. Moreover this was achieved despite vast increases in coastal population.

This book is not dedicated simply to tracing a century of progress in dealing with hurricanes affecting the United States, but rather to analyzing the reasons for such progress, and in some instances why more progress wasn't made. More specifically, it examines the interactive roles between technology and the sciences (physical and social), and their respective dependency upon each other. Finally, in a look ahead we examine where this may lead us, including potential roadblocks, the realistic limits of predictability, and of our ultimate abilities to cope with imminent hurricane threats.

In Part I, we trace the struggle to improve the accuracy of hurricane tracking and prediction of landfall, and the expansion of communications for acquiring weather data and for disseminating public warnings. This struggle drew heavily upon technologies generated during the two world wars. The explosion of scientific research, particularly in the U.S following WWII, presented challenges as well as demands for developing additional technology, some extensions of which produced only products in search of a mission rather than fulfilling an existing need. In meteorology, particularly tropical meteorology and its severe storms, this sequence of supply and demand peaked when scientists entered the digital age and rapid data processing and computing became available. Soon computer models for weather prediction appeared, and exploration of weather on a global scale made its debut. With these advantages scientists rapidly acquired the ability to frame more mature questions and hypotheses for extending understanding and prediction.

Part II focuses on societal problems in coping with hurricanes. Chapter 5 tells the gripping story of events leading up to and during the Galveston Disaster of 1900. Despite the lessons learned in Galveston, in-depth consideration of societal problems posed by hurricanes was slow in being addressed nationally until after hurricane Camille ravaged the U.S. Gulf Coast. At that time a collaborative effort between social and physical scientists in government, universities, and private institutions called attention to such needs as more direct communication of forecasts and warnings, and the ever-more-difficult problems of evacuating coastal residents to safe locations. It stressed the need to improve building standards and codes, but addressed as well important economic questions. At what point, for example, does the cost of research, new technology, and safety policy measures begin to produce diminishing returns.

Part III, which speaks from the present maturity of scientific understanding, reviews and contrasts progress in scientific understanding with improvements in warning skills. It defines the significant gaps in knowledge that remain, reframes the questions and approaches that new research should address, and suggests the technologies needed to obtain answers. New satellite technologies are described that will enable three-dimensional viewing of changes in motion and strength of a hurricane as well as thermodynamic properties that set the stage for hurricane seedlings to acquire the necessary structure to develop and retain hurricane strength.

Chapter 12 is an example of on-going research on one of the more intractable problems in predicting hurricane development; demonstrating how new technologies permit researchers to accomplish today what they could never have undertaken even a decade ago.

The final chapter is a personal perspective of how well a century of progress positions us to cope with hurricanes. Not everyone will agree. Nevertheless, it is a well reasoned account from which the author projects what science and technology may provide in the future; he then poses the unanswered question of whether or not this progress ensures that the United States will never suffer another hurricane disaster which rivals that of Galveston, 1900.

This book is based upon presentations at a historical symposium on hurricanes convened by the American Geophysical Union at its fall meeting in San Francisco, December 16, 2000. Here 10 eminent scientists presented invited papers on assigned topics. The effectiveness of the symposium suggested that these discussions be published in book form, not as a direct historical treatise on twentieth century progress, but rather, in the spirit of the symposium, as an analysis of factors which stimulate, or at times inhibited, a century of progress.

The authors of this volume represent widely diversified backgrounds and technical experience, including the physical sciences, engineering, technology, and the social and behavioral sciences. The opinions expressed, and at times the conclusions reached from their differing perspectives, provide the reader with interesting and instructive options, not only in their evaluations of a century of progress, but also in what the future may hold as we seek means of perfecting our abilities to cope with hurricanes.

The editors, while attempting to eliminate factual inconsistencies, have not sought a consensus opinion, and indeed have condoned some restatement of subjects or events where differing viewpoints may be useful to the reader. Despite these differences, each chapter survived the customary, multiple peer review.

While we have attempted to provide discourse that will persuade readers with broad ranges of interests and backgrounds to read on, we have also attempted to minimize difficulties for those with less experience in science. Inevitably the terminology in sections of several chapters will need clarification, and to that end we have included glossaries of technical terms, and definitions and conversion factors for units of measure. Rather than use International System of Units (SI) exclusively, as required in scientific journals, we have varied the units to accommodate the convenience and familiarity of a broad spectrum of readers.

Robert Simpson

Acknowledgments

Rarely are specialists from such different professional and experience backgrounds enticed to pool their views and concepts of how to confront and cope with a menace of common concern such as hurricanes, especially to collaborate in writing this kind of book. I admire and applaud the willingness of our authors who contributed their time and talents in preparing their assigned chapters, and their patience in responding to the needs imposed by editors to make these chapters meld into book format. I am even more grateful for the active support, wise counsel, and able professional contributions of my two colleagues and close friends of long standing, Rick Anthes and Mike Garstang who joined me as partners in guiding and coordinating the work of the authors, and other editorial tasks of producing this book. A number of others who made important contributions are identified elsewhere.

Despite this able assistance, this book probably would not have been completed without the relentless support, both emotional and professional, from one who has played this role so effectively throughout our 37 years of sharing – my wife Joanne Malkus Simpson. At the outset she opted to play a minor role as co-author of two chapters. However, during my serious illness midway through the production, she grabbed the reins and played a leading role in completing the editorial tasks. She is listed here as an assistant editor. But she did so much more. For this, I hope all who participated will join me in gratitude.

I want to thank Marvin Geller, who asked me to organize the symposium in San Francisco, and later helped to meld the contributions into a book. We also thank Fred Spilhaus for his expressions of interest and, on behalf of AGU, for having invited us to submit a proposal to AGU for publication of the book. AGU staff has provided outstanding support of our efforts, particularly Allan Graubard, our acquisitions editor, whose encouragement and constructive reviews were so helpful, and more recently from Terence Mulligan, our production editor, in readying the manuscript for publication. The contributing authors were supported by a total of six universities, four government agencies, and six corporations. We are particularly grateful, however, for the support of NASA's TRMM Project for its support of color graphics used in some chapters, as well as the artwork created by Barbara Summey. Finally we wish to express our deep appreciation for generous sustained support, though all stages of the book's production, from the University Corporation for Atmospheric Research (UCAR), Simpson Weather Associates, and Goddard Space Flight Center for editorial staff support and other assistance in processing the manuscript.

Robert Simpson

PART I

Struggles to Understand and Track Hurricanes

The "Struggles" which headline Part I relate primarily to the scientific and technical tasks of acquiring information and providing warnings for coastal residents when hurricanes threaten. The obstacles which limited the effectiveness of warnings at the time of the Galveston disaster of 1900 and years afterward were not the result of limitations in technology and scientific understanding alone, but also of organizational deficiencies in applying what was already known and in use elsewhere. After examining shortcomings of services during early years of the century, Part I identifies the explosive growth of technology during World Wars I and II as the driving force of scientific progress and rapid improvement in warning services during the middle and latter half of the century.

Chapter 1, an overview of stepping stones of progress throughout the century, includes a personal assessment of the author, whose research has contributed significantly to this progress. Chapter 2 traces the struggles and handicaps in developing warning services during the early years and the technology-driven improvements in services through mid-century up to the establishment of the National Hurricane Research Project (NHRP). Chapter 3 discusses the contributions of NHRP, particularly of hurricane explorations with research aircraft. Finally, Chapter 4 discusses the succession of objective prediction methods and computer models responsible for vastly improved accuracy of hurricane forecasts and warnings.

Hurricane! Coping with Disaster

10.1029/055SP01

Chapter 1

Twentieth Century Challenges and Milestones

William M. Gray

INTRODUCTION

Each year nearly 80 tropical cyclones, TCs, will form somewhere on the globe. These TCs have low-level sustained winds of 20 m/s or greater and about two-thirds of them attain hurricane intensity (33 m/s). Tropical cyclone activity in the Atlantic basin averages one-eighth of the global total with approximately ten TCs per year. Both population and property values in areas influenced by tropical cyclones are increasing at rates significantly greater than global population and property values as a whole. This is particularly true in the United States. Consequently, the potential impact of tropical cyclones in the United States has grown disproportionately and will probably continue to do so. Property losses of $25-50 billion or more from individual hurricanes will likely occur in coming decades. And public and scientific interest in the special characteristics of these storm systems will inevitably expand.

The World Meteorological Organization (WMO) estimates that during the 30-year period from 1963-1992, TCs caused almost three times as much damage globally as did earthquakes, while influencing the lives of almost five times as many people. Total deaths caused by tropical cyclones are about 50% greater than those from earthquakes. For the United States these figures are even more heavily weighted to tropical cyclones where, during the 20th century, ten times as many deaths and more than three times as much damage occurred as a result of tropical cyclones as compared with earthquakes.

Hurricane! Coping with Disaster

10.1029/055SP02

In coping with hurricanes, the human aspects are as critical and complex as the forecasts. A perfect forecast is of no use unless it reaches the people at risk. Even so, warnings are wasted if people are unwilling or unable to act to protect themselves and their property. Although the critical problem of how to safely evacuate threatened populations remains, the 20th century did see recognition of this problem with major steps taken to address the societal aspects of coping with hurricanes. Yet more needs to be done. The watershed of this process attended the aftermath of Hurricane Camille, which devastated the Gulf coast in 1969. Local, state and federal governments, and private sector entities at last began to cooperate as communication of warnings to those at risk began in earnest via television weather broadcasts. In 1992, some 23 years after Camille, Hurricane Andrew provided another wakeup call, particularly to the insurance industry, with nearly 25 billion dollars worth of damage done in South Florida. This disaster re-emphasized the need for improved coping strategies for dealing with landfalling hurricanes. As cooperation grew between the private sector and government, more programs were established to educate the public about the hurricane threat, so that they would undertake all necessary steps to protect themselves when warned. Each of these topics is discussed in detail in Chapters 5, 6, and 8.

At the time of the 1900 Galveston hurricane, which killed some 8,000 people, little was known about the physical processes associated with hurricanes. Why were the winds at the surface so much stronger than those of middle-latitude over-land cyclones? Why was hurricane formation generally restricted to late summer and early fall? How high did hurricanes extend in the atmosphere? Why did hurricanes form only over the warm oceans and what dictated their motion process? Why were there such differences in hurricane and tropical cyclone frequency from one year to the next and between different decadal periods?

These and many other basic issues concerning hurricanes could not be answered in 1900. Today, at least partial answers exist for most of these questions as a result of large gains in observational and communication technology. While there is much yet to be learned, a consistent and plausible conceptual model of hurricanes has begun to emerge.

If we did not know from observations that hurricanes actually do occur, however, we might never have thought their existence possible because their origins and energetics bear only marginal resemblance to those of similar sized mid-latitude cyclones. Hurricanes possess a unique combination of mechanical and energy interactions not present in typical fluid mechanical systems. We observe that tropical cyclones develop in regions of negligible horizontal gradients of temperature and vertical shear of horizontal wind. We now have a good understanding of the mean structure of the hurricane and of its basic mechanical processes. At the same time, we are still not in full agreement on the specific processes that accommodate the observed large differences in hurricane formation, structure, and intensity change characteristics. The full suite of physical processes involved with the hurricane life cycle and why there are so many differences in individual cyclone structure and intensity change features have yet to be well explained. In particular, details of the temperature-moisture and derivative thermodynamic processes of the hurricane need much more study. How small-scale cumulus convective processes interact with the surrounding larger-scale momentum fields also requires further analysis. Nevertheless, it is possible to make many new observational and physical inferences that, when synthesized into a conceptual framework, lead to a greatly improved understanding of how the typical hurricane forms, maintains itself, and moves – an understanding mostly developed since the end of World War II.

Trends in Major U.S. Hurricane Landfalls in Recent Decades

This is a particularly important time for developing improved coping and forecasting strategies, as well as improving our basic understanding of just how to deal with Atlantic basin hurricanes. As mentioned, a substantial increase in U.S. hurricane-spawned destruction is highly likely during the next few decades. Indeed, the outlook for an increase in major Atlantic activity appears ominous. Since 1995, we have seen an increased level of intense hurricane activity in the Atlantic basin. Within the same period, however, U.S. coastal populations and property values have skyrocketed.

Table 1 provides a breakdown of the number of U.S. landfalling named storms by intensity category during the 20th century for the 11 U.S. coastal regions shown in Figure 1. These coastal areas have been chosen by the frequency of landfalling major hurricanes. A total of 73 major hurricanes have struck the U.S. coastline during this 100-year period. Note also that there are about equal numbers of landfalling cyclones along the Gulf Coast (Regions 1-4 of Figure 1) and the Florida Peninsula and U.S. East Coast (Regions 5-11). As the total coastline length of regions 5-11 is twice as large as that of Regions 1-4 (the Gulf Coast), we see that landfall frequency per unit coastal distance is twice as great along the Gulf Coast as compared to the Florida Peninsula-East Coast domain.

Figure 2 shows the locations along the U.S. where these 73 major hurricanes have made landfall in the last century. Note that four U.S. coastal regions (4, 7, 9 and 11) have had no major hurricane landfalls during the last century, extending through 2001. There has been an unusual downturn in U.S. major hurricane landfall frequency since the mid-1960s, apparently a result of the combined trends in global atmospheric and ocean climate and just plain luck. This downturn has been most pronounced in peninsular Florida and the East Coast.

Hurricane destruction can result from wind, storm surge, and flooding due to heavy rains. Flooding is also especially hazardous in areas with complex, elevated topography. When normalized for coastal population, inflation, and wealth per capita (which doubles about every 35 years), we find that hurricane-spawned damage rises dramatically with cyclone (damage potential) category. Table 2 shows the

TABLE 1. Twentieth Century (1900-1999) total of landfalling tropical cyclone's of various Saffir/Simpson intensity categories by 11 U.S. coastal regions as shown in Fig. 1. The acronyms and abbreviations are defined as follows: NS - Named Storms, TS - Tropical Storms, HUR - Hurricanes, Categories 1-5.

Coastal Region	Coastal Distance km	NS	TS	HUR	Cat. 1-2	Cat. 1	Cat. 2	Cat. 3-4-5	Cat. 3	Cat. 4-5
1 (Gulf)	503	56	21	35	11	9	15	10	5	
2 (Gulf)	257	20	12	8	6	4	2	2	1	1
3 (Gulf)	666	75	36	40	21	16	5	19	16	3
4 (Gulf)	382	27	19	8	8	4	4	0	0	0
5 (FL+EC)	373	24	15	9	5	4	1	4	4	0
6 (FL+EC)	483	41	11	30	15	6	9	15	8	7
7 (FL+EC)	574	18	13	5	5	2	3	0	0	0
8 (FL+EC)	673	50	18	32	20	13	7	12	10	2
9 (FL+EC)	527	11	5	6	6	4	2	0	0	0
10 (FL+EC)	426	13	3	10	4	3	1	6	6	0
11 (FL+EC)	447	8	5	3	3	3	0	0	0	0
Total 1-11	5311	344	158	186	113	70	43	73	55	18
Total 1-4	1808	179	88	91	55	35	20	36	27	9
Total 5-11	3503	165	70	95	58	35	23	37	28	9

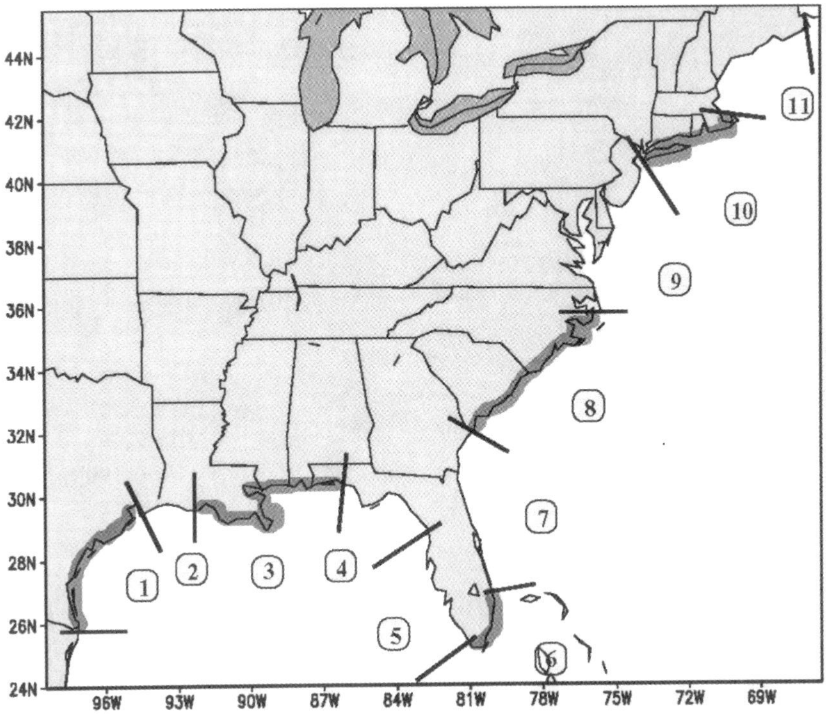

Figure 1. Designation of landfalling regions of the U.S. that are chosen by the frequency of 20th century (1900-1999) major (Category 3-4-5) hurricanes. Shaded coastal regions 1, 3, 6, 8 and 10 have had the most major landfalling events.

typical, normalized U.S. landfalling hurricane destruction expected by cyclone category in dollars for year 2000 for coastal and adjacent inland areas with nominal populations of 1 million per 100 km of coastline. Note that hurricane destruction goes up by a factor of four for each unit of increased Saffir/Simpson intensity category[1]. For example, the average landfalling tropical storm does about one-fourth the damage of a landfalling Category-1 hurricane; a landfalling Category-2 hurricane typically causes about four times the normalized damage of a Category-1 hurricane, and so forth. No distinction is made between landfalling Category-4 and -5 hurricanes of which there have been only 18 during the last century. These very intense Category-4 or -5 hurricanes should be expected to do about 16 times the destruction of a typical Category-2, and 64 and 256 times the destruction of a typical landfalling Category-1 hurricane and landfalling tropical storm, respectively.

It is important to note that these are average category damage estimates. Damage amounts from individual storms vary widely as a result of differences in

[1] Saffir/Simpson scale of hurricane intensity consists of five categories ranging from 73 miles/hour to above 156 miles/hour (see Chapter 7).

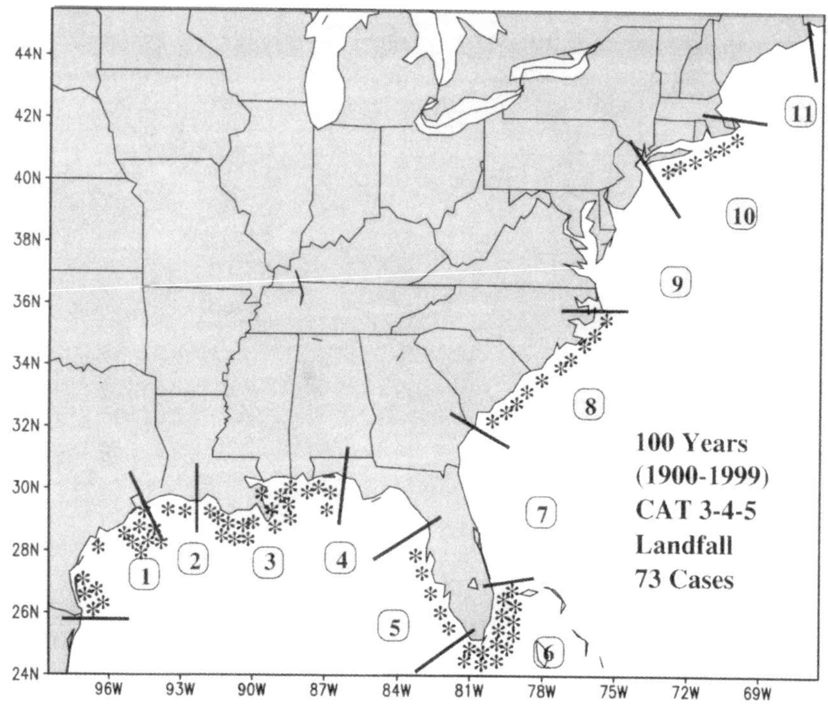

Figure 2. Location of 73 landfalling major (Category 3-4-5) hurricanes during 1900-1999. Note the lack of major hurricane landfall in coastal regions 4, 7, 9 and 11.

coastal shape, terrain, population, wealth per capita, the direction and speed of the landfalling cyclone, tidal stage at landfall, and rainfall amounts. These comparative landfall destruction values are thus representative of normalized average

TABLE 2. Typical rates of normalized destruction by tropical storms and hurricanes of various Saffir/Simpson category.

Cyclone Category	Typical Sustained Max. Wind MPH	Typical Relative Wind Destruction	Typical Relative Area of Wind Destruction	Typical Relative Destruction	Typical Destruction (in 2000 $) for 1 million per 100 km of Coast
Tropical Storm	60	.25	.25	.06	60 Million
Category 1	85	.50	.50	.25	250 Million
Category 2	103	1	1	1	1 Billion
Category 3	120	2	2	4	4 Billion
Category 4-5	150	4	4	16	16 Billion

damage by all Category-2 hurricanes during the last century, for example, versus all normalized average Category-3 landfalling hurricanes.

Typical hurricane destruction values rise dramatically with increases in category number, resulting from the combined influences of the following:

1. The dynamic force of the wind, which rises with the square of wind speed (V^2), not wind speed alone.
2. The increase of the area of damaging winds, which typically expands with cyclone category.
3. The sharp increase in flying debris with increased wind speed, where damage comes from wind and wind-driven missiles.
4. Very intense Category 4-5 cyclones, which frequently generate small concentrated sub-vortices inside their radius of maximum wind. In sub-vortices wind speeds over small regions of a few hundred meters can be substantially greater than the cyclone's maximum wind as a whole. This occurred with Andrew in 1992 and was responsible for the large irregular swaths of small-scale destruction around the cyclone's center.
5. The more intense a hurricane is at landfall, the greater the probability of hurricane-spawned tornadoes in outer reaches of the storm. Such tornadoes can also cause significant damage.

Although major hurricanes make up about one-fourth of all named tropical cyclones, the data in Table 3 shows that they cause about 85 percent of all cyclone-spawned damage when normalized by population, inflation, and wealth per capita. These landfalling major hurricanes are of primary concern.

Major hurricane activity in the Atlantic basin experiences large multi-decadal (20-50 years) excursions that appear to be linked to the effects of variations in the Atlantic Ocean thermohaline circulation [Gray et al., 1997]. For example, major hurricane activity in the Atlantic was suppressed during 1900-1925 and

TABLE 3. Typical increase of normalized hurricane destruction with rise in Saffir/Simpson category for 1900-1999. Note that major (category 3-4-5) hurricanes cause about 86 percent of all TC damage.

Category	Typical Relative Damage	Number	Percent of Total Named Storms	Total Normalized Damage	Relative Percent of Total Damage
Named Storm	.0625	157	(39%) 46	20	4%
Category 1	.25	71	(21%) 21	18	3%
Category 2	1	41	(14%) 12	41	7%
Category 3	4	55	(19%) 16	220	37%
Category 4-5	16	18	(7%) 5	288	49%
Total		342	(100%)	587	100%

1970-1994 but enhanced during 1926-1969 and since 1995. If future trends approximate the past, then the recent upswing in major Atlantic hurricanes since 1995 will probably last another two to three decades or longer.

Official records indicate that during the 20th century approximately 215 major hurricanes developed in the Atlantic basin with approximately one-third (73) making landfall on the U.S. coastline (Figure 2). This ratio also applies for the other categories of weaker TCs. From 1995 to 2001, the incidence of Atlantic basin hurricanes has been the highest of any seven-year period on record. Twenty-seven major hurricanes developed in the Atlantic basin during the last seven-year period, but only three hurricanes came ashore – Opal in 1995, Fran in 1996, and Bret in 1999.

If the previously noted long-term average of one landfalling major hurricane for every three that form held during the last seven years, then we should have experienced nine U.S. major hurricane landfall events rather than the three that actually came ashore. We owe our good fortune to a persistent upper-air trough that has been located along the U.S. East Coast for a high percentage of the active portions of the last seven seasons. This fortuitous trough location caused an unusually large fraction of the major hurricanes moving west-northwesterly within the trade winds to recurve to the north and northeast before reaching the U.S. coastline. However much we may wish it to, we cannot expect this seven-year scenario to continue.

Our good fortune in having fewer major hurricane landfalls has also persisted longer than the last seven years. The period between the mid-1960s and the mid-1990s (1966-1994) saw greatly reduced major hurricane activity compared with last century's average. During the most recent 36-year period (1966-2001), the Florida Peninsula and East Coast (Regions 5-11 of Figure 1) experienced only six major hurricane landfall events (Diana 1984, Gloria 1985, Hugo 1989, Andrew 1992, Emily 1993, and Fran 1996) — an average of 0.17 per year. During the prior 62 years (1899-1960), there were 34 major hurricane landfalls along this same coastal area (0.55 per year) or an annual normalized rate that was more than three times the rate of incidence during the recent 36-year period.

The Gulf Coast (Regions 1-4 of Figure 1) has also experienced a large reduction in major hurricane activity during the last 21 years (1981-2001) wherein only five Category-3 hurricane landfalls occurred and, more importantly, no Category 4-5 landfalls. During the prior 81 years (i.e., 1900-1980), there were 31 major hurricane landfalls on the Gulf Coast of which nine were Category 4-5. Considering the normalized damage estimates shown in Table 3, we infer that during the last 21 years, the annual potential hurricane damage threat in the Gulf region was only one-third as much as during the prior 81 years.

The situation has been even more dramatic along the southern portion of the Florida Peninsula (Figure 3). During the last 35 years (1966-2001), only one major hurricane, Andrew 1992, made landfall on the Florida Peninsula. During the two 25-year periods (1916-1940 and 1941-1965), there were seven and nine

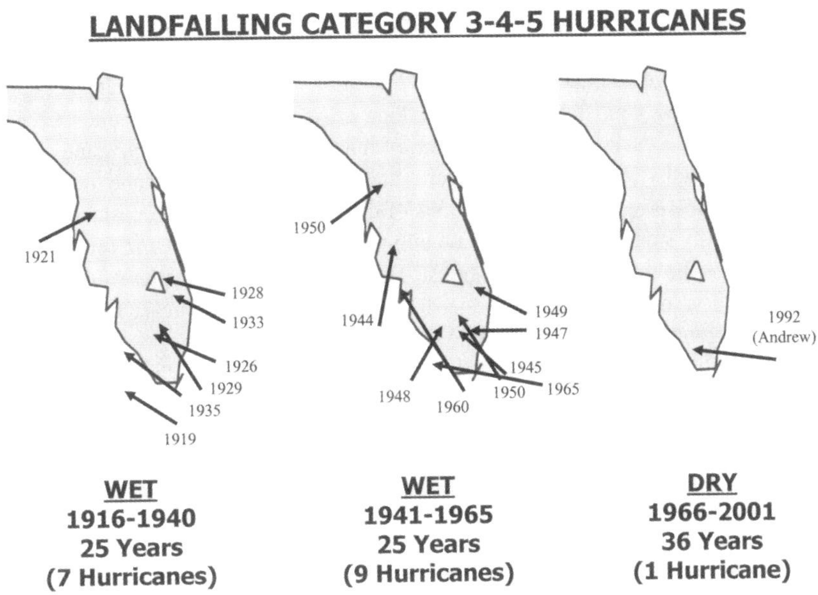

Figure 3. Location of major hurricanes making landfall on the Florida Peninsula during three multi-decadal periods.

major hurricane landfall events, respectively. As noted previously, this large downturn in East Coast and Peninsular Florida (major) landfalls appears to be associated with a rare combination of reduced production of major hurricanes and a multi-decadal variation in thermohaline circulation in the Atlantic Ocean, together with unusual good luck. Again, we cannot expect such fortunate conditions to continue. A large increase in major landfalling hurricanes in Florida must inevitably resume in the years ahead. With exploding coastal populations and increasing property values, we must prepare for hurricane damages much greater than in the past. Probably very few residents of the southeastern U.S. coastline are aware of how fortunate they have been over the last few decades. New and improved means for coping with this large, expected increase in hurricane landfall are urgently needed.

UNIQUE CHARACTERISTICS

Tropical cyclones are a special class of large-scale rotating wind systems. They occur over selective portions of the tropical and sub-tropical oceans and develop distinctive characteristics:

1. A warm central core that differs markedly from the cold-core structure of mid-latitude cyclones. Because of this warm-core structure, the strongest

winds in hurricanes occur at low levels. As the force of the wind increases with the square of the wind speed (V^2), hurricanes have the potential to cause massive amounts of damage. By contrast, middle-latitude cyclones typically have their strongest winds in long jet streaks in the upper atmosphere, far above the surface layer domain of human habitation, though some oceanic middle-latitude cyclone systems are able to generate hurricane force winds.

2. They develop in those portions of the tropics where, on occasion, there are only small variations of wind direction and speed through the vertical extent of the troposphere. In contrast, middle-latitude cyclones develop in conditions of large temperature gradients and where large vertical shear of the horizontal tangential wind is present.
3. Their low pressure centers do not slope much with height. Middle latitude cyclone centers typically have a large slope.
4. The distribution of precipitating clouds in tropical cyclones is usually organized into spiral rainbands (Figures 4 and 5) and the release of condensation (latent) heating from rain clouds is fundamental for the storm's formation and maintenance. In contrast, the rain producing processes of middle latitude cyclones, although very important for some aspects of their structure, are not an essential component.
5. Tropical cyclones can be viewed as naturally occurring thermoydnamic heat engines that draw thermal and latent energy from the warm tropical oceans and convert this thermal energy into kinetic energy at an efficiency of about 2-4 percent [Riehl 1954]. Although the purely mechanical (i.e., physical) processes of hurricanes are rather well understood, the complex and non-linear interactions between the cyclone's mechanical effects and the diabatic heating and moisture processes (including condensation-evaporation, radiation, cumulus cloud convection, sea to air energy exchange) have not yet been fully resolved. More observational and theoretical studies are needed to better determine how these interacting processes combine to form and maintain hurricanes and those special processes that account for such large differences in cyclone structural features and intensity change processes.
6. Typically they develop an intense inner core with extreme winds occurring in a closed or broken cloud band known as the eye-wall. This feature is usually located within a 10-40 km radius from the center. A circular to broken eyewall cloud typically surrounds a very warm center or "eye" of the hurricane (Figure 6). Cyclone inner-core winds strengthen as angular momentum is imported into the system at a rate that is greater than that at which momentum is being dissipated through friction in the surface layer. This momentum spin-up process within the inner core occurs primarily by virtue of the cyclone having low-level inflowing air of higher velocity than the speed of the compensating outflowing air at upper-levels. Additional momentum inflow occurs as a result of horizontal eddy processes whereby the azimuthal varying exchange of air at an individual level will have an inflow greater

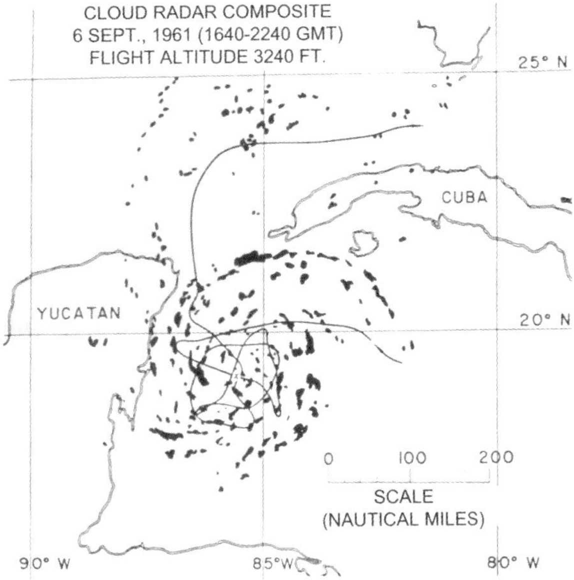

Figure 4. Radar composite of cumulus rain radar echoes during the early intensification of pre-hurricane Carla on September 6, 1961, as measured by a DC-6 aircraft at ~ 1 km elevation. Maximum surface winds and minimum sea-level pressure at this time were 60 knots (70 miles/hour) and 999 mb, respectively. The flight track of the aircraft, from which the radar echoes were obtained, is the solid line. Note the strong rainfall banding features.

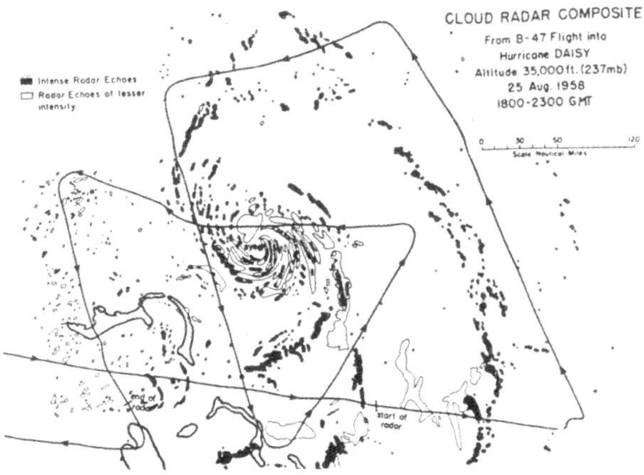

Figure 5. Radar composite surrounding Hurricane Daisy's rain areas during its intensification stage on 25 August 1958 when it was located off the Florida coast. The solid line is the B-47 aircraft flight track at 237 mb altitude (~ 10 km) from which the radar echoes were taken. Central pressure was about 980 mb. Note the heavy rainfall line far to the east of Daisy's center or eye.

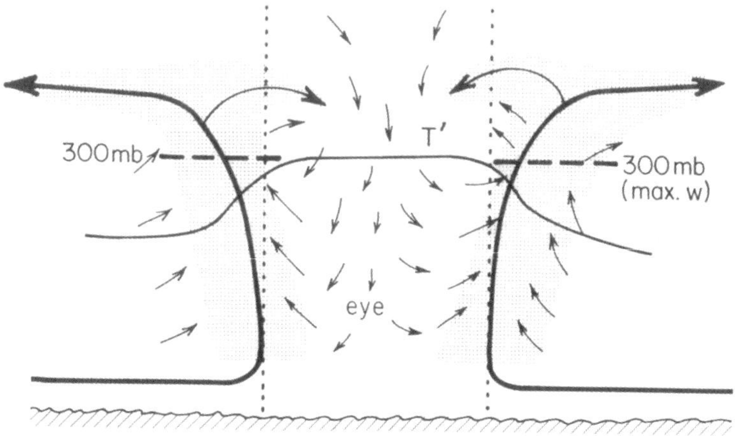

Figure 6. Schematic illustrating how entrainment into strong sloping eye-wall convection drives compensating eye-wall subsidence. This subsidence causes hydrostatic warming along the vertical dotted lines, which allows for a larger pressure drop at the low-level radius of maximum winds than would occur if a vertical eye-wall cloud were present. This illustration follows the conceptual picture proposed by Malkus [1958] and has been generally verified by inner-core aircraft measurements.

than the compensating outflow at the same level. This exchange, known as the "eddy flux process," also contributes to balancing surface friction. The value represented by V_R' is the radial wind deviation (positive outward) from the azimuthal mean and V_T' is the tangential wind deviation from the azimuthal mean. Inward eddy flux occurs when $V_R'V_T'$ is negative. The needed inward horizontal eddy momentum flux requires that the cyclone have favorable asymmetric environmental winds such as occur with an upper-level trough to the northwest and/or favorable azimuthal sectors of stronger than normal inflowing air at lower levels. It also requires that there be a large release by latent heat from rainfall to allow upward vertical motion to take place around the cyclone's center in an atmosphere that is stable for non-condensing vertical motion. As low-level air flows inward to lower pressure near the center of the TC, it absorbs heat and moisture from the ocean surface and, thereby, experiences less expansion-related cooling than would occur if no ocean energy source were present. This sea-to-air energy flux is the hurricane's primary energy source and the principal reason that hurricanes form only over the ocean and rapidly weaken when they move over land where this energy source is no longer available.

7. The cyclone's in-up-and-out radial circulation is observed to withstand (i.e., overcome) the stabilization effects of the interior upper-level warming associated with the cyclone's large surface pressure decrease. Upper-level warming near the center by itself is not the limiting factor to a hurri-

cane's maximum intensity, as has been hypothesized in the past. Rather, the circulation in tropical cyclones is able to withstand large upper-level inner-core warming and apparent lapse rate stabilization by forming an outward sloping eye-wall cloud (Figure 7). The boundary layer energy gain from the ocean also helps vertical motion to continue at inner radius despite upper-level thermal stabilization. The stronger a hurricane becomes, the greater the inner-core convection and vertical motion must become, in order to balance the cyclone's ever-increasing rates of frictional dissipation. But there is an upper limit beyond which the rate of spinup through increased vertical motion (a linear process proportional to V) is able to overcome the increased rate of frictional dissipation (a process proportional to the square or higher power of V). If all physical conditions for increased cyclone intensity are favorable (i.e., background climatological features including sea-surface temperature, lapse rates, etc.), there is always a limit to which the hurricane's central surface pressure can be reduced. This is about 10-12 percent of the surrounding environmental surface pressure.

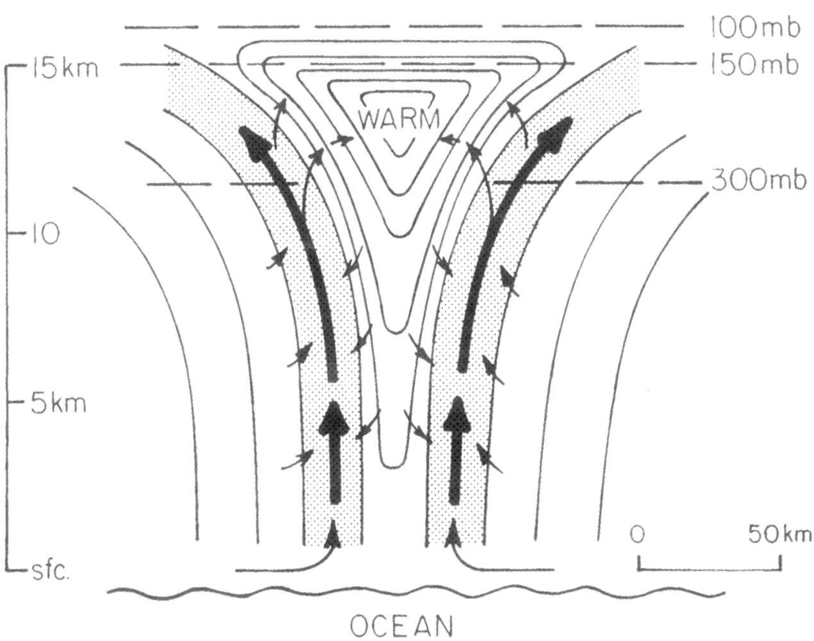

Figure 7. Idealized cross-section of eye and eye-wall cloud temperature anomalies relative to the outward sloping eye-wall updraft. Note the large eye to eye-wall temperature gradient and how the sloping cloud wall can maintain its thermal buoyancy by skirting the edge of the stabilizing warming in the upper level of the eye.

Formation

Tropical cyclones tend to develop in environments which are experiencing positive deviations (for TC development) from climatic values of summer and early fall. Tropical cyclone development requires warm ocean surface temperatures (> 80°F or 26.5°C), potentially unstable moist-lapse rates to allow deep cumulonimbus (Cb) convection, minimal vertical shear of the tropospheric horizontal wind, and positive low-level rotation (or vorticity). Some expression of all four of these favorable environmental factors must be simultaneously present for hurricane formation. This is why TC formation is a relatively rare event while organized cloud areas of tropical disturbances and easterly waves are quite numerous. Most organized cloud systems can persist for only a day or less, although some last for 2-3 days before dying. Only about one organized tropical disturbance in 10 or 20 (Figure 8) is able to maintain itself for 3-4 days or longer.

The early stages of tropical cyclone development and intensification typically occur from rapid and intermittent bursts rather than a gradual and progressive process. Thus, early stages of formation also often depend on the chance occurrence of specific low level, time varying asymmetrical environmental wind surges that are associated with the passage of large-scale weather systems on either the poleward or equatorial side. My hypothesis is as follows: An initial wind surge

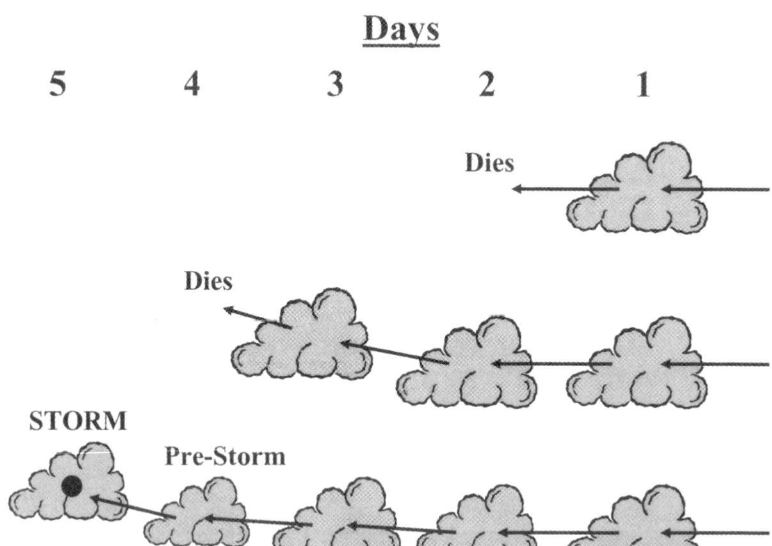

Figure 8. Conceptual illustration of typical cloud cluster systems as they move westward in the Atlantic trade winds. Whereas most cloud clusters live for less than one day, some persist for 2-3 days. If environmental conditions are right, then some longer living cluster systems are able to transform themselves into named tropical storms.

(typically lasting about 12-18 hours) causes the formation of a cloud cluster of perhaps 300 km diameter. Within this cloud cluster, a small middle level (~ 3-5 km height) vortex often forms which may be 50-100 km in diameter (see Figure 9). As I discussed previously [Gray, 1998], this small scale Stage 1 vortex does not immediately intensify further even though, once formed, it is able to maintain itself for several days. Most of these pre-cyclone, Stage 1 systems require the arrival of a second wind surge 1-3 days later to further concentrate vertical motion within the already present small-scale vortex. It is this second (Stage 2) wind surge that typically initiates an unstable wind growth process, which leads to the formation of a circulation of named storm intensity (sustained winds > 40 miles/hour). If this sequence of wind surges does not occur or, if the second wind surge does not sufficiently concentrate convection within the original small-scale vortex, TC formation will not occur. Hence, for this reason, fewer than one in ten or 20 cloud cluster areas or easterly waves ever become named storms.

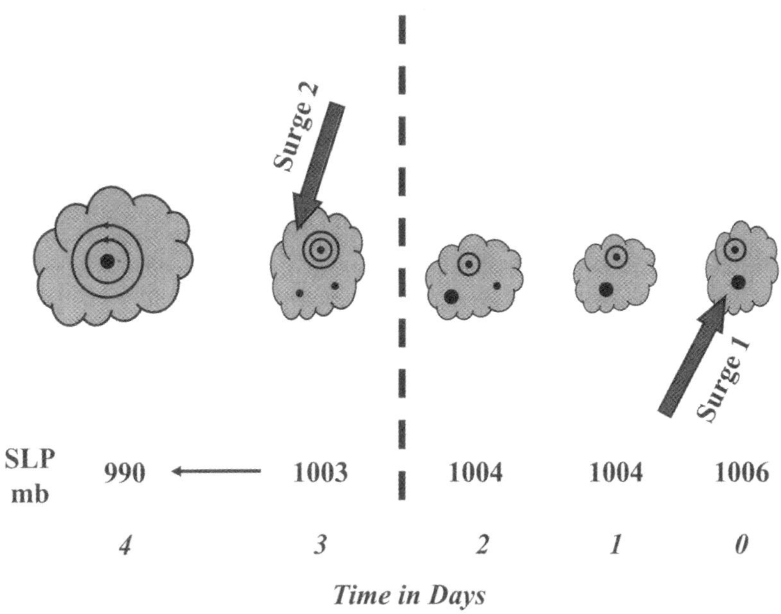

Figure 9. Illustration of the day-to-day (right to left) westward propagating sequence of the two stages typical of tropical cyclone formation. In Stage 1, a transient environment wind surge establishes a cloud cluster, within which a small middle-level vortex (~ 100 km on a side) is formed in association with the larger cluster development. This small convective vortex (seen in the north portion of the cluster) persists for a few days until a second external wind surge provides a second packet of converging momentum to the cloud cluster. This second packet of momentum, if it leads to convergence within the original small vortex, is typically able to initiate an unstable growth process where the central area of lowest pressure commences a sustained decrease to the named-storm intensity.

Figure 10 shows a cross-section diagram of an idealized hurricane's in-up-and-out radial circulation. Once an incipient tropical cyclone has grown into a well-developed vortex, its deep outer layer rotational circulation (often termed the "inertial circulation") acts to isolate and protect the interior of the storm from the environment (except within the upper-level outflow region). At this stage, the major contribution that the environment can make to the maintenance and intensification of the cyclone is one of non-interference. This involves allowing the cyclone to be isolated from strong cross-vortex winds and vertical wind shear patterns.

Environmental conditions vary greatly from storm to storm and can dramatically change during the lifetime of individual cyclones. These conditions also vary with season and between the different global tropical cyclone basins. For example, the trade winds of the western tropical Atlantic present a different type of low-level environment than do the monsoon trough environments of the tropical Pacific and Indian Oceans (Figure 11). In most of the global tropical cyclone basins, formation occurs within a low-level monsoon trough environment wherein the monsoon trough provides favorable low-level background rotation and sustained deep convection. In the Atlantic basin, TCs more often develop from organized cloud clusters that are totally embedded within the trade winds. In 1954 Herbert Riehl explained why westward moving "easterly waves" carry the

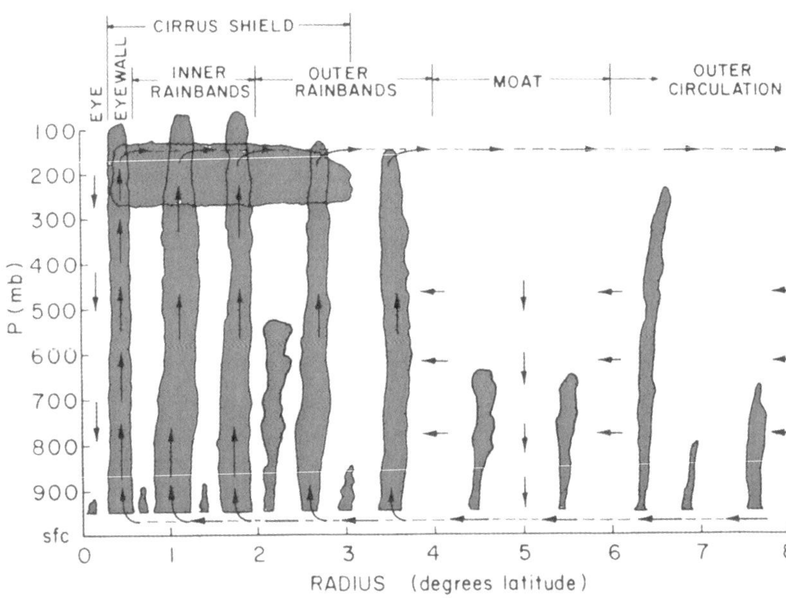

Figure 10. Schematic diagram of the radius-height cross-section typical of the hurricane's cloudiness and radial and vertical winds. Note the concentrated rain areas near the center. Arrows denote the expected radial and vertical motion within and surrounding the clouds [from Frank, 1977].

essential environmental rotation to develop into TCs. They have wavelengths of 1000-2500 km and occur at four- to five-day intervals, typically moving from Africa long distances westward in the trade-wind belt before the masses of concentrated cloudiness they generate find a sufficiently favorable environment to spin up into a TC. During the mid-1970s, Neil Frank showed that about 60 of these easterly wave disturbances come out of West Africa each Atlantic hurricane season. In some years, as few as 1-2 (or none) of these westward-moving waves

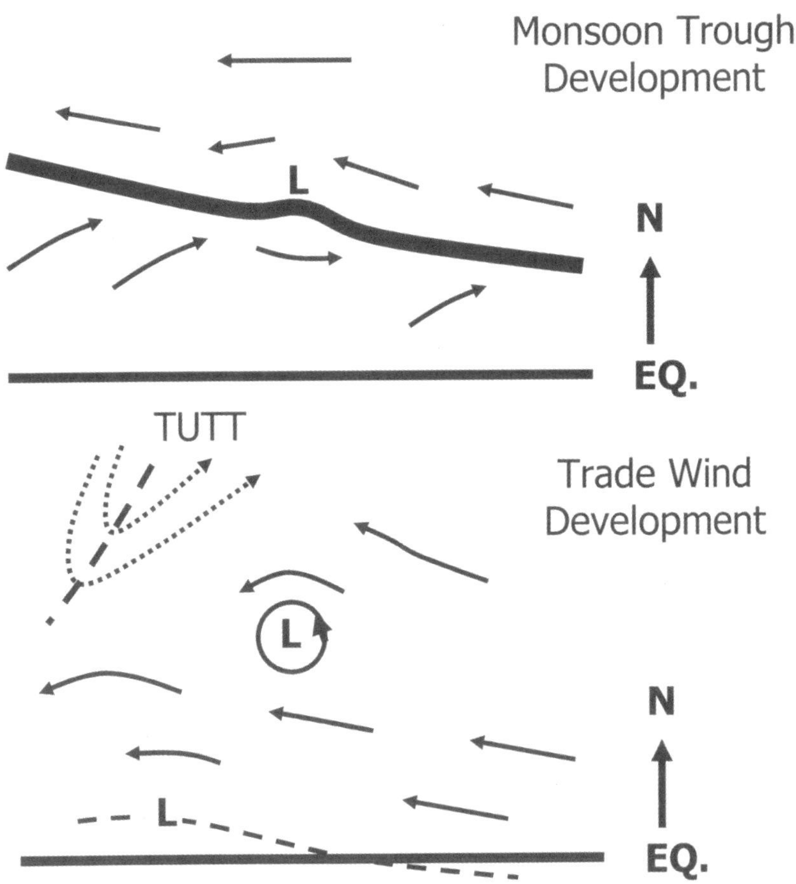

Figure 11. Illustration of the contrast between tropical cyclone early stage development within a monsoon trough (top diagram) versus the bottom diagram, which shows typical TC development characteristics in the Atlantic. Here the early-stage cyclone development occurs entirely within the trade winds in association with an easterly wave. A Tropical Upper-Tropospheric Trough (TUTT) is often located 300-500 km on the northwest side of developing Atlantic systems and often assists in its development.

will develop into named storms. In other years, 8-10 (or more) such waves may develop. Small differences in the large-scale vertical and horizontal wind shear and thermodynamic fields affecting these waves often determine the difference between an active or inactive hurricane season.

Measurement Difficulties and Cyclone Variability

Tropical cyclones spend most of their lifetimes over the tropical oceans where traditional data sources are generally sparse. Observations from ocean vessels in the vicinity of hurricanes are also sparse for obvious reasons. As a result, tropical weather analyses are not always reliable. Consequently, it is rarely possible to acquire enough observations from all available sources to conduct energy budget studies and analyses of the flow of mass through an individual hurricane at one point in its life cycle. This deficiency led to a rawinsonde compositing method, which many associates and I applied extensively during the 1970s and 1980s. Much quantitative documentation of the structure and energy budgets of the "average" hurricane and typhoon and their average formation characteristics was obtained using these compositing procedures. These analyses contributed greatly to the understanding of hurricane behavior, despite the inclusion in the database of individual tropical cyclones with significant differences in size and strength.

Propeller driven aircraft have provided the best information on the structure of the hurricane's central core region, nominally 300 km in diameter. Logistical considerations, however, have limited the ability of prop-aircraft to provide information on the hurricane's outer circulation regions. Aircraft observations are nominally restricted to no more than three flight levels per mission. The duration of these missions, typically less than 4 to 6 hours within the storm core, limits our ability to study temporal variations in cyclones. Further details of aircraft measurements and associated research are discussed in Chapter 3.

Variability in Hurricane Intensity, Size and Outer Wind Strength

Figure 12 demonstrates the large range of surrounding wind speeds that can occur within the cyclone's 1-2.5° latitude (111 to 278 km) circulation for storms of similar Mean Sea-Level Pressure (MSLP) or, conversely, the large range of MSLP for cyclones with similar 111 to 278 km radius winds. The stronger the storm, the greater is this variability of center versus outer radius circulation. The strength of a tropical cyclone's surrounding circulation is often not well related to its intensity. Figure 13 shows a comparison of the radius of 50-knot winds versus MSLP where there is also a very large variability. Figure 14 reiterates this wide variability of cyclone size versus intensity. Size is expressed as the mean radius of the greatest closed isobar versus maximum sustained winds. Note that only about 10 percent of the variance of maximum wind is explained by cyclone size. Large hurricanes may have only modest intensity while some small cyclones may be very intense. These

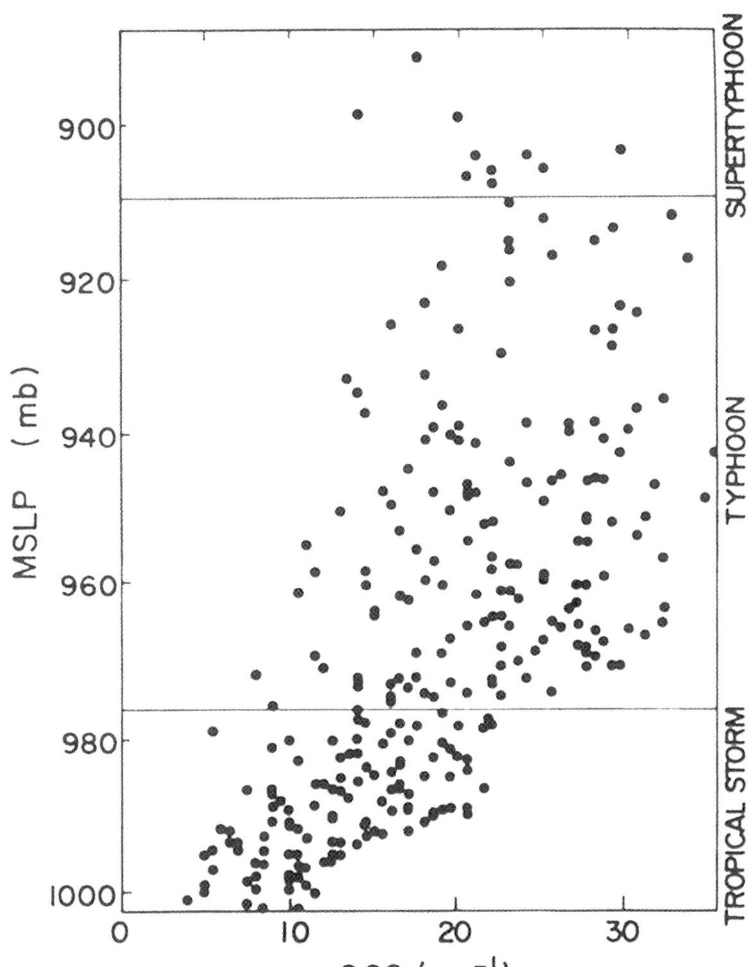

Figure 12. Scatter diagram of cyclone minimum sea-level pressure (MSLP) versus Outer-Core wind Strength (OCS) for typhoons of the northwest Pacific, as measured by reconnaissance aircraft [from Weatherford and Gray, 1988]. OCS is defined as the mean tangential wind between 1-2.5° latitude (111 and 278 km) around the center of the cyclone. Note the wide scatter of values. These observations are believed to be very similar for Atlantic hurricanes.

individual case variations are very large indeed. No wonder it is so difficult to forecast the impact of hurricanes; each tropical cyclone has peculiar characteristics that, to this day, can be difficult to observe and especially difficult to understand and forecast. Even with all the new measurement technology that has come forth in recent years, it is difficult to adequately monitor or explain these large structural variations. This is an important topic for further research.

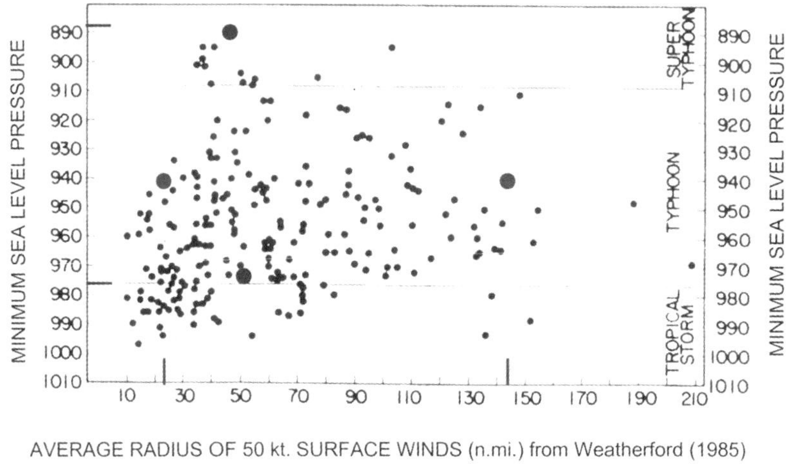

Figure 13. Intensity (minimum sea-level pressure) versus the radial extent of 50 kt (60 miles/hour) winds in nautical miles (n mi) for northwest Pacific typhoons as measured by reconnaissance aircraft [Weatherford and Gray 1988]. One n mi is equal to 1.85 km. Large dots illustrate the large differences that can exist for cyclones of the same minimum sea-level pressure (MSLP) of 940 mb, 23 kt versus 144 kt. Note also that the cyclones of similar outer radius of 50 knot winds can have MSLP differences as great as 890 versus 975 mb. These measurements are also typical of the Atlantic.

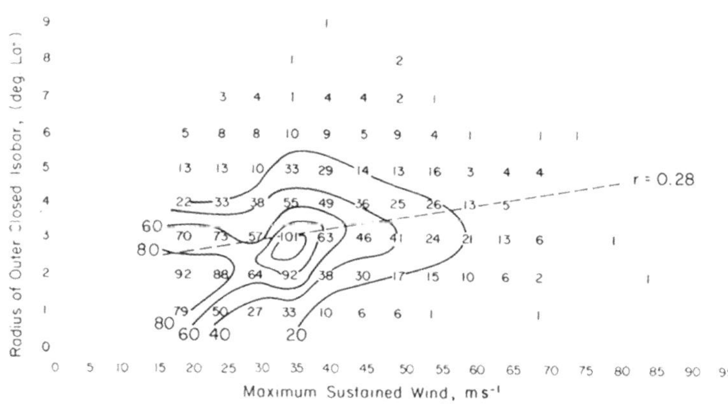

Figure 14. Tropical cyclone size as a function of maximum sustained wind for Atlantic tropical cyclones. Observations are tabulated in classes of 1 degree latitude (111 km) and 10 kt (12 miles/hour). The least square regression line fitted to the raw data is shown by the dashed line. Size is determined from the mean radius of the outer closed isobar on surface weather maps. Note the very wide spread of values. On a statistical basis, tropical cyclone size is little related to cyclone maximum winds [from Merrill 1984].

HISTORICAL MILESTONES

Observations and Forecasting in the First Half Century

Much of the observational database used to advance our forecasting and physical understanding of hurricanes was not available prior to World War II. Observations we now take for granted — including those from satellites, aircraft reconnaissance, coastal radar, and GPS dropwindsondes — along with new capacities for instant communication, were scarcely imagined in the early decades of the 20th century, when only scattered surface observations were available. The fact that there were great problems in predicting the onset and landfall of hurricanes at the turn of the 20th century is not surprising.

In those years forecasters often tracked approaching hurricanes by analysis of changes observed in the period, height, and direction of ocean swell, and the movement of high clouds, following the earlier work of Father Benito Viñes, a meteorologist and hurricane forecaster of some renown. Wireless (radio) did not become available on ships at sea until the early years of the 20th century. While marine weather reports were first transmitted to ships at sea early in the new century, the first warning of a hurricane from a merchant ship at sea did not occur until 1909. Even the existence of an atmospheric layer wherein temperature increased with height (the stratosphere) was not known until the early 20th century. Before the advent of radio communication, it was impossible to receive timely weather information from distant islands or from ships at sea.

Most current hurricane practitioners have already accepted and ingested the last 40 years of measurement gains into their routine procedures. Consequently, it may be difficult for the current group of hurricane forecasters to fully appreciate the challenges of making hurricane forecasts without the aid of continuous satellite products, aircraft reconnaissance, coastal radar, and the guidance of computer prediction models. It thus may be instructive to consider how hurricane forecasters did their job with the limited set of tools at their disposal during the first half of the 20th century.

Given the primitive state of their technology, it is amazing how well forecasters were able to monitor the approach of hurricanes and issue effective warnings. I personally observed Gordon Dunn and his staff at Miami making hurricane forecasts through much of the early 1960s, and was impressed with the insights they gleaned from the sparse set of ship reports, upper-air balloon soundings, and commercial aircraft reports they received.

Twentieth Century Milestones Prior to the Weather Satellite Era

Knowledge of the physical processes that explain hurricanes has slowly grown over the last century. In my view, the more important knowledge gains include the following:

1. The analysis of air temperature and pressure data from western Pacific ships (which were caught within the circulations of typhoons) revealed reduced pressure near the centers of typhoons [Horigute, 1928] but surface air temperature values which were not significantly lower. This condition implied quasi-isothermal pressure decrease within the typhoon's boundary layer. Byers [1944] explained that this observation shows that energy transfers (warming) from the ocean to the air substantially diminished the expected 1°C adiabatic cooling per 10 mb surface pressure decrease. This result, plus the massive amounts of rainfall in these storms, verified that the ocean was the primary energy source for the formation and maintenance of hurricanes. This result also explains why hurricanes form only over warm ocean areas and rapidly weaken as they move over land where this strong lower energy source is not available.
2. Much controversy existed during the early decades of the 20th century about how far up into the atmosphere hurricanes extended. Some thought they were confined to the lower troposphere. Haurwitz [1935] used fragmentary high altitude balloon observations to make hydrostatic calculations that verified that hurricanes had to extend through the full depth of the troposphere, which, in tropical environments, typically extends close to 100 mb (~ 16 km).
3. The quantitative demonstration that hurricanes move in the general direction and with the approximate speed of their surrounding environmental winds was fundamental to cyclone track prediction. This so-called "steering current" concept came forth as upper-air observation networks were established during the 1930s and 1940s. Most recent rawinsonde compositing research shows that tropical cyclones typically move 1-2 m/s faster and 10-20° to the left of the surrounding deep-layer mean circulation out to 500-600 km radius.
4. Palmén [1948] showed that only late summer-early fall ocean temperatures and lapse rates were sufficiently warm and moist to easily sustain deep cumulonimbus (Cb) convection in the West Indies. This observation provided the thermodynamic link explaining why Atlantic hurricane activity is confined primarily to the late summer and early fall periods.
5. Working with Herbert Riehl at the University of Chicago, Hughes [1952] verified from western Pacific low-level aircraft reconnaissance data that the boundary (or Ekman) layer of typhoons had substantial frictionally driven radial flow towards the storm center. This finding indicated that the primary source of the hurricane's in-up-and-out mass circulation occurred as a consequence of low-level frictionally forced convergence.

6. Besides these thermodynamic requirements, Riehl and Schafer [1944] and Gray [1968, 1979] verified that TC formation required Cb convection to occur and be centered in areas of low shear of the horizontal wind through the depth of the troposphere. These requirements limit TC formation to summer and fall periods in regions when the 850 mb (~ 1.5 km) to 200 mb (~ 12 km) wind shear vector becomes comparatively small and changes its sign across the cyclone center. The geographical and seasonal variations in these two factors (i.e., conditionally unstable lapse rates and vertical shear) plus the need for low-level rotation lead to an understanding as to why hurricane formation is limited to the locations and seasons where these factors co-exist.
7. In early aircraft flights through hurricanes during the late 1940s and early 1950s, Robert Simpson (see Chapter 2) observed surprisingly strong temperature concentrations within the hurricane's eye and eyewall clouds at both lower and middle tropospheric levels. This condition was fundamental for interpreting and explaining the concentration of strong winds near the hurricane's center.
8. Joanne Malkus [1958] proposed that the very warm temperatures within the eye were a consequence of mechanically forced subsidence that, in turn, was driven by intense eyewall cloud deep convection. This remains our current view on how it is possible to have mechanically forced subsiding warming air that acts against buoyancy in the warmest portions of the hurricane. Figure 6 shows an idealized view of this process.
9. The inner-core measurements from aircraft during the 1950s and 1960s showed that condensation heating from cumulus convection does not directly warm the troposphere at the location of its release, but rather in the surrounding areas of return mass flow subsidence. This subsidence may occur nearby (as with the eye) and/or at large distances from the convection. With small tropospheric wind shear, subsidence warming can be more easily concentrated near the source of the condensation release. This concentration of warming will be ineffective, however, unless the convection occurs in a region with cyclonic wind rotation, where a horizontal pressure gradient is able to develop an equivalent wind that is able to balance such warming-induced pressure variations. Unbalanced warming causes gravity waves that propagate out of the system and act to inhibit temperature concentration.

Major Research Developments in Decades Following World War II

The formation of the U.S. Weather Bureau's National Hurricane Research Project (NHRP) in 1956, with Robert Simpson as its first director, was a major landmark for increased hurricane understanding and forecasting. Aircraft dedicated to research became available with capability to record flight data (on punch cards) that were almost immediately available for analysis. The simultaneous development of Doppler wind capability allowed winds over the ocean to be accurately measured for the first time; and precision navigation over water was now possible. Doppler navigation—a major technical accomplishment at that time—made a significant impact on tropical meteorology. Quantitative analyses of the balance of forces controlling the wind circulation in hurricanes could be dependably made. The data derived from these NHRP research flights led directly to new findings by many prominent cyclone researchers of the 1950s through the early 1970s. They include Herbert Riehl, Joanne and Robert Simpson, Noel LaSeur, Cecil Gentry, Banner Miller, Harry Hawkins, Charles Jordan, and many others.

From these flights Joanne Malkus (Simpson) and Herbert Riehl developed the so-called "hot tower" hypothesis, whereby it was shown that the majority of the hurricane's vertical motion occurred in individual convective towers and not on a larger meso- or synoptic scale. The formation of NHRP also led to the first development of objective statistical track forecast methods by H. Riehl, B. Miller, P. Moore, and K. Veigus and to the many other statistical track forecast schemes that soon followed. Of special note is the more than 30 years of effort by Charles Neumann and colleagues who developed the most skillful of these statistical track forecast models. Significant advancement in TC track forecast skill has also been more recently achieved from initial value numerical track forecast models. Unfortunately, there has not been much advancement in the prediction of cyclone intensity change. In 1999, however, DeMaria and Kaplan combined statistical and numerical model products to obtain the best skill so far for both track and intensity forecasts for the Atlantic basin. Chapter 4 provides a very thorough and detailed discussion of the current suite of TC guidance models.

Most early theoretical analyses of hurricane circulation assumed vortex circulation symmetry. This symmetry assumption appears to be quasi-valid for the hurricane's inner-core but invalid for radii beyond 100-150 km where asymmetric vortex physics and horizontal eddy processes ($-V_R'V_T'$) play a fundamental role. Momentum maintenance in the outer cyclone region requires that horizontal eddy flux processes act to import large amounts of tangential momentum to compensate for the substantial outer radius boundary-layer frictional dissipation. The cyclone's symmetric in-up-and-out radial circulation is never sufficient to balance frictional dissipation at outer radii. The outer tangential circulation can be maintained, as previously discussed, only through processes whereby azimuthal variations of inflowing air at individual levels are stronger than are the azimuthal variations of outflowing air at the same level. In the early 1970's, Rick Anthes did the first fully three-dimensional numerical hurricane models, as described in his 1982 book. His results helped verify the importance of these horizontal eddy-flow processes.

With the advent of the satellite/computer age during the late 1960s, considerable advancement has occurred in our understanding of the physical processes by which hurricanes develop and maintain their circulation. These many advances and their associated forecast implementation are too numerous to discuss in detail here but are covered in other chapters of this volume and in the references.

Advances in Technology and Its Implementation

The evolution of techniques for hurricane location and track-structure-intensity prediction has closely followed the basic technical developments during the 20th century, as shown in Figure 15. These technical advancements have been truly phenomenal. To this list must also be added the large advances in weather radar and, in particular, the dual-Doppler radar developments that have allowed the upper-level winds of hurricanes to be measured wherever there are hydrometeors available for tracking. These dual Doppler radar techniques are available on the NOAA/HRD P-3 research aircraft. Much new information about the hurricane wind structure has been obtained from these flights. Airborne dual-Doppler radar measurements can be taken over the oceans far from land. The entire southeast

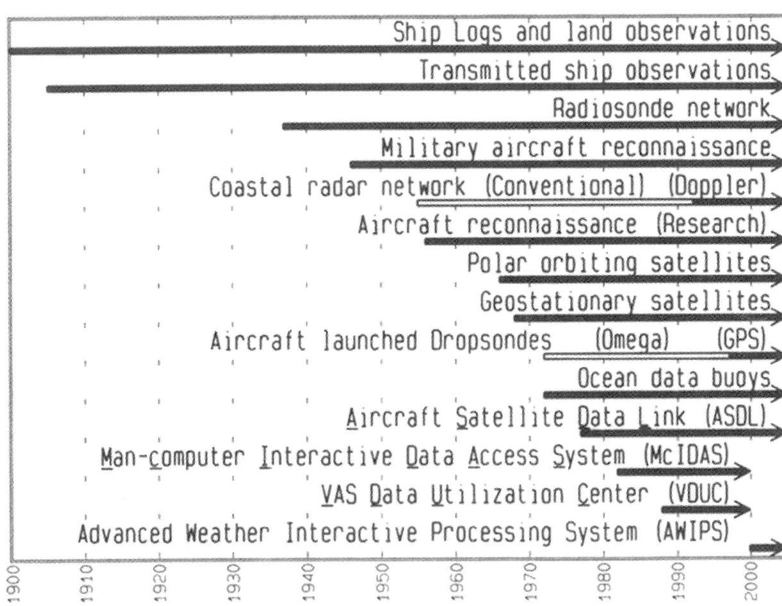

Figure 15. Advancement in observing technology during the 20th century. Note the large advance in technology since the beginning of the satellite age in the mid-1960s. To this list should also be added developments in radar, particularly the dual-Doppler radar, and the new technology of the GPS dropwindsonde.

U.S. coastline is now ringed by land-based, overlapping dual-Doppler coastal radar networks. These provide excellent immediate information on the wind and cloud patterns of hurricanes approaching landfall. Unfortunately, land-based radar cannot monitor storms that are more than 300-400 km from the coastline.

The deployment of the Global Positioning Systems (GPS) aircraft dispensing dropwindsondes also needs to be added to the technical developments of Figure 15. GPS dropwindsondes, commonly available for the first time in the 1990s, now provide reliable near surface wind measurements within the hurricane's inner core as well as accurate vertical gradients of wind, temperature, moisture, and pressure height in the lowest few kilometers of the hurricane. These GPS sondes have put us into a new era with regards to the measurement of accurate *in-situ* low-level wind and thermodynamic features within the hurricane core region (see Chapter 3).

The new technology has also led to significant improvements in 2-5 day track prediction from the 6-8 numerical weather prediction (NWP) models that are now making real time extended-range track predictions. The upgrading of the dropwindsondes from the Omega to the GPS instrumentation has been an improvement for hurricanes that have the potential to hit the U.S. coastline. Upper-level dispensing of GPS dropwindsondes from the NOAA G-4 Gulf Stream and NASA ER-2 aircraft now provide crucial ambient steering wind information, as illustrated in Figure 16. When this GPS dropwindsonde information is collected in the (heretofore) data-void areas surrounding hurricanes, distinct improvements in track prediction skill have been observed (Figure 16).

Weather Satellites. Of all improvements in data acquisition technology, weather satellites have certainly led the way. In recent decades, the NOAA and NASA research laboratories have developed many new techniques for better monitoring and prediction of cyclone position, formation, structure, and intensity change as a result of our ever-improving satellite technology, which has greatly transformed tropical cyclone prediction, (particularly in the non-Atlantic basins where aircraft reconnaissance is not flown). Although most satellite applications are discussed at length in later chapters, one early area of satellite research must be noted here: the unique work of Vern Dvorak. Starting in the early 1970s, Dvorak [1975] led the development of a pattern recognition system for using satellite visual and infrared (IR) data to specify cyclone intensity. Dvorak's intensity scale, now used universally, is the fundamental technique for early-stage detection of developing cyclones. In lieu of aircraft reconnaissance, it is the basic tool for hurricane location and intensity determination for cyclones that are beyond the range of land-based measurement tools.

Improved observational technology and physical insight in recent decades have led to slow but significant improvements in the crucial 1-2 day cyclone center track predictions. Forecasters value new technical advances, since observational redundancy is helpful in diagnosing the varying position and intensity of

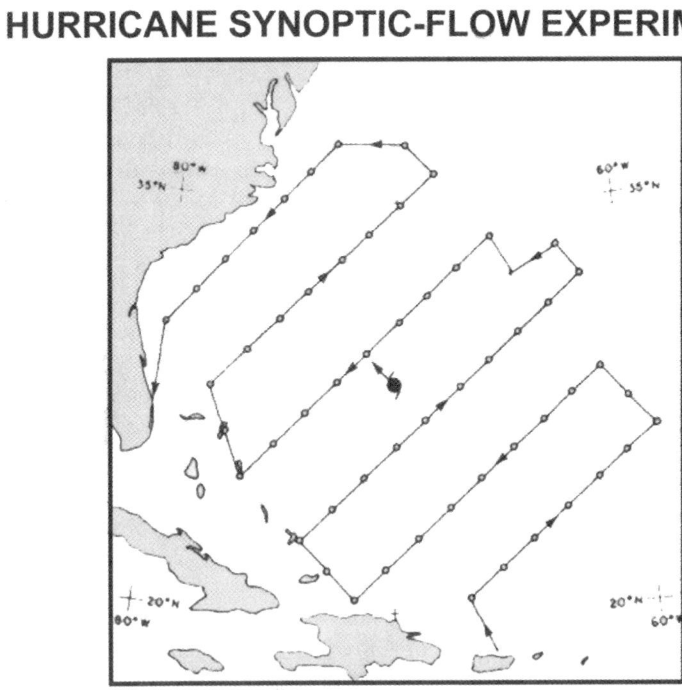

Figure 16. Typical hurricane synoptic mission by the NOAA Gulfstream-4 jet aircraft upper-tropospheric flight from San Juan to Miami. The circles show where GPS dropsondes are dispensed. These dropsondes fill in the Atlantic data void areas. These data are ingested into the numerical track models and improve and give more confidence to the hurricane track forecasts.

hurricanes. Also each observational tool has weaknesses so that greater confidence accrues if independent observations – such as aircraft, radar, and satellite measurements – yield a consistent picture.

Other important late 20th century computer-related developments include facilities that make satellite products available on the Internet in near real time as well as retaining such data in an accessible archive. The website of the Naval Research Laboratory (NRL) is a perfect example here. Tailored to forecaster needs, it gives most of the new microwave data as well as Visible and IR data from satellites in geosynchronous orbits. Details of new satellite instruments such as scatterometers for surface winds and passive microwave (which help in detecting warm cores) are discussed in Chapters 11 and 12.

Leadership at the National Hurricane Center. Robert Simpson, who was Director of the National Hurricane Center (NHC) from 1968-1973, worked hard for the implementation of new technology in hurricane forecasting. His tenure began with the emergence of reliable daily satellite images. Simpson reorganized the National Hurricane Center to meet the challenges of this and other new technological developments. Earlier he had led in the establishment of the U.S. Weather Bureau's National Hurricane Research Project (NHRP) in 1955 and was its director until 1959. The NHRP and its successor, the NOAA/AOML/ Hurricane Research Division (HRD), have made many hundreds of flights into the centers of hurricanes to study inner-core structure and rainband features. This information still provides the basis for most of what we presently know about the hurricane's inner-core structure.

Neil Frank (NHC Director 1974-86) gave major attention to raising hurricane awareness in the vulnerable community and to developing hurricane emergency management procedures. He brought the media into the Center when significant hurricane threats occurred, as he felt that they (the media) could better inform the public if they had direct contact with the current forecast situation. This action was partially based on the assumption that improved technology and forecasting skill are of little benefit if they are not adequately communicated to the public. Frank was also one of the founders of the annual Interagency National Hurricane Conference meetings from which multiple strategies for coping with hurricanes are being continually developed and upgraded. These meetings bring together people from a variety of federal, state and local governments, private industry, insurance companies, research and engineering groups to cooperate in the development of better coping techniques.

Robert Sheets (NHC Director 1987-1994) and the subsequent directors have followed this same "open door" media approach. Table 4 lists the directors of the Miami Hurricane Forecast Office since its establishment in 1943—subsequently designated the National Hurricane Center (NHC). Each NHC director and his staff have made special efforts to implement the new technological changes that are continually becoming available, leading to a substantial improvement in hurricane observing and forecasting capability.

Hurricane Modification

A conceptually attractive method for coping with hurricanes would allow us to moderate their destructiveness. Although many ideas have been suggested for

Table 4. Directors of the U.S. Hurricane Forecast Center (Miami)

Grady Norton (1943-54)	Robert Sheets (1986-95)
Gordon Dunn (1955-67)	Robert Burpee (1995-96)
Robert Simpson (1968-73)	Jerry Jarrell (1997-99)
Neil Frank (1974-86)	Max Mayfield (2000-present)

accomplishing this objective – some ridiculous, others well founded — so far at least they are logistically and/or economically unfeasible. A recurring suggestion since the end of World War II, for example, has been to "blow the storm apart" using a nuclear bomb. A mature hurricane of moderate strength and size releases as much condensation heat energy through its cloud systems in a day as the nuclear fusion energy of about 400 20-megaton hydrogen bombs. On the basis of energy comparisons alone, this idea must be discarded immediately, without even considering the inevitable degradation of the environment from radioactivity.

Another approach involves cutting off the hurricane's surface energy supply. If the oceanic source of heat and moisture could be shut down, the hurricane would rapidly weaken, as shown in nature when a storm encounters colder water or it moves over land. Since mid-century, several logistically unfeasible methods have been suggested to artificially cut off the hurricane's surface energy source. One scheme to artificially suppress surface evaporation involves the placement of a chemical component on the ocean surface that contains evaporation and heat flux, a method used with some success on reservoirs. However, to suppress a sufficient area of ocean surface, the chemical retardant film would have to be very thin — and ocean wind and wave action would prevent its maintenance.

Several colleagues and I have proposed another scheme. We envisaged dispensing very small, sub-micron carbon black particles into the boundary layer around the hurricane's cloud shield. These particles would absorb solar radiation and warm the boundary layer, enhancing the cyclone's outer radius cumulus convection while reducing the cumulus convection near the cyclone center – finally weakening the cyclone's strongest winds. The small carbon particles would be obtained by burning petroleum products in an oxygen deficient environment. We had originally envisaged using large military (C5A) aircraft for dispersing the particles, but this turned out to be unfeasible. Further efforts with this approach, however, might be tried using surface ships for burning the dispensing agent. Appropriate cautionary measures would be required. Only after demonstration at sea should this operation be attempted, and only when a hurricane threatened a densely populated area such as southeast Florida. This idea may eventually be re-examined. It must be assessed whether the cost of precipitation loss and other negative pollution effects would be outweighed by the reduction in hurricane damage, if real.

In the 20th century, only one type of experiment on hurricane modification has actually been attempted: the strategic seeding of towering storm clouds near storm center. By inducing the super-cooled water in the towering clouds at storm center to freeze, it was felt that a modest reduction of maximum wind strength would ensue, and hence a reduction in damage, without interfering with the rainfall that many regions rely on for agriculture. This was a very attractive idea in the 1960s and 1970s. The difficulties of carrying out this experiment, called STORMFURY are discussed at greater length in Chapter 3.

Inter-decadal Variability and Individual Seasonal Prediction

Another notable advancement during the last two decades has been the development of an understanding of slowly varying global atmospheric and oceanic conditions that promote comparatively active versus inactive hurricane seasons. Of seven major tropical cyclone basins, the Atlantic has the greatest inter-seasonal variability of tropical cyclone activity. As such, the Atlantic may be viewed as a marginal basin for hurricanes. When seasonal climate trends are favorable, many cyclones develop. When climate conditions are unfavorable, very few cyclones are able to form. Activity in the Atlantic basin is vulnerable to the effects of several strong and persistent climate factors that can inhibit or enhance TC formation. For example, Atlantic basin hurricane activity is typically reduced during El Niño and easterly tropical stratospheric (or QBO) years and in years of western African drought. Activity is enhanced during La Niña years and when Atlantic Sea Surface Temperatures (SSTs) are above average and sea level pressure below average. There is also substantial multi-decadal variability in major (Category 3-4-5) hurricane activity that appears to be largely governed by multi-decadal variations of the Atlantic oceanic thermohaline circulation. These new seasonal to multi-decadal global climate-related insights have been made possible by advancements in data processing technology that provide global data sets and analysis tools not available just a few decades ago.

EPILOGUE – PERSONAL REFLECTIONS FROM 45 YEARS OF HURRICANE RESEARCH

Nearly all the evolutionary milestones in the understanding of hurricanes — and most other meteorological advancements as well — occurred as quantitative deductions derived from observations. Comparatively few insights have come from non-observationally based theoretical speculations. Many of the physical processes of hurricanes are much too complex to be imagined or deduced directly from an understanding of the laws of physics alone. The following quote by Jacob Bronowski [Ascent of Man, 1973] concerning his general insight on science is relevant, I believe, to aspects of our developing understanding of tropical cyclone structure and intensity change:

> *A science that orders its thought too early is stifled. Until a science has passed through a long stage of observation and trial, it cannot develop a system of ordering its observations; and it is mere presumption to try to fit it with so ambitious an order as Newton's.*

Concerning the large differences in individual tropical cyclone structure features and the wide variety of cyclone intensity and outer circulation modes (as illustrated in Figures 12-14), I believe we are still in Bronowski's "observational and trial" stage. It is the measurement and understanding of these individual cyclone structure and intensity features that are so crucial for predicting and understanding hurricane-spawned destruction.

Progress in the physical understanding of hurricanes derived directly from numerical modeling over the last few decades has been controversial. Several of the following chapters provide illustrations of valuable results from model applications. This author's view is that the rush to tropical cyclone numerical simulations in the 1960s and 1970s may have been premature in terms of improved understanding of these storms. Some of the resources expended in these modeling ventures probably limited opportunities to make great advances from observations already enabled from new technology and data accessibility. With hurricanes having so many physical variables interacting in such a complex and nonlinear manner, there is an almost infinite number of wrong ways to obtain right-looking answers from numerical simulations. Having a numerical model that is able to simulate a vortex with characteristics of the hurricane does not constitute understanding. Greg Holland in 1987 stated it quite well:

> *There is, in particular, an inclination to assume that because one set of mechanisms produce a hurricane-like vortex that the other processes are unimportant. For example, each of the axisymmetric analytic and numerical models described by Ooyama [1969], Carrier et al. [1971], Yamasaki [1977], Chala and Pfeffer [1980], and Emmanuel [1986] contains quite different explicit physics from the others, yet each also produces "realistic" tropical cyclone structures. Further, observational studies have noted consistent environmental interactions associated with tropical cyclone structure changes, yet most numerical models develop tropical cyclones without including such environmental interactions.*

The rapid advances in computer technology we have seen and our successes in global and middle-latitude numerical prediction, as far as 5-10 days in the future, have led many to believe that similar successes can be achieved in hurricane simulation and prediction. This is certainly true of tropical cyclone track prediction (out to 3-5 days), but it is not necessarily true for the prediction of cyclone structure and intensity change or for understanding the real internal physics of hurricanes. Much faith exists that ever more complicated numerical models and sophisticated theories will produce improved basic understanding of hurricanes. This is not necessarily true.

Hurricane structure involves complex interactions between processes occurring on a huge range of spatial scales and under extreme conditions. There may well be significant details of the hurricane structure that are just too complex and too nonlinear to be reliably simulated by computer models in the foreseeable future. Two major physical problems that models must continue to address by judicious simplification are how to handle the complex cumulus convective processes (often in long banding features) of hurricanes and the internal viscosity of the hurricane. In fact, there is probably a relation between the convective processes and the internal viscosity, which is not easy to formulate. How does one write numerical code to realistically deal with cloud scale condensation-evaporation, cloud-scale vertical momentum exchange, convective downdraft influences, ice particle habits that influence fall velocities, and realistic air-sea interaction with sea spray driven by turbulence?

These important and complex interactive processes mainly occur on scales below that which current computer grids can resolve. And as greater computer resolution becomes possible, new hierarchies of still smaller-scale physical processes may need to be modeled, and thus also measured and understood. The speed of computers is not the primary constraint on understanding hurricanes (and probably other geophysical systems), as so many model-oriented scientists and uninformed government officials have often implied. We should not expect more powerful computers, by themselves, to necessarily bring improved hurricane understanding and intensity-structure prediction. In contrast, improvements in computer technology are proving to be an invaluable benefit for observational hurricane study, using many different types of observations together (see Chapter 12). Future capabilities for improved data storage, retrieval, and processing of hurricane observations appear almost unlimited. As such, we should expect large observational-based hurricane knowledge gains in the coming years. As these many types of observations are wisely assimilated into models, and the model results tested against many different observations, the combination should optimize predictions and reveal their degree of uncertainty.

Traditional Science Versus Needed Research

My personal view is that certain methodological barriers have inhibited the advancement of tropical cyclone research. This involves the strong cultural legacy of traditional science, academic prestige, and peer pressure to follow popular, current themes and current methodology. The prestige of the so- called "pure science" subjects, such as fluid mechanics, turbulence, mathematics, and numerical modeling, engenders a certain cachet versus the less glamorous research subjects, which require "slugging it out in the trenches" of the tedious and dirty world of observational data analyses. The labor-intensive observational efforts are necessary before really improved hurricane understanding and prediction can be achieved. In particular, hurricane research and forecasting need more people like Vern Dvorak and Charles Neumann. The latter individuals have contributed so much to our improved forecasting of tropical cyclones over the last quarter century in their innovative applications of statistical analyses. The main concern of Vern Dvorak and Charles Neumann was to make needed contributions to improvement in tropical cyclone forecasting. They did not worry whether their contributions would be judged as good science or not. But basic science gains have resulted from their efforts. The most profitable scientific insights often occur from individuals attempting to solve practical problems. Dvorak's empirical pattern recognition for estimating the intensity of tropical cyclones from satellite images is used the world over. His success was largely due to his being unencumbered by any theory or preconception on how tropical cyclones are supposed to function. Neumann's statistical tropical cyclone track forecasts for the Atlantic have been operationally tested for many years and (where properly normalized) show forecast skill at one to two days comparable to the better initial value-numerical prediction schemes. But fewer resources are now being directed toward implementation of empirical track forecasting methodology. Nearly all track forecasting efforts are now going to the development of improved methods of initial value numerical prediction. Statistical forecast schemes use information from many past years. Initial value numerical forecast models, by contrast, introduce only current information. In my view, it is unfortunate that statistical track prediction has fallen out of favor. The simultaneous development of both techniques would offer many advantages, although the so-called "ensemble models" approach to initial value numerical prediction does embrace some aspects of this philosophy.

OVERVIEW

There is much to be learned about tropical cyclones from the wide variety of new observational data sets that are now becoming available. Although each type of data is inadequate as a distinct entity, an improved underlying synthesis of the internal physics of cyclones and their environmental interactions becomes possible if one is able to piece together the interlocking associations between many different data sources. We need to integrate satellite, radar, aircraft, buoys, surface ships, rawinsondes, and related data sets better. There are vast amounts of valuable tropical cyclone data gathering dust on shelves or stored unstudied on computer tapes and discs. Currently, it is easier to secure government funding for the acquisition of new data than to drudge at the necessary analyses of existing data. What we now require are persistent tropical cyclone "generalists" who insist on getting their hands dirty to analyze and interpret many types of data, even at first if by "back of the envelope" calculations. We need men and women who are able, through a wide variety of different data platforms, to see the complicated physical linkages that operate within the tropical cyclone's internal thermodynamics and its environment. Even if such physical linkages may at first be only qualitative in nature, new physically based ideas are much in demand at this time. It is important that hurricane data be used interactively and very carefully with numerical models to verify or reject physical assumptions from the data. At this stage of our understanding, it is also important that computer models be used primarily as an adjunct to observationally based physical hypotheses. Data should not be viewed primarily as a source of verification of model simulations.

Recent advances in computer technology have opened the door to many new possibilities for the processing and analysis of massive quantities of tropical cyclone-related data sets.

Herbert Riehl, who studied hurricanes from the 1940s through the 1980s, is considered by many to be the father of tropical meteorology. He is perhaps the best example of one who could consult a variety of data sources and come forth with synthesized physical hypotheses of the most likely relevant physics. His 1954 book, *Tropical Meteorology*, is a classic in terms of what he was able to conceptualize from tropical observations available to him before 1954.

REFERENCES

Anthes, R. A., Tropical Cyclones: Their Evolution, Structure and Effects. *Meteor. Monogr.*, 19, Boston, MA: Amer. Meteoro. Soc. (ISBN 0-033876-54-8), 208 pp., 1982.

Bronowski, J., *Ascent to Man.* Popular TV series accompanied by a book. 1973.

Byers, H., *General Meteorology*. New York: McGraw-Hill, 461 pp., 1944.

Dvorak, V. F., Tropical Cyclone Intensity Analysis and Forecasting From Satellite Imagery. *Mon. Wea. Rev.*, 103, 420-430, 1975.

Frank, W. M., The Structure and Energetics of the Tropical Cyclone, II: Dynamics and Energetics. *Mon. Wea. Rev.*, 105, 1136-1150, 1977.

Gray, W. M., Global View of the Origin of Tropical Disturbances and Storms. *Mon. Wea. Rev.*, 96, 669-700, 1968.

Gray, W. M., Hurricanes: Their Formation, Structure, and Likely Role in the Tropical Circulation. *Meteorology Over the Tropical Oceans*. D. B. Shaw, ed., Royal Meteor. Soc., 155-218. 1979.

Gray, W. M., Recent Advances in Tropical Cyclone Research from Rawinsonde Compositing Analysis. Geneva, Switzerland: World Meteorological Organization, 407 pp., 1981.

Gray, W. M., The Formation of Tropical Cyclones. *Meteor. and Atmos. Physics*, 67, 37-69, 1998.

Haurwitz, B., The Height of Tropical Cyclones and the "Eye" of the Storm. *Mon. Wea. Rev.*, 63, 45-49, 1935.

Horigute, Y., *Mem. Imp. Obs. Kobe*, 3, 23-33, 1928.

Hughes, L. A., On the Low Level Structure of Tropical Cyclones. *J. Meteor.*, 9, 422-428, 1952.

Malkus, J. S., On the Structure and Maintenance of the Mature Hurricane Eye. *J. Meteor.*, 15, 337-349, 1958.

Merrill, R. T., A Comparison of Large and Small Tropical Cyclones. *Mon. Wea. Rev.,* 112, 1404-1418, 1984.

Neumann, C. J., B. R. Jarvinen, C. J. McAdie and G. R. Hammer, *Tropical Cyclones of the North Atlantic Ocean, 1871-1992*, Dept. of Commerce, NOAA, 206 pp., 1999.

Palmén, E., On the Formation and Structure of Tropical Hurricanes. *Geophysica, III, 26-3.* 1948.

Riehl, H., *Tropical Meteorology*. New York: McGraw Hill, 392 pp., 1954.

Riehl, H. and H. Schafer, The Recurvature of Tropical Cyclones. *J. Meteor.*, 1, 42-54, 1944.

Weatherford, C. and W. M. Gray, Typhoon Structure as Revealed by Aircraft Reconnaissance: Part I: Data Analysis and Climatology. *Mon. Wea. Rev.*, 116, 1032-1043, 1988a.

Weatherford, C. and W. M. Gray, Typhoon Structure as Revealed by Aircraft Reconnaissance: Part II: Structural Variability. *Mon. Wea. Rev.*, 116, 1044-1056, 1988b.

Chapter 2

Impact of Technologies from Two World Wars

Edward N. Rappaport and Robert H. Simpson

INTRODUCTION

For nearly half a century after the Galveston disaster of 1900, the struggle to learn how to cope with hurricanes was slow and painful. Much of the progress can be attributed to the availability of new technologies developed in support of military operations in World Wars I and II—developed mainly for purposes having little to do with hurricanes, their prediction, or the ensuing social consequences. Here we will examine how these technologies were put to work, not only to improve the ability to detect, track, and warn of hurricanes, but also to demonstrate and justify the need to expand scientific understanding of hurricanes and their behavior. The net result will be seen in later chapters describing the science-driven acceleration of progress during the last half century.

BACKGROUND

On their own terms and time lines, weather and war rank as two great threats to society. When a significant meteorological event or prolonged weather siege coincided with a major military operation, history was often made. The sinking of the Spanish Armada, and the fall of the Third Reich upon encountering the frozen grounds of a harsh Russian winter, are well-known examples. Tropical cyclones, too, have had an impact. In the Far East, a thirteenth century typhoon destroyed Kubla Khan's invading forces, sparing Japan from the Mongol invaders. In the West, great tropical cyclones changed the course of colonial

Hurricane! Coping with Disaster

10.1029/055SP03

expansion and supremacy in the New World, including Florida's early political fate — events described in more detail by Rappaport and Fernandez-Partagas [1997]. Yet, the relationship between these two threats has sometimes worked to society's broad advantage. Some of the benefits to meteorology will be evident in the discussions and analyses here.

Formative Years of Weather Services in the United States

As a military necessity in 1814, James Tilton, physician and Surgeon General of the Army, ordered his staff to begin recording daily weather during the War of 1812. Military benefits from these records contributed to the establishment and development of weather services in 1870 when the U.S. Congress directed the Secretary of War,

> *...to provide meteorological observations at military stations in the interior of the continent and at other points in the States and Territories...and for giving notice on the northern (Great) Lakes and on the seacoast by magnetic telegraph and marine signals, of the approach and force of storms.*

This task became the responsibility of the U.S. Army Signal Service (later "The Signal Corps"). The nation's first advisory on a tropical weather system was issued in 1873 by this service. In 1890, Congress transferred the weather function to the Department of Agriculture, becoming a civilian service known as the Weather Bureau.

In 1900, weather forecasts were prepared by the Weather Bureau's central office in Washington, twice daily, at 10 AM (EST) covering the next 36 hours, and at 10 PM for the next 48 hours. According to G. K. Grice [Appendix A], all severe weather advice, including hurricane warnings, were prepared and disseminated from Washington, the responsibility for issuance of all warnings residing with a single forecaster.

Meteorological science and understanding of storm systems and their behavior advanced slowly in the early years, in part inhibited by inept management and decision-making driven more by political than scientific considerations [Whitnah, 1961; Bates and Fuller, 1986].

Data and communication systems during this era were limited. Each of the numerous options was deficient in some way. Delivering data to the meteorologist and forecasts and warnings to the public often stretched the wits of the dispatcher to their limit. Among the more novel approaches for weather warnings was the use of electric searchlights and whistles. Fire alarms, bombs, rockets, and flags flown from moving express trains foretold approaching cold waves.

From its inception, the telegraph was seen as a communication tool with great potential for weather forecasting. In 1845, Joseph Henry, then Secretary of the Smithsonian Institution, envisioned the telegraph would provide

> *...a system of observation which shall extend as far as possible over the North American continent...[and] will furnish a ready means of warning the more northern and eastern observers to be on the watch from the first appearance of an advancing storm.*

In the early 1900s, telegraph offices were closed at night, and each alternative system came up short in some way. While the mail service provided daily forecasts on rural-free-delivery routes, hurricane information was too perishable for this form of dissemination. Kiosks displayed weather information to the public, but were available only in the largest cities. On request, telephone operators would read weather forecasts to subscribers. But phone service then was far from universal.

Weather advice for mariners posed additional challenges. A system of hurricane flags and lights gave warnings to those in visual range and were applicable for eight hours within 100 miles of the display. In 1905, wireless communications were introduced to the marine weather community. It paid off dramatically for the first time in 1909 when a ship at sea transmitted its report on a hurricane in time for warnings to be effective.

Several years later, in 1912, arrangements were completed to acquire regular observations from ships at sea. Thirty ships plying the New York to New Orleans route began sending weather observations twice daily by wireless telegraph. This program proved difficult to maintain, however, and was unattractive to shippers who received only 10 percent of their telegraph costs plus $0.50 per observation.

With the exception of weather reports from ships at sea, few changes were made in operational tools and procedures for weather services during the new century until World War I, when more remote observing stations were added. These extended into tropical regions including Swan Island, British Honduras, Nicaragua, Cuba, the Canal Zone, and the Virgin Islands. As we shall see, war stimulated the development of a succession of new devices, few of which were in direct support of meteorology. Nevertheless, they were critical to the scientific progress made during the first half century.

Early Scientific Understanding of Storm Systems and Hurricanes

Whatever progress followed the Galveston disaster of 1900, a remarkable understanding of global scale meteorology, climatology and wind circulations had pre-existed for decades. What was lacking was the ability to understand hurricane inception, and predict the movement of storm systems with sufficient timeliness to minimize loss of life and property. Pertinent to this discussion are two quite different roots or cornerstones of knowledge upon which progress fed, albeit slowly — one rooted in Europe, the other in Cuba. Interestingly, both grew in prominence and public credibility during, or in association with, wartime pressures, although the United States was slow to take advantage of scientific progress made abroad. It will be useful to examine this background and question why the United States seemed to be slow in taking advantage of it.

The Bergen School and its Influence

Carl Anton Bjerknes, a brilliant and resourceful Norwegian physicist, spent most of his lonely career during the latter half of the 19th century near Kristiania (now Oslo), Norway, trying to expand the knowledge of hydrodynamics into a universal theory of motion (including that associated with electromagnetism). In this he failed, a saga described in fascinating detail by William McPeak [1995]. Nevertheless, Carl's unrelenting enthusiasm, primarily in search of a universal theorem, spread to his son Vilhelm, and later his grandson Jacob. Vilhelm and Jacob's research interests, however, soon turned to a more restrictive application of hydrodynamics: atmospheric circulation and storm development in temperate latitudes. The collective efforts of this triumvirate, sometimes referred to as the Bjerknes dynasty, led to the founding of the "Bergen School." From the collaboration there, Jacob published his revolutionary three-dimensional concepts of air mass and (warm and cold) frontal analysis and the theory of mid-latitude storm development. It ultimately found uniquely effective applications in defining and predicting aviation weather.

Jacob's concepts evolved during World War I when Scandinavia found itself isolated, unable to access previously available weather reports, or to provide weather warnings for their all-important fishing industry. To remedy this, the Norwegians established a dense network of cooperative weather reporting stations. Detailed analyses of these closely spaced observations by Jacob and his colleagues revealed distinct discontinuities between weather regimes. These led Jacob to the concept of warm and cold fronts and the theory of frontal cyclones, including the basic energy transformations that generated these storm systems.

The influence of the Bjerknes dynasty did not find immediate or direct applicability in the tropics, or to hurricane development. Nevertheless, the emphasis it placed on three-dimensional analysis in understanding storm development, and the need for more closely spaced observations to understand and predict storm behavior, made an impact globally. It challenged technology to find a means of describing weather systems in greater detail, especially over broad ocean expanses. Not until almost mid-century, however, was it possible to describe weather systems in sufficient detail to understand the importance of interactions between weather in the tropics and temperate latitudes.

Cuban Leadership in Hurricane Prediction

Father Benito Viñes, a native of Havana, Cuba, gained international recognition as a scholarly Jesuit scientist during the latter decades of the 19th century for his contributions to tropical meteorology and, more particularly, for his skills in predicting and warning of hurricanes. Figure 1 is a pen sketch of his profile. J. J. Fernandez-Partagas, in an unpublished manuscript [1989], presents an enlightening account of Viñes' career and fame as a hurricane forecaster. While his avid interest in tropical meteorology and concern with hurricanes claimed most of his professional effort, Viñes acquired a remarkable contemporary understanding of the basic processes in tropical meteorology. Nevertheless he was more widely known and remembered as a keen observer of weather, who applied sound physical reasoning to understand the implications of what he observed. As early as the 1880s, Viñes made a simple, yet profound, statement about hurricanes:

> *The (steering mechanism) of cyclones is, in my opinion, the upper current...*

Figure 1. Portrait of pioneering tropical meteorologist Father Benito Viñes, S. J. [courtesy Mrs. Berta Randín].

Using time sequences of high cloud movement, 24-h changes in barometric pressure, and changes in direction and frequency of ocean swells as tools of prediction, he was credited both at home and in the Caribbean area with making remarkably accurate predictions of location and timing of hurricane land falls.

When the Spanish-American War loomed on the horizon, with the future of Cuba at stake, the United States rushed to emulate the hurricane prediction methods of Cuba's master meteorologist, Benito Viñes. Ivan Ray Tannehill in his book *Hurricanes* [1940] identifies the esteem in which Viñes' hurricane warning prowess was regarded in Washington when he quoted President McKinley, presiding over preparations for war:

> *...I am more afraid of a West Indian hurricane than I am of the entire Spanish Navy. Get this new hurricane warning service inaugurated at the earliest possible moment...*

The United States promptly set about establishing a Caribbean warning facility located in Jamaica, though ironically the war ended before the service was inaugurated. The facility was then moved to Havana, enabling collaboration with Viñes and his colleagues, before finally being relocated to San Juan, Puerto Rico. The war also led to an expansion of the Caribbean weather-observing network as a defensive measure. Ironically, the legacy of this expansion was a first small step which eventually led the World Meteorological Organization to designate the United States as the area's Regional Specialized Meteorological Centre for tropical meteorology, supported by an international network of surface and upper-air observing sites.

Viñes' reputation and skill in predicting hurricane movement was known and admired by intellectuals elsewhere in the United States, although his methods were slow in gaining recognition. Isaac Monroe Cline, head of the Galveston Weather Bureau Office during the disaster of 1900, had studied and become impressed with the success of Viñes' methods while at Galveston; although at that stage of his career, Cline lacked authority to apply them or make forecast decisions. When Cline was transferred to New Orleans in 1901 in charge of the forecast office there, it was no surprise that he promptly introduced procedures similar to those of Viñes for hurricane predictions. Some of these procedures (including analyses of changes in upper-level cloud motions, ocean swell direction and frequency, and 24-h pressure changes from all observing stations in the potential path of the hurricane) were still in use at New Orleans as late as 1942. In the wartime absence of weather reports from ships at sea, New Orleans forecasters made a remarkably accurate 48-h prediction of hurricane landfall, relying substantially on the same variations in trends of cloud movements, surface pressures, and ocean swell used by Viñes, observed from one coastal station to another around the Gulf of Mexico and plotted as an x-t diagram[1].

[1] A diagram in which x (space between stations) on the abscissa is plotted against t (time) on the ordinate.

From this perspective of science and technology at the turn of the century, we shall examine in this and later chapters the struggle to improve our abilities to cope with hurricanes. But first, to set the stage for understanding what happened, we will recount some of the handicaps encountered along the way.

OBSTACLES ENCOUNTERED

Little progress was made during the first four decades of the 20th Century in helping coastal residents cope with the hurricane menace in the United States. Contributions by the Bergen School, applicable at the time primarily in temperate latitudes, and the concepts of tropical meteorology and procedures for predicting hurricane landfall, so effectively employed by Viñes, were slow to take root in the United States.

Why were U.S. weather services so far behind? The answers probably reside more in the internal management of the Weather Bureau than the limitations of science and technology. For this highly visible government agency, political pressures compromised the adoption and orderly implementation of plans to help society cope with threats of severe weather.

Politically Driven Obstacles

Foremost in compromising the orderly growth and progress of the Weather Bureau (as well as other Federal agencies with science or technology-based missions) was what has been described as Congress' proclivity for *administration by crisis*; that is, the delay in funding public programs essential to the long-term welfare of the nation until a major crisis generates enough public outcry to mandate it [Simpson, 1998]. Various justifications for delay can be given. For example, at any individual locality hurricanes are rare events. Even with the motivation of a Galveston-type disaster, Congress tended to view the problem as a local one, requiring only a "band-aid" solution. In the Galveston case, the government response was to support the local construction of a 17-foot seawall and to increase the base level of the city area by five feet — an essential, effective, but costly measure to protect Galveston. Those enhancements, however, offered no protection against such disasters elsewhere.

Not until 1954, when three successive hurricanes (Carol, Edna, and Hazel) ravaged the coastal constituencies of 24 senators and a multitude of congressmen from New England and the Mid-Atlantic States, was a massive infusion of government funds supplied to expand observing and hurricane detection facilities, as well as research programs to help understand and cope with hurricanes.

Internal Procedural Problems

Politically, the broad mission of the Weather Bureau, from its early years as a civilian service, centered on the well being and protection of the nation's economic life blood: agriculture and transport of manufactured products. Most of the Weather Bureau's fiscal resources were required for maintaining a network of observing stations used for those objectives. Not surprisingly, it suited both convenience and political advantage to concentrate the professional tasks of forecasting and warning, and the small amount of research that could be supported, at the seat of government in Washington. This turned out to be a mistake that was perpetuated for decades: the remotely located forecaster often failed to account for local peculiarities of the coastal environment that modulates the hurricane risk, and this neglect prevented him from anticipating the responses to his advice and warnings by the coastal population-at-risk with whom he had little or no contact.

In 1900, the Bureau had an annual budget of about $1 million to cover operations at about one hundred offices. Interestingly, the National Weather Service (formerly Weather Bureau) a century later maintains a comparable number of offices. Its budget is now 700 times larger, and it employs about nine times the 546 employees of 1900.

Internal Management

The whims and idiosyncrasies of early Weather Bureau administrations sometimes severely hampered scientific progress and the credibility and effectiveness of weather prediction and warning. For example, Bureau Chief Willis Moore pointedly instructed the Galveston office that only headquarters could issue storm warnings. Eric Larson [1999] describes Moore's obsession with control and public image, as one that guaranteed no one in the Galveston office "would even whisper the word *hurricane* without a formal authorization from Moore himself."

Staff members during these early years often worked under difficult conditions, leading to disgruntled pleas to a mostly unsympathetic Bureau leadership. Moore testified with pride to Congress that Weather Bureau employees had to be committed to insane asylums more often than employees of any other federal agency.

Moore soon got his due; he was fired in 1913 for using Bureau officials and government time to campaign for the position of Secretary of Agriculture. This was not the first trouble for the country's weather leadership. A scandal in the 1880s had led to the arrest of Henry Howgate, the Bureau's chief financial manager, for embezzlement. In retaliation, Congress reduced the Bureau's annual appropriation by the embezzled amount, estimated to be one-sixth of the annual budget. All this would have made little more than interesting anecdotes if not for their cumulative effect — a tightened budget responsible for a near total curtailment of research for years.

Then, in the 1920s, "hurricane season" and the Bureau's customary warning services were reduced by six weeks, a victim of the Coolidge economy.

Employee lassitude also took its toll on the Bureau's reputation. Gordon Dunn, Director of the National Hurricane Center from 1955-66, [Appendix A] recalls an incident from 1934:

> *...on a Sunday forenoon the Washington (based) forecaster issued a hurricane warning for the Upper Texas coast...by mid-afternoon the anxious (Galveston) Chamber of Commerce wired the Washington Weather Bureau for the latest information. The map plotter on duty honestly but indiscreetly wired back: 'Forecaster on golf course — unable to contact'. In Galveston, the weather remained quiet but temperatures in the Chamber of Commerce rose rapidly.*

The widespread criticism that followed reached Congress and the President. A review of hurricane service by a Presidential Science Advisory Board led to a major reorganization of the Bureau. Hurricane forecasting was decentralized, with responsibility transferred from Washington to regional operations in Jacksonville, New Orleans, San Juan, and reconstituted in Washington. A 24-h teletype circuit was proposed for hurricane-prone states (but was initially terminated in Jacksonville due to insufficient funding).

The changes provided for ship and land observations at six-hour intervals. Advisories were also to be issued every six hours. The upper-air network got a boost with an increase in observations. Still, meteorological research activities got very little attention within the Weather Bureau until 1936, when it established its meteorological research division.

While the Washington Bureau staff included a number of scholarly meteorologists who published credible papers and books, the Bureau, in its service operations, failed to take advantage of modern scientific advances and concepts in meteorology, even though doing so would not have imposed a significant fiscal burden. For example, even in forecasts for temperate latitudes, it was more than two decades after the Bergen School had published its treatise on air mass and frontal analysis before these concepts were adopted by the Weather Bureau. Finally, the remoteness of the Washington forecast office from most hurricane-prone coastal areas made it difficult for forecasters to establish a bond of credibility with the public for hurricane forecasts and warnings.

ROLE OF NEW TECHNOLOGY

At the turn of the century, research had already made it clear that the largest gap in understanding, as well as prediction of storm development and movement, was the lack of wind and thermal data from the middle and upper layers of the troposphere. Viñes sought understanding from eyeball observations of cloud movements, and the ever-changing character of ocean swell reaching the shores of an array of islands. Vilhelm Bjerknes devoted considerable time devising procedures for measuring upper-air winds with kites, as was later done in the United

States. Aside from Viñes' procedures, which required little instrumentation, attempts to install enough observing stations to describe the character of upper-air circulations met with limited success in the face of insurmountable operational difficulties and inadequate resources prior to WW I.

Upper Air Soundings

From the fertile and innovative mind of Charles F. Marvin, destined to become Chief of the Weather Bureau, came the development, before and during World War I, of a wide array of instruments for measuring atmospheric properties, including the first aerosonde. Mounted on the wing of an aircraft, the aerosonde recorded changes in atmospheric structure from the surface to 15 or 20 thousand feet. C.C. Bates and J. F. Fuller [1986] give an interesting account of the first such flight in 1918 by Navy Lieutenant Reed using an R6 aircraft.

In rapid succession, additional new technologies appeared, motivated by military requirements of World Wars I and II. These included the pibal, apob (aerosonde measurements of atmospheric properties during the spiral ascent of an airplane), radiosonde, rawinsonde, and radar. Created to serve the urgent needs of wartime operations, these marvelous devices rapidly became available to the meteorologist, not only in support of weather observing and forecasts, but to explore puzzling atmospheric questions previously beyond reach. Moreover, the procurement of the new devices for civilian use was often funded from military budgets.

Among the more important of these devices deployed during World War I was the pilot balloon, or pibal, an inexpensive means for measuring upper-level winds using a theodolite to visually map trajectories of balloons during their ascent. The apob (airplane observation) flights were eventually contracted to a most-hearty breed —pilots who were not paid unless they reached 13,500 feet — for flights begun in the dark and conducted often in sub-freezing and oxygen-scarce surroundings in an open cockpit. For this effort, they were paid around $25. For each additional 1000 feet of altitude attained, they earned a 10 percent bonus, a somewhat limiting performance incentive for those willing to attempt, and able to survive, the hazardous duty. At those altitudes, pilots could be overcome by lack of oxygen. Some attempted to guard against this danger by taking along small pets whose heads would droop from oxygen exhaustion before the pilot was adversely affected. Moreover, no consideration was given for the weather as a hazard, and if there was no flight, there was no pay. In 1938, after about two dozen major crashes and at least ten deaths, this form of data acquisition was discontinued.

Soon thereafter, as an extension of WW I technology, came the balloon-borne radiosonde. It transmitted electronically successive values of temperature, humidity and pressure during its ascent. Winds aloft were computed as before using a theodolite to track the balloon trajectory. In a further electronic refinement, during WW II, winds from the ascending radiosonde were computed by radio direction-finding devices, making it possible to measure wind and thermal structure well into the stratosphere, day or night, irrespective of cloudiness or adverse weather. It became known as a "rawinsonde."

Role of the Rawinsonde in Hurricane Prediction. In 1944, Grady Norton, Head of the Miami Hurricane Forecast Office, provided a remarkable demonstration of the value of the rawinsonde network when his analysis of high-level winds helped predict accurately the movement of a hurricane almost directly toward the center of a surface anticyclone. Figure 2 is an advisory message issued from Miami on October 10th, based upon upper air analysis of winds over the area of the hurricane in the Caribbean Sea northward following Norton's steering concepts.

Miami, Fla., October 16, 1944 2230 BWT

No. 15.

Miami Advisory 2230 BWT. The Caribbean Hurricane continuing its listless drift is now central about 19 to 19.5 degrees North 83 degrees West. Heavy squalls and gale winds prevail over a very large area with winds up to 100 MPH near center.

The hesitancy of this severe storm to assume a definite course makes the threatened area uncertain. However, there is now some evidence that the storm may be turning slowly northwest or northward. Such a course, accompanied by gradual acceleration, will seriously affect western Cuba, the southeastern Gulf of Mexico and possibly the Florida Peninsula within 24-36 hours. Nevertheless this storm continues to offer a very serious threat to the Yucatan Peninsula and Honduras. Every precaution should be continued by all interests west of Longitude 80 within a six hundred mile radius of this storm. Stand by for further advices.

Storm warnings are displayed on the Florida Keys and small craft on Florida West and East Coasts should remain in port.

Radio broadcast authorized.

Weather Bureau - Simpson

Figure 2. Teletype copy of Weather Bureau advisory issued by Robert Simpson on October 17, 1944, forecasting hurricane to move toward Florida, into an area of surface high pressure.

In Norton's summary of this experience, he indicated:

> *The first really large, fully developed hurricane for which Rawin data were available...was in October 1944, when we surprised ourselves by being able to predict the (cyclone's) movement right through a surface high pressure area... In September 1945 another hurricane...gave such a good account of itself following the wind-Coriolis rule, that for the first time in my career I ordered full hurricane warnings on a definite and limited stretch of the Florida coast when the storm was nearly 400 miles away, and fully 24-hours ahead of its arrival.*

His summary of 1944 storm activity described further the impact of upper-level wind data for hurricane prediction:

> *The two major hurricanes of the 1944 season, the September hurricane, and the one covered by this report, had circulation depth well above the top of ordinary pilot balloon observations. It was the RAWIN data reaching to much greater height that told the story of future movements, and without them future movements could not have been indicated with as much certainty in the forecast.*
>
> *(This office) has made copies of our RAWIN charts —(which) will emphasize the urgent need for extending the RAWIN network, especially in the tropical areas, as an invaluable aid in hurricane forecasting. It is urgently recommended that the Weather Bureau lend all possible support to the establishment of additional RAWIN stations in the Caribbean and Gulf area to further implement the hurricane warning service.*

Wartime Training

World War II provided meteorology with another crucial resource for the decades ahead — *people*. Demands of the war, particularly in support of tactical and transport aircraft operations, required an enormous increase in trained meteorologists including, for the first time, many women. When the United States entered the war, the Air Corps had 400 weather officers and 2000 enlisted weathermen, while the navy had 90 aerologists and 600 aerographic "ratings." By the end of hostilities, those numbers had exceeded 25,000.

To meet these wartime needs, special training programs were quickly established at a number of major universities, where young recruits were taught air mass and frontal analysis and prediction methods. They were then sent overseas to generate air route and terminal forecasts and severe weather warnings. Many ended up in advance bases in the tropical Pacific where fronts were rarely a factor in prediction. Worse, their training had provided little guidance in detecting, tracking, and predicting tropical cyclones and the weather accompanying the intertropical convergence zone.

In the struggle to remedy this situation, the short-term solution for operational meteorologists was retraining, to which we will return shortly. The longer-term problem facing the research scientist was the need to understand the connecting links between weather processes dominant in temperate latitudes and those dominant in tropical and equatorial latitudes. These would have to be much better understood if significant progress was to be made in predicting the development and movement of hurricanes.

After the attack on Pearl Harbor spread hostilities across the tropical Pacific Ocean, the military response was to establish two centers for the study of tropical meteorology and the retraining of military weather officers. These officers, having been hurriedly schooled in the application of air mass and frontal analysis, encountered only frustration in understanding the behavior of typhoons and hurricanes. To remedy this, the Navy established the Institute for Tropical

Meteorology (ITM) in Puerto Rico in the early 1940s. The Air Force, somewhat later, established the Air Force School of Tropical Meteorology (TWS) at Howard Air Force Base in Panama. The ITM was headed by a talented New Zealand tropical meteorologist, Clarence Palmer[2], with staff and research support during protracted visits by Herbert Riehl, George Cressman, and Horace Byers from the University of Chicago, and Gordon Dunn of the U.S. Weather Bureau, all with concerns for, or emerging prominence in, tropical meteorology. The TWS seconded two weather bureau forecasters, Robert Simpson of the Hurricane Center in Miami, and Robert Fletcher, Official in Charge at Los Angeles.

Impact of RADAR

Without question, the two most important gifts to meteorology as fallout from wartime technology are radar and weather reconnaissance aircraft.

Radio detection and ranging, commonly known as *RADAR*, was developed in Great Britain in direct response to the boast of Adolph Hitler in 1933 that he had developed a "death ray" capable of widespread destruction of life and property. Britain turned to Robert Alexander Watson-Watt[3] of the National Physical Research Laboratory for advice. His conclusions denied the practicability of developing a death ray as claimed. However, he did propose the likelihood of devising means of identifying and tracking the movement of aircraft by radio waves. His proposal was funded, and the first operationally successful radar was developed before the start of WW II in 1938. Many credit radar with turning the tide in the Luftwaffe's aerial attack on Britain.

Some meteorologists may surmise that radar's effectiveness as a defensive device was little more startling than its multi-faceted benefits and applications to meteorology, a continuing boon not only to weather operations and warning, but to weather research as well. Figure 3 obtained from a U.S. Navy ship, is one of the first radar images ever acquired of a hurricane and its characteristic spiral rain band structure. The progress of half a century is well reflected by comparison with satellite imagery of severe storms in Chapters 11 and 12.

After WW II, the military gave the Weather Bureau a large number of surplus APS-2F aircraft radars. At the instigation of Delbert Little, the Bureau's Assistant Chief for Operations, the Bureau's Instrument Division embarked on an experimental program to convert these obsolete systems for ground-based weather detection. The 19-inch diameter original antenna was replaced with a six-foot dish for higher resolution of weather imagery, and propagation wavelength was increased from approximately 3 cm to 10 cm (X-band to S-band) to reduce attenuation in heavy precipitation.

---

[2] Palmer went on to head a post-war Air Force research and training group in Hawaii concerned with typhoons and disturbances in the tropical Pacific.

[3] Knighted in 1942 for his services to Britain in developing radar.

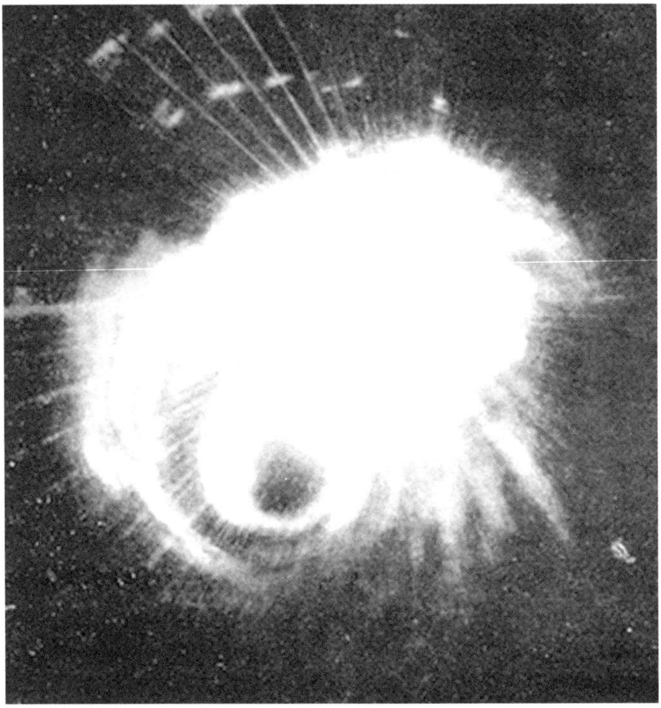

Figure 3. Second tropical cyclone photographed on radar screen, taken aboard a U.S. Navy ship, December 18, 1944. [after Maynard, 1945].

The prototype new system had undergone only a few weeks of testing when a remarkable sequence of events transpired, ultimately setting in motion Congressional approval of plans for a national radar network. Near midday, May 11, 1953, a tornado devastated the heart of downtown Waco, Texas, killing 114 and causing $14 million in damage. A few hours later the same day, San Angelo, Texas suffered a similar devastation leaving 11 dead, 130 injured, with $3.5 million in damages. Both events occurred without official warning. State authorities responded quickly to the outrage streaming in from the devastated communities—outrage that the Air Force, having established an experimental tornado-warning program at Tinker Air Force base in Oklahoma to warn air bases of the potential for tornadoes, refused to disseminate these warnings to the public.

This was not an arbitrary refusal, but one negotiated in a formal interagency agreement requested by the Weather Bureau. The Bureau's position—a correct one as it turned out —was that the processes that led to tornado formation could not be sufficiently described, within the existing network of observations and scientific understanding, to determine the location and timing of occurrence. Once a tornado had formed, its signature could be detected and tracked by radar within viewing range. Predictions of occurrence, however, on a national scale of need would have to warn too large an area and too many communities to escape local charges of "crying wolf" and the consequent lowering of warning credibility.

A meeting was called in Austin of the responsible State and local officials to explore the options for averting or minimizing such tragedies in the future. A representative of the Weather Bureau and Military were also invited to attend.[4] One option was the establishment of a separate State Weather Warning program. After reviewing the circumstances of the two events and the options available, the State Officials accepted the offer presented by the Weather Bureau: to install enough copies of the new prototype weather radar in Texas communities for detection of tornadoes, to man these systems, and establish communications for relaying warnings of tornadoes to communities in their path as soon as sighted by radar, providing the State or individual communities paid the cost (approximately $10,000 each) of modifying and installing these radars. The result was the establishment of a Texas radar network that soon proved its operational worthiness. As a pilot program, this network demonstrated the importance and justification of establishing a national network of radars. Congress appropriated the necessary funds for this purpose in 1955. This, in turn, led to the design and production of sophisticated new state-of-the-art weather and storm detection radar, the S-band WR-57. The first system was installed in Washington, D.C., the second at the hurricane forecast office in Miami in 1957, forerunners of an elite national network of radars.

Aircraft Reconnaissance of Hurricanes

One of the primary landmarks in the struggle to understand the behavior and improve the prediction of hurricanes occurred in July 1943 when Col. Joseph B. Duckworth became the first to deliberately fly an aircraft into the eye of a hurricane. Duckworth, accompanied solely by his navigator Lt. Ralph O'Hair, took off from a military air base at Bryan, Texas in a single engine Air Force AT-6. An unauthorized mission Duckworth later characterized as "a lark to satisfy my curiosity," the flight was made at an altitude of 7000 feet, during which they entered and explored the eye of a small hurricane nearing Galveston. It may have been "a lark" to Duckworth, but upon review it earned him an Air Medal. More importantly, it demonstrated the feasibility of maneuvering an aircraft into the core of a hurricane. After three additional "curiosity" but confirmatory flights were made into hurricanes in 1943 by military personnel[5], a formal program of daily reconnaissance of Atlantic hurricanes was inaugurated the following hurricane season by both the Air Force and Navy.

---

[4] Captain Howard "Shorty" Orville, Director of Navy Weather Service, represented the Military; Robert Simpson represented the Weather Bureau.

[5] A second flight by Duckworth on July 27, one from Antigua on August 3, and one in September from Tinker AFB, Oklahoma are described by Tannehill [1955] in his popular book, *The Hurricane Hunters*, p. 118 [See Appendix A.]

The success of these early operational flights presented a scientific challenge to seize any opportunity available to pursue such research objectives as exploring the structure and energy processes of hurricanes. The Military, with many concerns about hurricanes of their own, was willing to cooperate, and the first research mission was flown in 1944. A second occurred in 1945.

Fifteen months after the Duckworth flights, (then Major) Harry Wexler, a brilliant young scientist destined to become Weather Bureau Director of Research, together with two Air Corps meteorologists, flew in an A30 aircraft into a major hurricane off Cape Henry, Virginia. On September 14, 1944, the A30 departed Bolling Field in Washington D.C. on the first exploratory research mission into a hurricane. The plane entered the western sectors of the recurving hurricane while the storm center moved swiftly northward, paralleling the coast less than 150 km offshore. Flying at an elevation of 3000 feet, the pilot encountered what was described as unique patterns of turbulence and heavy rain in transiting what was probably the eyewall, although it appears uncertain whether the plane entered the eye before turning back. During the transit, however, the pilot reported an "average wind of nearly 120 mph" at flight level. From a reconstruction and analysis of cloud and rain structure, Wexler [1946] reached some remarkable hypotheses concerning the role of penetrative cumulus clouds as sources of energy for the hurricane, foreshadowing the in-depth research on this subject generated from data later obtained by the National Hurricane Research Project (NHRP).

On August 4, 1945, a C-47 (military transport version of the civilian DC-3), used in forecaster training at the TWS in Panama, reconnoitered a small tropical cyclone moving west-northwestward across the Caribbean Sea. Well equipped with navigation and (relatively crude) meteorological sensors of the era, it also carried radar repeater scopes at each of 11 training stations. As the C-47 approached the storm from the south, it encountered the storm's cloud system several hundred kilometers south of Haiti. Flying at 6000 feet, it approached the hurricane core along a spiral flight path to the east, north, and west of center and entered what turned out to be a well-formed eye located approximately 80 miles south-southwest of Port au Prince. After circling and describing the eye, approximately 15 miles in diameter, the aircraft departed through the same corridor it entered and landed without incident in Santo Domingo (Ciudad Trujillo in that era). The research flight, planned and directed by Robert Simpson, (then) of the TWS staff, was noteworthy for at least one reason. In transiting the spiral path from the cloud-free environment to the eye center at 7000 feet, a rise in temperature of 5ºC was observed. The eye was free of turbulence but not of clouds (patches of fractocumulus were observed at numerous levels above and below flight level). This small tropical cyclone had likely reached hurricane force briefly, judging from the temperature increase at 7000 feet and the frequent glimpses of the sea state viewed through the drift meter as the plane approached the eyewall. The storm apparently lost strength rapidly during the following 24 hours. On August 5th there was no evidence of a low-pressure center from weather reports available.[6]

While knowledge gained from these early explorations was less than earth shaking, the experience, opportunities, challenges, and new frontiers of research opened up by exploiting the technology for hurricane reconnaissance drew the attention of many scientists previously unimpressed with research possibilities in tropical meteorology. As mentioned earlier, however, the resources to support these opportunities would not soon become available without motivations beyond the promises of rewarding results from scientific research.

"PIGGY-BACKING" RESEARCH MISSIONS ON HURRICANE RECONNAISSANCE FLIGHTS

Grady Norton, whose early leadership role in the hurricane warning service is referenced above, came to Miami in 1943 as head of the newly established Miami Hurricane Forecast Office. A forecaster whose skill and credibility in communicating hurricane forecasts and warnings to the public has rarely been equaled, Norton was widely regarded as a "wise old head" in dealing with hurricanes long before he arrived in Miami. He not only learned from his mistakes, but he usually reasoned his way to hypotheses that, if correct, assured the mistake would not be repeated.

In preparing this portion of the chapter (as second author), I (Simpson) recall well my early realization of how fortunate I was to have been transferred to Miami to work with Norton. Arriving there in May, 1944, after a year as forecaster in New Orleans and one as Weather Bureau scholarship student at the University of Chicago, I was immediately impressed, though at times puzzled, with Norton's insights and hypotheses regarding hurricane movement and development.

One of Norton's principal concerns was predicting hurricane landfall positions and in identifying the steering current that would carry the storm there. Like Viñes half a century earlier, Norton believed the hurricane was steered by winds in the upper troposphere, more specifically, by winds at the top of the hurricane "where its vortex disappeared." Unlike Viñes, however, Norton, as indicated earlier, had available a network of radiosonde and rawinsonde observations with twice daily reports of wind and pressure distributions usually extending into the stratosphere. While these reports were, for the most part, widely spaced, the analysis of circulations at successive levels, bounded by the assumption that a hurricane vortex disappeared at some level in the upper troposphere, provided

[6] Weather reports from the TWS' C-47 were neither transmitted, nor made available during the war. In the absence of evidence otherwise, official published analyses concluded the storm was only of depression strength and had broken up on the mountains of Haiti's southern peninsula. Some reports did indicate heavy damage and loss of life from rains and mud slides in that area.

the rationale for locating the steering current. Norton's remarkable record with hurricane forecasts was based on this approach to steering. He raised poignant questions of hurricane structure, upper-level circulations in the hurricane, and interactions with the environment. With curiosity fully aroused, we devised plans to seek answers to the interesting questions brought into focus by Norton's concepts. We hoped to obtain use of one of the Air Force's new WB-29 weather reconnaissance planes to explore circulation structure in upper levels of the hurricane. Early in 1947, Francis W. Reichelderfer, then Chief of the Weather Bureau, requested the Army Air Corps to support this effort with one or more research flights dedicated exclusively to the description of circulations in upper layers of the hurricane.

The positive response to this request set in motion plans for a dedicated research flight during 1947. Also, a series of missions combining operational and research objectives was to be flown from Kindley Air Force Base in Bermuda by the Air Corps' 63rd Weather Reconnaissance Squadron. After operational information had been obtained, the remaining flight hours were to be devoted to research plans to investigate the three-dimensional structure of the hurricane.

First Over-the-top Research Flight

The research mission in 1947 proved to be not only a remarkable and exciting adventure, but exceeded scientific expectations. Serendipity played an unusually rewarding role. Figure 4 shows an analysis of data from the first of two missions flown into the Great Atlantic Hurricane of September 15th. At that time the hurricane was located about 300 miles south-southwest of Bermuda. It had paused on its path of recurvature before sweeping westward just north of the Bahamas and across Florida to a final landfall near New Orleans. The WB-29, with special modifications to enable flight at altitudes above its normal service ceiling, left Kindley AFB in early morning and proceeded toward the hurricane, climbing slowly through the warm tropical air of the hurricane environment to a pressure altitude of 36,500 feet (more than 38,000 feet above the water surface). As the aircraft flew through the tops of clouds in the eyewall, it completed three circumnavigations of the eye and two diametric transits before departing for Kindley. Figure 4 from a paper in *Scientific American* [Simpson, 1954] shows an analysis of "D" values—differences between pressure altimeter elevation and actual (radar) height above the water surface, representing the pressure gradient. While no direct measure of winds at flight level could be made, a good approximation of tail wind speeds during the three encirclements of the eye was derived from successive measurements of plane position by Loran. Within probable observational error, a vigorous vortex clearly existed at the top of this hurricane's inner core cloud system, surrounded by several small anticyclonic pressure cells apparently associated with hurricane outflow circulations. Of greater and unexpected interest was the finding that tail wind speeds at cloud tops averaged

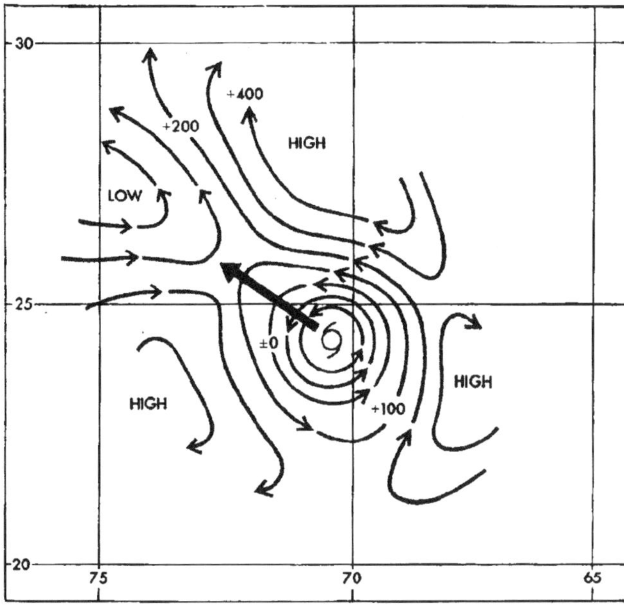

Figure 4. Mass circulation at eyewall cloud tops in Hurricane George, September 15, 1947. Isopleths are "D"-values (true altitude minus pressure altitude in the mean tropical atmosphere).

approximately 120 knots—just a little less than the maximum expected sustained winds at the surface below. Considering the reduction in horizontal pressure gradient with height mandated by the warm core, a much greater reduction in maximum wind speeds with height would have been anticipated from hydrodynamic considerations. This result, startling at the time, is probably the first clear evidence of massive vertical transport of momentum by convective clouds, a process that Joanne Malkus [1958] defined as a dynamic necessity in her analysis of energy processes sustaining the eye of the hurricane.

Research Mission in Great Typhoon Marge

On August 15, 1951, under a cooperative arrangement between the Weather Bureau and Air Weather Service's 2143rd Air Weather Wing in Tokyo, a landmark research mission was flown into Typhoon Marge when its central pressure was 895 mb—at that time the lowest central pressure measured from aircraft. During a 15-h flight out of Guam, the RB-29 spent more than 11 h in Marge, completing a three-dimensional investigation of the nearly cloud-free 40-mile diameter eye, and its surrounding storm core, flying at elevations extending from 1500 to

20,000 feet [Simpson, 1952]. Among other remarkable findings, the temperature in the eye was virtually isothermal to midtroposphere, where it was nearly 20°C warmer than Marge's environment 260 miles southeast of the hurricane eye and 22°C warmer than at the same altitude in the mean West Indies sounding for July to October (see Figure 5).

Final Piggyback Research Missions

From 1952-1954, combined operational and research missions, or "piggy-back" research flights, continued from Bermuda. Each provided building blocks

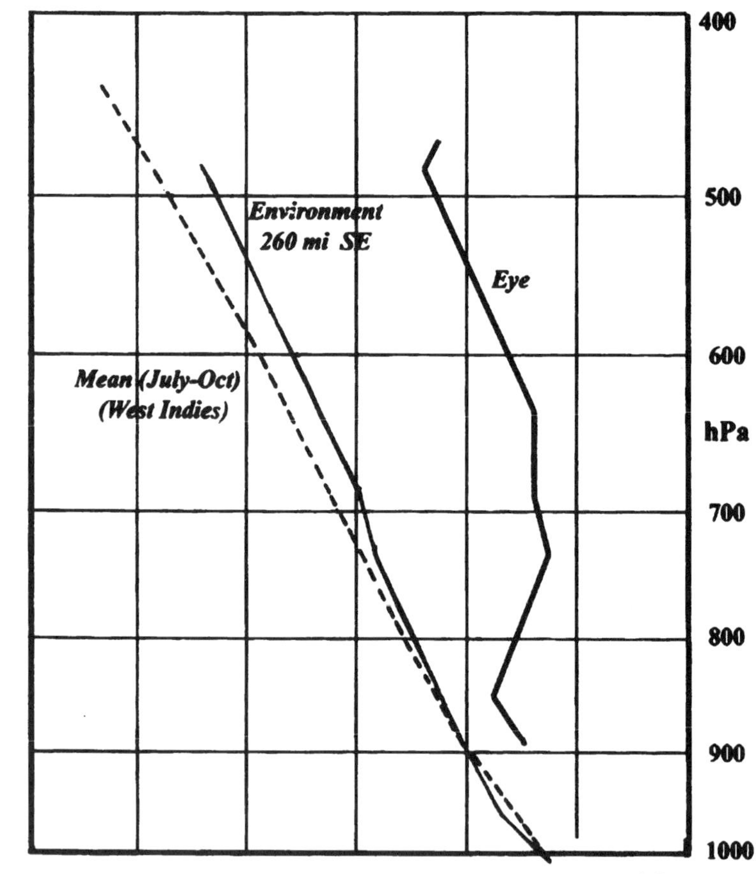

Figure 5. Comparison of the Marge eye sounding with that of the environment 425 km southeast of center and the West Indies Tropical Mean Sounding (July to October).

of new knowledge about hurricanes, contributing to the ever-expanding plans, proposals, and oft-frustrated hopes of the Weather Bureau to attract resources for a hurricane research program to improve the accuracy of forecasts and the ability to cope with hurricanes.

An example of interesting, if controversial, findings from a piggyback research flight appears in Figure 6. From the eye center of Hurricane Edna, south of Bermuda on September 10, 1954, this is a view looking upward from a piggyback flight at 7,000 feet, showing the transport of cirrus clouds in a cyclonic spiral from the top of the eyewall to the eye center [Simpson and Starrett, 1954]. This process is of dynamical interest and concern in understanding energy processes influencing the central pressure and strength of the hurricane.

The National Hurricane Research Project (NHRP)

Finally, as mentioned earlier, the year 1954 produced three significantly damaging hurricanes whose impact created a sufficient crisis to spur Congress to appropriate funds to launch the NHRP. Established at West Palm Beach in April 1956, the NHRP was generously supported by the Air Force with the loan of three high performance aircraft, two WB-50s and a WB-47. It also provided flight staff and maintenance crews dedicated to on-site support of research flights into every hurricane that came within flight range of the Palm Beach facility (Morrison Air Force Base). This project will be discussed more fully in Chapter 3.

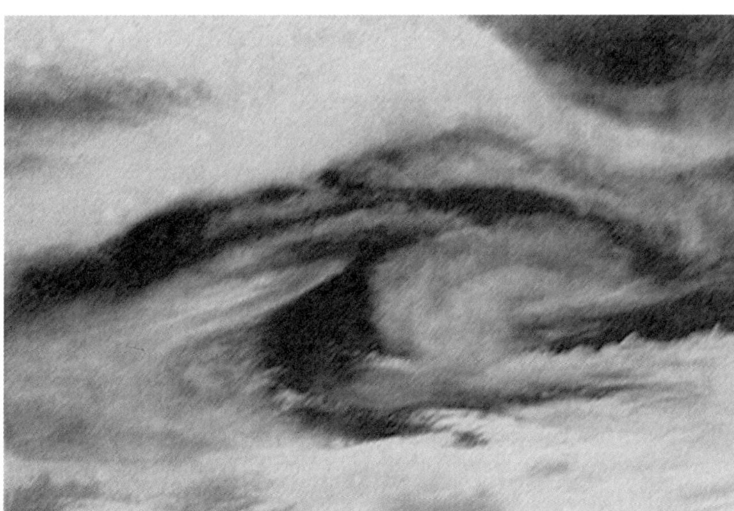

Figure 6. Cyclonic inflow of cirrus clouds from the top of the eyewall, as viewed (overhead) from an aircraft at 7000 ft in the eye of Hurricane Edna, September 10, 1954.

The three aircraft were instrumented with state-of-the-art meteorological sensors in accordance with specifications generated by the Weather Bureau in collaboration with university scientists and engineers. One of the more important components of this new age technology was the data recording system, an icon of the digital age in meteorological research, permitting direct recording and rapid processing of research data. Digital recorders on the NHRP aircraft permitted data from a flight of nominally 8 h, recording 12 channels of information as frequently as once per second, to be processed and ready for analysis within 24 h after the crew returned home. By contrast, in earlier years the data from analogue or more primitive recorders from several hours of rapid sampling could take months or even years to process before significant research could make headway. So far as we are aware, NHRP was the first to collect and directly record meteorological data digitally from aircraft.

At last, the framework and resources for expansion of scientific research in tropical meteorology were available independent of stimulation from appearance of new technology. And, gratifyingly, the roles began to reverse, as we shall see, with new research results providing motivation for development of new technology.

Grady Norton would find eminent gratification with this outcome, recalling his summary comment in an unpublished essay entitled "Soliloquy":

> *...Men with better equipment and better training should build on the foundation we have laid in "blood, sweat, and tears" until the ultimate of perfection has been reached. They will have to find a better method than we have (for analyzing and applying) the winds aloft, however, if better hurricane forecasting is done.*

REFERENCES

Bates, C. C., and J. F. Fuller, *America's Weather Warriors 1814-1985*, College Station, TX: Texas A & M University Press, 360 pp., 1986.

Larson, E., *Isaac's Storm*, New York: Random House, Inc., 323 pp., 1999.

Malkus, J., On the Structure and Maintenance of the Mature Hurricane Eye. *J. Meteor.*, 15, 337-349, 1958.

Maynard, R. H., Radar and Weather, *J. Meteor.*, 2 (4), 214-226, 1945.

McPeak, W. J., Hooked on Hydrodynamics. *The World and I: Scientists Past and Present*, American Geophysical Union, 168-176, 1995.

Rappaport, E. N., and J. J. Fernandez-Partagas, History of the Deadliest Atlantic Tropical Cyclones Since the Discovery of the New World. Chapter 5 of *Hurricanes: Climate Change and Socioeconomic Impacts*, Edited by H. F. Diaz and R. S. Pulwarty, 93-108, New York: Springer-Verlag, 1997.

Simpson, R. H., Exploring the eye of Typhoon Marge. *Bull. Amer. Met. Soc.*, 36, 286-298, 1952.

Simpson, R. H., Hurricanes. *Sci. Amer.*, 32-37, 1954.

Simpson, R. H., Stepping Stones in the Evolution of a National Hurricane Policy. *Wea. Forecasting*, 13, 617-620, 1998.

Simpson, R. H., and L. G. Starrett, Further Studies of Hurricane Structure by Aircraft Reconnaissance. *Bull. Amer. Met. Soc.*, 36, 459-468, 1955.

Tannehill, I. R., *Hurricanes*, Princeton, NJ: Princeton University Press, 308 pp., 1940.

Whitnah, D. R., *A History of the United States Weather Bureau*, Champaign-Urbana, IL: University of Illinois Press, 267 pp., 1961.

Wood, F. B., and H. Wexler. Flight into the September 1944 Hurricane off Cape Henry, Virginia, *Bull. Amer. Met. Soc.*, 5, 153-159, 1946.

Chapter 3

Hurricane Surveillance by Specially Instrumented Aircraft

Robert C. Sheets

INTRODUCTION

As discussed in Chapter 2, the first deliberate flight into a hurricane, piloted by Colonel Joseph B. Duckworth and accompanied by Lt. Ralph O'Hair, occurred on July 27, 1943. Apparently, no observations were recorded on that flight, but when Col. Duckworth repeated the flight, now accompanied by the base weather officer, Lt. William Jones-Burdick, they made written observations of static on the radio, turbulence, cloud conditions, and temperature. Those transcribed observations of temperature would later indicate the likelihood of errors caused by a "wet bulb" effect – which lowers the sensed temperature below that of the air by the evaporation of water on the temperature probe. These two historic flights paved the way for routine reconnaissance flights into hurricanes and piqued the scientific curiosity of people such as Harry Wexler and Robert Simpson. Wexler, then a Major in the Army Air Corps, "...participated in what may be called the first research reconnaissance flight into a hurricane" [LaSeur and Hawkins, 1963]. The data collected were "...largely visual observations and qualitative physical impressions..."

Some of Simpson's early "piggy-back" flights into typhoons and hurricanes and their important early findings have been previously described. Although the military flew regular reconnaissance missions to locate hurricanes and estimate their intensity, these aircraft had no recording systems or special instrumentation. Therefore, no organized effort had been made to use their limited potential to

Hurricane! Coping with Disaster

10.1029/055SP04

help improve forecasts and to consider ways humans might weaken hurricanes, although weather modification with other purposes was a hot topic in the period of rapid technological advances following WWII.

It became clear to Simpson and others that a much more dedicated effort was needed than the "piggy-back" operations then current — if significant advances in understanding hurricanes and in making substantial improvements in forecasts were to occur. This enthusiastic, self-confident and bold, middle-level meteorologist took his ideas to upper management in the Weather Bureau. He convinced them at least to put a modest proposal forward in their proposed budget to the Department of Commerce and finally to the Bureau of the Budget, where it was routinely turned down by the fiscally conservative Eisenhower administration.

THE NATIONAL HURRICANE RESEARCH PROJECT (NHRP)

As is frequently the case, it took major disasters to get the attention of the Congress and the Administration. Action was finally triggered by three major hurricanes during the 1954 hurricane season that caused large property losses in several very populous states from North Carolina northward through the mid-Atlantic and New England states.

The result was a special appropriation to start the National Hurricane Research Project (NHRP); arguably the single greatest step forward in improving our knowledge of hurricane structure and how these systems work to date! The funding was considerably more than the modest program originally proposed. Distinguished meteorologists from all over the world (Carl Rossby, Sverre Petterssen, Jak Bjerknes, Horace Byers, Herbert Riehl, Clarence Palmer, Tor Bergeron, C. Bleeker, Columbus Iselin, Joanne Malkus (Simpson), Victor Starr, J.S. Sawyer, Eric Palmén and T. C. Yeh) were invited to the Weather Bureau to help plan the NHRP. Robert Simpson, along with the top staff of NHRP (Robert Cecil Gentry, Art Johnson and Noel LaSeur), also visited most universities in the United States that had departments of meteorology, seeking additional comments and suggestions. Research proposals started pouring in from scientists from those universities, the results of which greatly increased the number of scientists and institutions now doing serious research on hurricanes and related topics in tropical meteorology.

The NHRP was a cooperative effort, under the sponsorship of the U.S. Weather Bureau, involving such government entities as the U.S. Air Force, Navy, and the Army Corps of Engineers as well as leading tropical meteorologists at various universities. The plan called for an "...intensive data collection program, beginning in May 1956, to continue for approximately 30 months. The research and development program is expected to require a minimum of four to five years" [Staff, NHRP, 1956]. The data collection and aircraft investigation programs were designed to accomplish the following objectives:

1. To investigate the mechanisms of hurricane formation.
2. To examine the details of hurricane structure and to determine the contribution to movement and changes of intensity, respectively, by the environment and the internal mechanisms of the storm.
3. To study the formation of clouds and rain in hurricanes and to investigate possible thermodynamic imbalances which may permit modification of the structure and movement of hurricanes.
4. To determine important parameters in hurricane forecasting and investigate how these can best be observed by the reconnaissance aircraft and other means.

The data sources included a special dense network of rawinsonde stations in the West Indies, with eight stations operated by the U.S. Air Force, three by the Navy, two by Great Britain, two by France, two by Cuba, and one each by the Netherlands, the Dominican Republic, Mexico, and Columbia. Thereby, the regional interest and cooperative nature of this program were emphasized.

The specially instrumented aircraft assigned to the project and operated by the Air Weather Service included one B-47 and two B-50s. During hurricane events, every attempt would be made to put all three aircraft into the hurricane simultaneously, with data collection levels of about 1,000 ft, 8,000 ft, 15,000 ft, 25,000 ft, 30,000 to 35,000 ft, and 40,000 ft. In addition, in these pre-satellite days, there was a plan by the Office of Naval Research to launch rockets that would attempt to photograph the entire hurricane cloud system and its immediate environment from altitudes above 80 km (50 miles).

Through yeoman efforts and the cooperation of several key individuals within the Department of Commerce and the Department of Defense, a pioneering project was functioning within twelve months time. This group, directed by R. H. Simpson, comprised 21 employees, three research aircraft with meteorological probes, and the first airborne digital systems for recording meteorological data. The operating base was Morrison Field at West Palm Beach, Florida [Staff, NHRP, 1956].

Robert Rados, Donald Hilleary, and Frank Christensen [1957] monitored the instrumentation of these aircraft by the General Precision Laboratory and conducted extensive calibration testing to assure reliable measurements of wind speed, direction, temperature, humidity, and pressure inside the extreme environment of a hurricane. This task was unprecedented in difficulty and success. Also, computing derived quantities and recording of the data was a formidable problem to be solved in these days of vacuum tubes and "punch" cards. For instance, the wind speed and direction computations depended upon the latitude and longitude values continuously computed by radar navigation, which measured ground speed and drift angle. This was done by evaluation of the Doppler effect on the microwave signal reflected from the ground (sea surface) coupled with the measured true airspeed and heading of the aircraft. The true airspeed computer used the static pressure and air temperature information. The density ratio and compressibility corrections to the indicated airspeeds were measured from the pitot-static pressure differences. All of these steps could introduce errors. Great efforts were taken to "calibrate" and "recalibrate" systems to get the most accurate data possible, no small task!

The initial research flight into a hurricane occurred in July 1956. Essentially, all meteorological sensor and data systems failed on the flight. Fortunately, most systems worked reasonably well on the second flight day. However, the recorded data were contained on more than 2,000 punch cards, which were scattered over the wet tarmac, when the technician removing them from the plane stumbled on the ladder. Noel LaSeur, Research Flight Director on this flight, pointed out that the cards were not numbered sequentially, and furthermore, noise problems entered random numbers in other elements, resulting in some of the cards never being properly sorted. However, there was a backup photo panel where the instrument dials were photographed at 5-sec intervals throughout the flight. Although quite laborious, missing punch-card, and later magnetic-tape recorded data, were checked and/or corrected using this information.

Rather extensive data sets were obtained during flights into Hurricanes Carrie and Frieda in 1957 and Hurricanes Cleo, Daisy, and Fifi in 1958. Much was learned from these flights, with the prime example being the first detailed (by standards of that time) vertical cross section of the inner core of Hurricane Cleo, a relatively weak but mature hurricane [LaSeur and Hawkins, 1963]. The minimum pressure ranged from 971 mb to 974 mb during the data collection period. Figure 1 shows the flight patterns flown at various levels in Cleo on August 18, 1958. Figure 2 shows a "vertical cross-section of cloud, precipitation, and strong radar echoes observed on (the) first traverse of hurricane Cleo" constructed from "...radar film, time lapse cloud film and observers' notes..." Figures 3 and 4 show vertical cross sections through the hurricane, revealing the extremely warm core in the middle to upper levels of the eye of the hurricane and the large pressure gradients associated with the strong winds in the eyewall. Over the next few years, several other studies were completed using the data sets collected during the 1957 and 1958 hurricane seasons. Although much was learned from the data collected in the three-year period of the project, it soon became obvious that the original objectives of the project could not be met with the data sets already collected. While there were common characteristics in the structure of the various hurricanes investigated, wide variations were found between the structure of one hurricane to another and, in fact, from a given hurricane from one day to the next. The hurricane was much more complex than originally thought.

On November 22, 1958, at what was later designated as the "First Technical Conference on Hurricanes," sponsored by the American Meteorological Society and held at Miami Beach, Florida, Robert Simpson summed up the project's status before the NHRP Technical Advisory Board: "...it is clear from the papers reported here that, while shedding light on many aspects of the storm structure

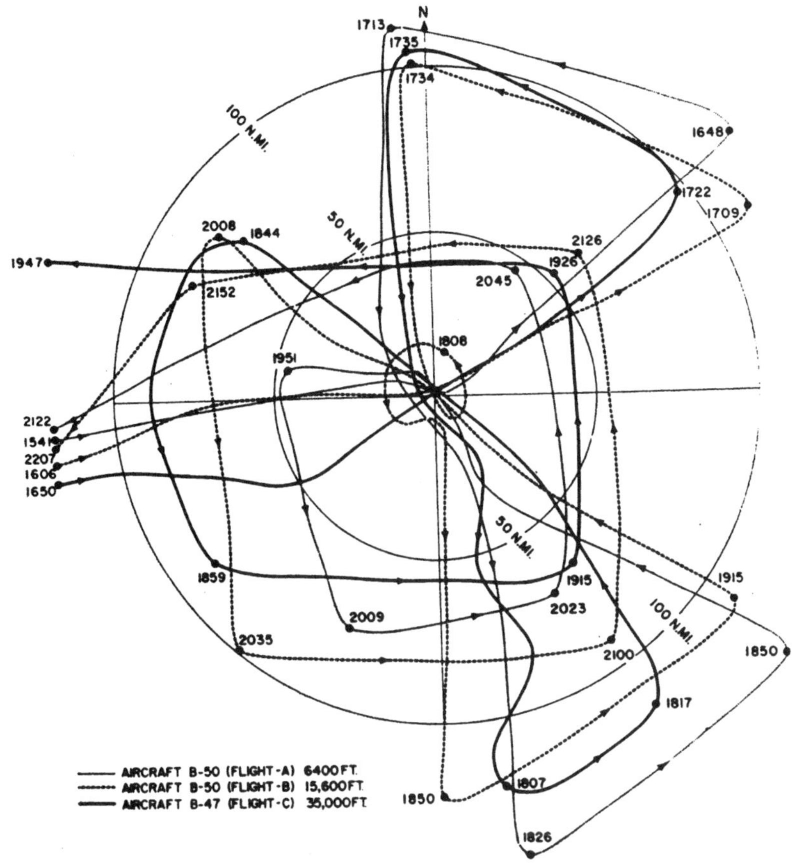

Figure 1. The tracks of each aircraft relative to the center of Hurricane Cleo on August 18, 1958 with time (UTC) indicated at various points along the tracks [after LaSeur and Hawkins, 1963].

and its energy problems, new and important questions have been raised which demand answers". He then stated that the Weather Bureau had developed plans "...for an additional three years of aircraft operation...which will not only continue to probe hurricanes...but...will extend the use of aircraft facilities to the investigation of tornadoes, equatorial jet streams, and extratropical cyclones". Thus began spin-off programs that continue today. Many other atmospheric and oceanographic research programs benefit from the efforts and resources generated by the hurricane research program. Simpson went on to report that the military services would not continue to provide the aircraft for the project. The two B-50s and one B-47 aircraft were to be replaced by two DC-6Bs and one B-57 operated by civilian contractors.

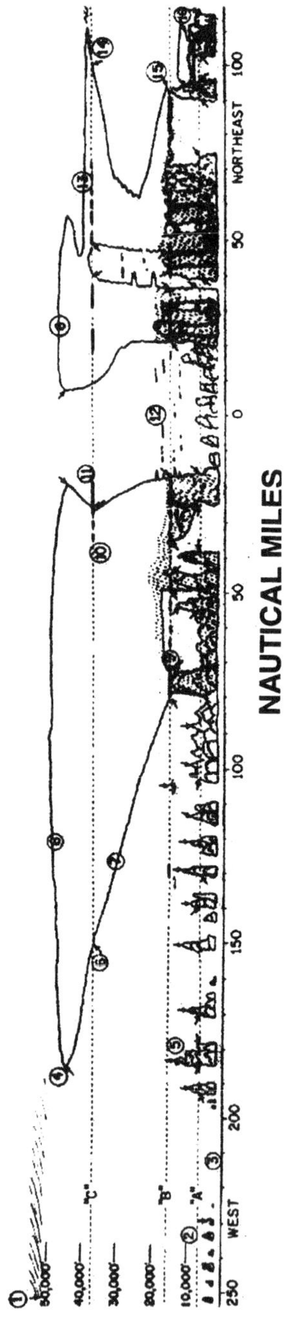

Figure 2. Vertical cross-section of cloud, precipitation and strong radar echoes observed on the first traverse of Hurricane Cleo [after LaSeur and Hawkins, 1963].

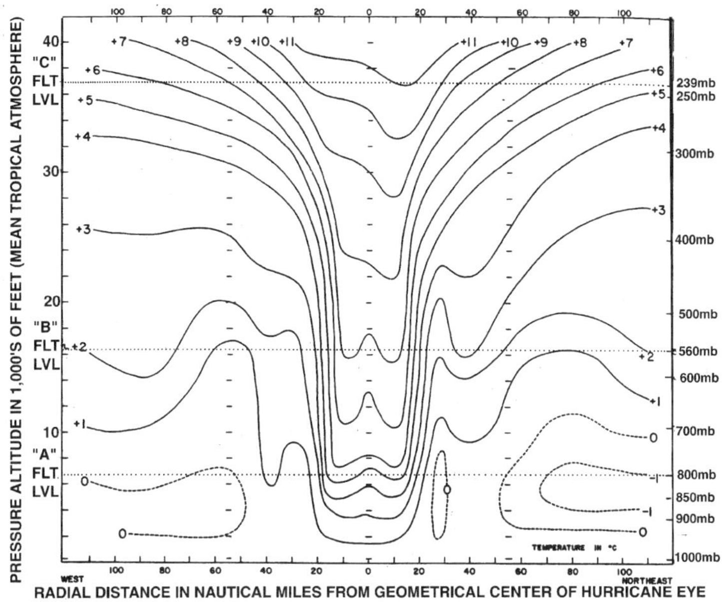

Figure 3. Vertical cross-section of temperature anomaly from the mean tropical atmosphere along the first traverse of Hurricane Cleo [after LaSeur and Hawkins, 1963].

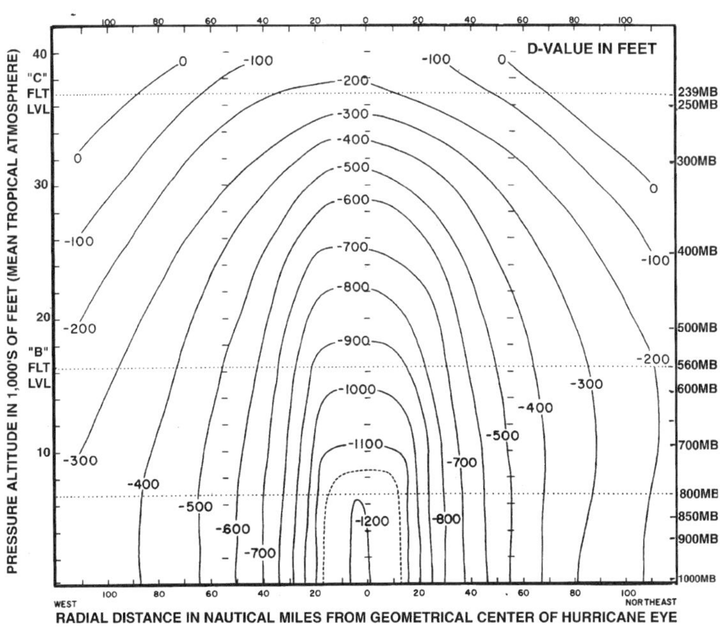

Figure 4. Vertical cross-section of "D" values (from mean tropical atmosphere) along first traverse of Hurricane Cleo [after LaSeur and Hawkins, 1963].

THE NATIONAL HURRICANE RESEARCH PROJECT EVOLVES INTO A NOAA LABORATORY

The next spring, operations were moved to Miami with aircraft based at Miami International Airport. Aircraft operations, which were administratively separated from the research staff, would later form the Research Flight Facility. The NHRP became the National Hurricane Research Laboratory (NHRL) and the research staff with its new Director, Robert Cecil Gentry, was co-located with the newly designated National Hurricane Center (NHC) led by Gordon E. Dunn. This now brought researchers and operational forecasters in daily contact with each other. That highly beneficial co-location continued through 1983. Unfortunately, at that time, due to budgetary reasons, the Hurricane Research Division (HRD – successor of NHRL) was physically separated from NHC and located with its parent organization (Atlantic Oceanographic and Meteorological Laboratory) on Virginia Key in Dade County, Florida. I am not alone in believing that the separation remains a major mistake.

In addition to the more standard hurricane reconnaissance flights conducted by the U.S. Navy and Air Force, specialized research flights into hurricanes continued, utilizing the two DC-6s and the WB-57 jet, for several more years. Among many publications on the structure of hurricanes, two are worth special note. These two were similar to the LaSeur and Hawkins paper on Hurricane Cleo of 1958. In one paper, Hawkins and Rubsam [1968] described the structure of a moderate strength 1964 Hurricane Hilda while it was in the Gulf of Mexico. The second study comprised in-depth analyses[1] of small, but very intense Hurricane Inez of 1966. Inez had a minimum pressure of 927 mb on September 28th. Hawkins and Imbembo [1976] used data collected at multiple levels in Inez, utilizing "butterfly" flight patterns (Figure 5). Horizontal analyses of winds relative to the moving center (Figures 6 and 7) showed the inward cyclonic spiraling air at low levels (950 mb), rapidly increasing in speed with wind speeds of approximately 80 m/s recorded in the eyewall. Vertical cross sections of wind speeds (Figure 8) and temperature anomalies (Figure 9) depicted a small, intense vortex.

In 1967, Banner I. Miller summarized the primary accomplishments of the NHRP/NHRL:

> *Through an intensive research program, either conducted by the Laboratory's staff or sponsored by the Laboratory under contracts with, or grants to, universities and other research organizations, NHRL has produced an almost complete description of the circulation and thermal structure of hurricanes, better understanding of the transformations of energy that take place in hurricanes, improved methods of forecasting, and additional basic knowledge about the meteorology of the lower latitudes.*

[1] These studies were limited by the types of instrumentation available at that time; i.e., no airborne Doppler radar or high resolution Global Positioning System (GPS) dropwindsondes, as we have at the beginning of the 21st century.

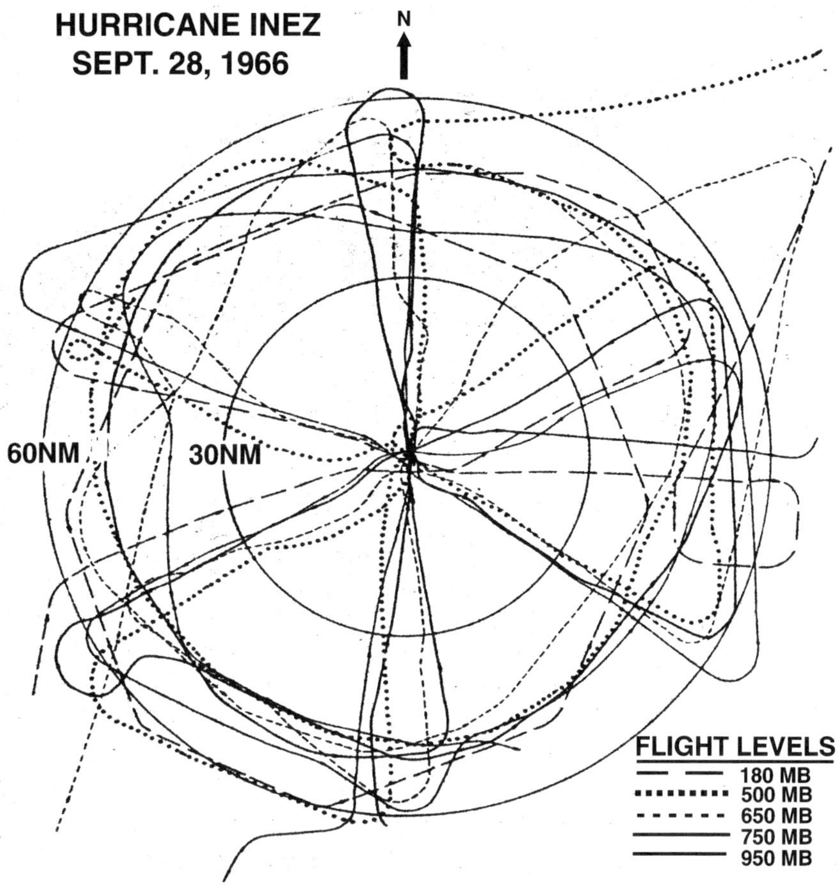

Figure 5. The tracks of each aircraft relative to the center of Hurricane Inez on September 28, 1966 [after Hawkins and Imbembo, 1976].

Note that Miller used the term "better understanding…" not "complete understanding." Hawkins recognized many shortcomings in the data collected to date and, thus, in the analyses performed. For instance, these analyses were generally based upon data collected at flight level and only along the flight path at a few discrete levels in the hurricane. The data then had to be "smoothed" both in time and space over a period of many minutes or perhaps a few hours to produce the analyses presented and the conclusions drawn. Hawkins and Imbembo [1976] recognized these and other limitations. They made the following recommendations at the end of their Inez paper:

"…Hurricanes of a reasonable range of intensities have been examined with the same instrumentation. What remains to be done?"

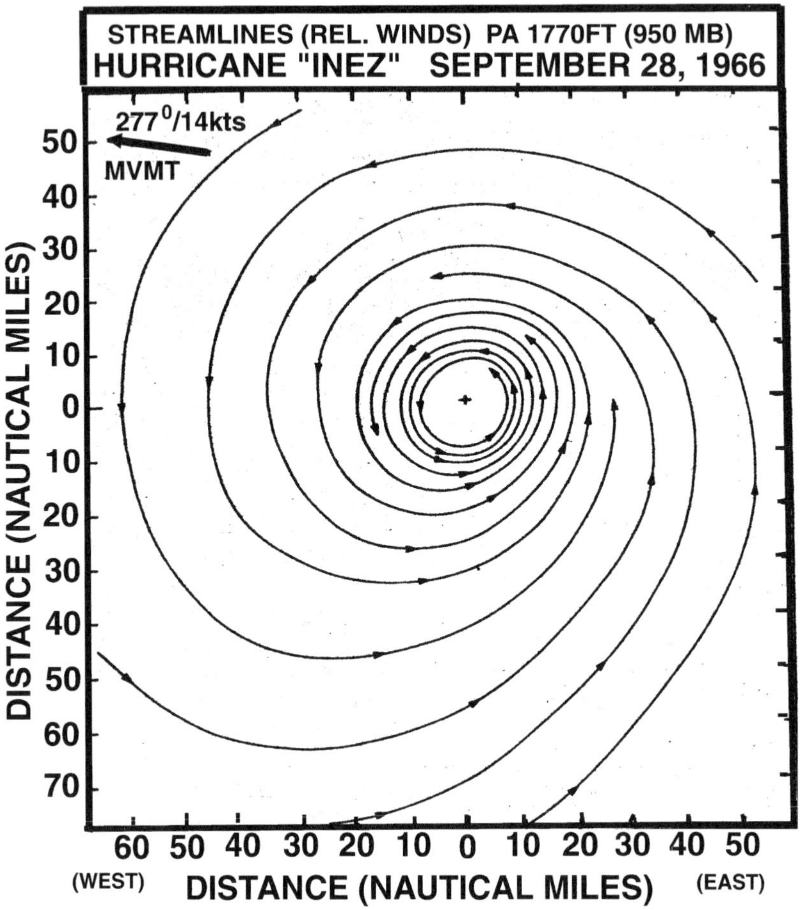

Figure 6. A streamline analysis of wind flow relative to the moving center of Hurricane Inez as recorded at 950 mb on September 28, 1966 [after Hawkins and Imbembo, 1976].

1. There should be a systematic study of the winds in hurricanes so that the Doppler winds are related to the more accurate winds derived from the inertial navigation system[2]. The Doppler radar computation system for navigation assumed a stationary reference surface. Over ocean, particularly in the high wind area (eyewall) of a very strong hurricane like Inez, the surface water, particularly sea spray could be moving very fast. This surface motion could introduce large errors into Doppler-derived positions and thus, the wind calculations.

[2] Here, Hawkins was referring to the Doppler navigation system, not the airborne and land-based Doppler radar systems of today that are used to measure the motion of precipitation particles and to infer from these, the wind speeds and direction.

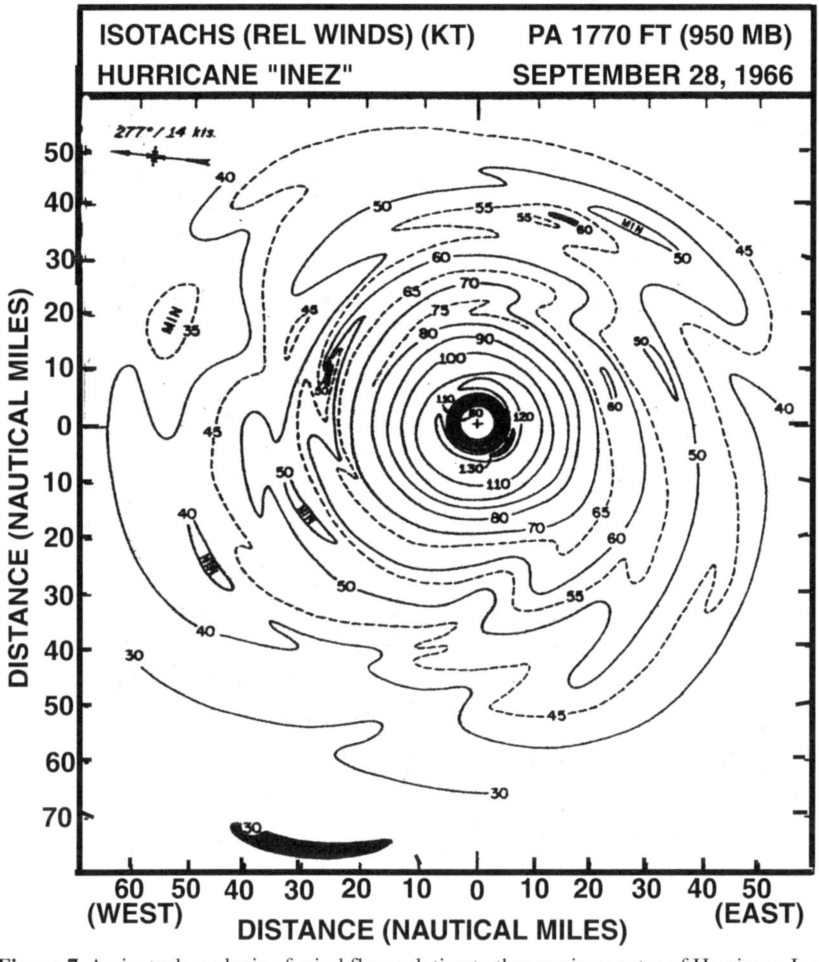

Figure 7. An isotach analysis of wind flow relative to the moving center of Hurricane Inez as recorded at 950 mb on September 28, 1966 [after Hawkins and Imbembo, 1976].

Shea and Gray [1973] suggested that Doppler derived wind speeds in high wind areas should be increased by 5% to 10%. These corrections would be applied to data where the wind speed and direction seemed reasonable. However, in some small portions of the high wind area of the storm, computed wind speeds and directions were highly spurious, and were generally subjectively discarded for that sector of the flight data. If systematic differences are discovered, then their implications need to be made clear.

2. When opportunity presents itself, a multi-plane mission should be executed with modern instrumentation (including cloud physics and turbulence gear) on modern aircraft with the ability to gather data in the mid-troposphere (around

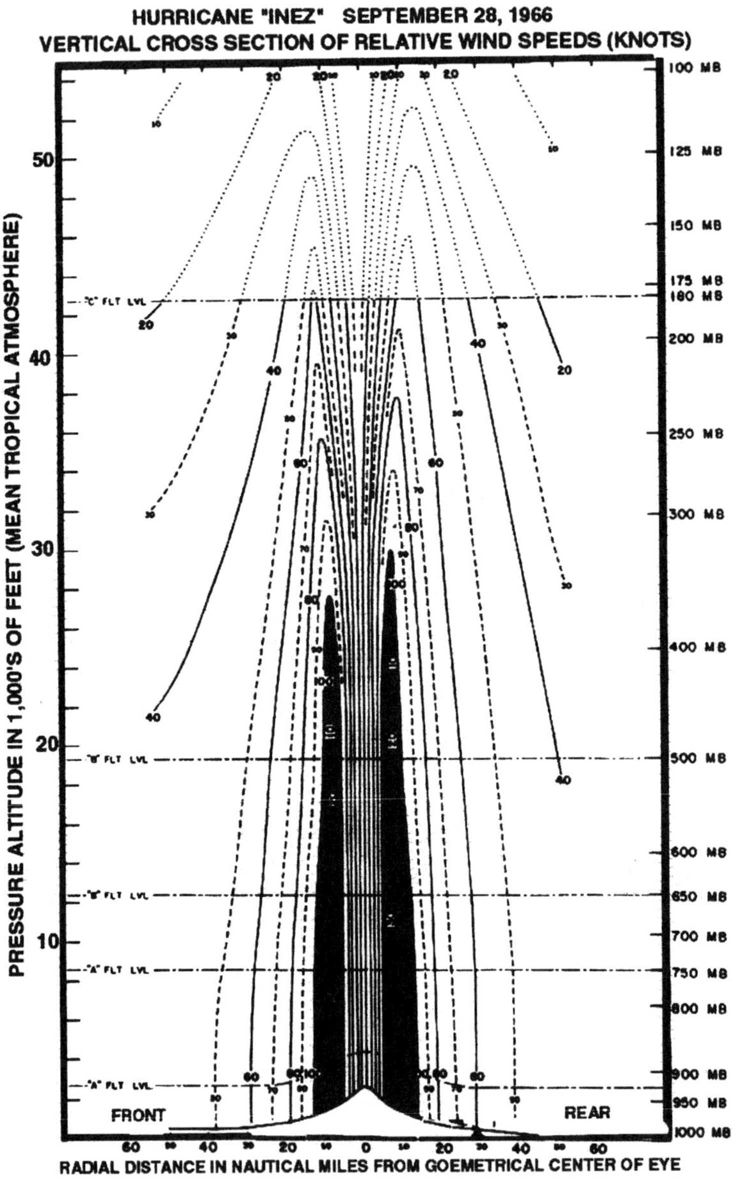

Figure 8. A vertical cross-section of wind speeds relative to the moving center of Hurricane Inez as analyzed from data collected at the indicated flight levels on September 28, 1966 [after Hawkins and Imbembo, 1976].

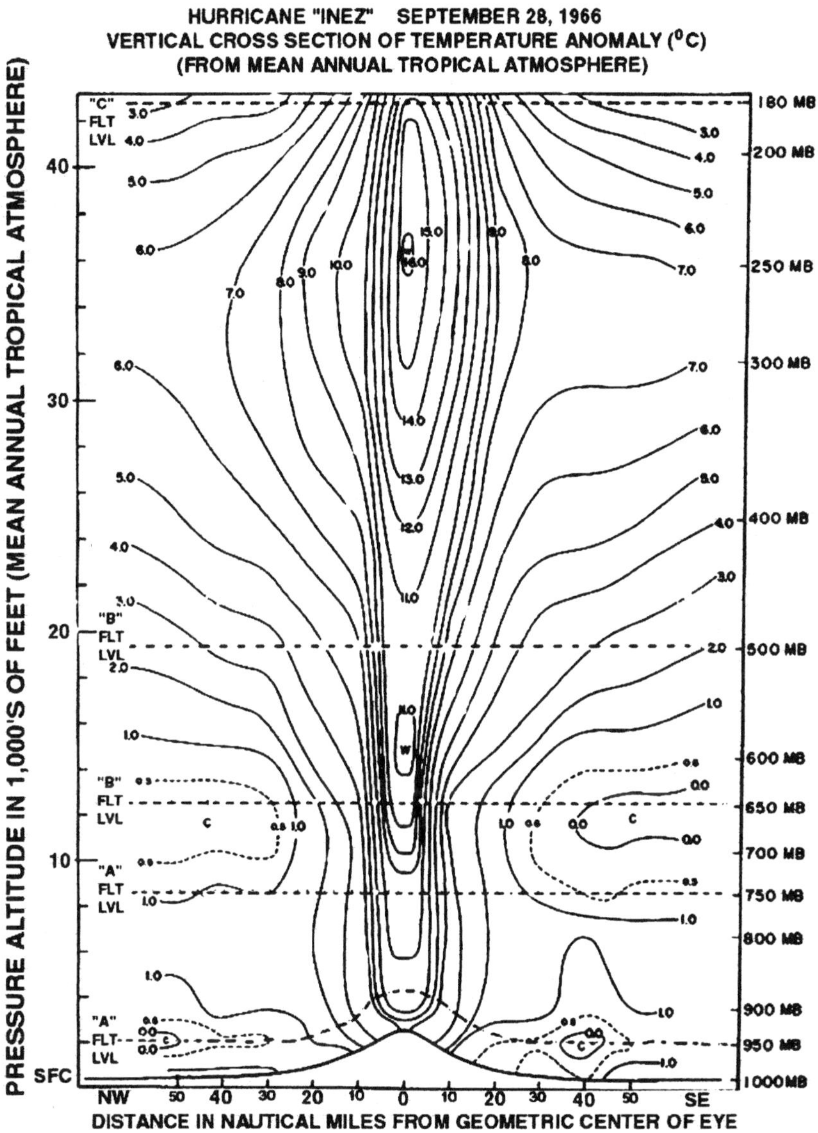

Figure 9. A vertical cross-section of temperature anomalies from the mean annual tropical atmosphere as analyzed from data collected in Hurricane Inez at the indicated flight levels on September 28, 1966 [after Hawkins and Imbembo, 1976].

25,000 ft. This was a void layer in essentially all of the earlier studies of the thermodynamic structure of hurricanes. The DC-6 did not have the operational capability to get and stay at those altitudes for an extended period, and the WB-57 would probably have had severe icing and turbulence problems at those altitudes.

3. There should be a surveillance of a storm, or storms, in which omegasondes are used extensively to determine what wind shears exist with height in the various parts of the storm[3]. The authors of these earlier studies had to considerably "smooth" wind speeds/directions with height. Knowing wind shears, particularly in the boundary layer, would not only be important for operational forecasting and warning purposes, but would help to determine the characteristics of energy transfer from the ocean to the atmosphere for various sectors of the storm.

WEATHER MODIFICATION IN HURRICANES: PROJECT STORMFURY

One of the objectives of the NHRP had been to study the formation of clouds and rain in hurricanes and to investigate possible thermodynamic imbalances that might permit modification of the structure and movement of hurricanes. That research led to Project STORMFURY, which, in turn, uncovered resources that could more than meet the recommendations of Hawkins.

Robert Simpson's observations on early flights into the upper levels of hurricanes revealed unexpected cloud characteristics. On one of these flights out of Bermuda onboard a B-29 at near its operational ceiling of 40,000 ft, he recalled that as they headed toward the center of the hurricane:

> *...through this fog in which we were traveling at 250 miles per hour, there loomed from time to time ghost-like structures rising like huge white marble monuments through the cirrostratus fog. Actually, these were shafts of supercooled water, which rose vertically and passed out of sight overhead...Each time we passed through one of these shafts, the leading edge of the wing accumulated an amazing extra coating of rime ice...*

Dr. Joanne (Malkus) Simpson later referred to these elements as "hot" towers. Those observations led to the initial STORMFURY hypothesis as developed by Drs. Robert Simpson and Joanne Malkus [1964] and later slightly modified by Dr. Robert Cecil Gentry [1970]. Briefly, the hypothesis was that large amounts of supercooled liquid water existed (above the freezing level) in clouds within and just outside of the "eyewall." Introduction of silver iodide crystals into those

[3] Dropwindsondes based upon the omega navigation system had been developed and were being used at the time of this publication.

clouds on the outer edge or just outside of the eyewall would cause the liquid water to freeze, releasing latent heat. The latent heat of fusion released was postulated to increase the buoyancy of the upper portion of the cloud, resulting in increased ascent in the cloud and then increased convergence at the cloud base. Those clouds would now grow to the outflow level, providing a major new conduit for the vertical mass transport of the warm moist air spiraling in toward the storm's center. This new conduit would intercept flow that would normally "feed" the original eyewall, and form a new eyewall at a larger radial distance from the storm's center (Figure 10). The maximum wind speeds would then be reduced because of the outward shift of the radius of maximum winds and the principles of partial conservation of angular momentum. The pressure field would then adjust to this new wind field.

Based upon model simulations [Rosenthal, 1971], several hours after the seeding ended, the storm was expected to return to its "natural" state. These simulations gave some degree of confidence to researchers and advisors that any artificially induced changes to the storm by the experiments would likely be of short duration (a few hours) and certainly masked by natural variations after several hours. Thus, conducting the experiments over open ocean areas well away from land would minimize chances that any effect of the experiments would be felt if the storm later made landfall. Of course, in an effort to develop technology that could possibly be used at some future time to mitigate damage at landfall, much more needed to be learned about the actual changes that might be induced by the modification experiments in addition to model simulations. One other factor that had to be considered for any possible future operational modification efforts was that the hypothesized larger radius of the maximum wind could result in a larger storm surge. However, Jelesnianski and Taylor [1973] used their Storm Surge model to show that these increases would be relatively small. The lessened damage from the significantly reduced maximum surge caused by the peak wind speed reduction would be greatly beneficial.

Figure 11 shows the tracks of all hurricanes ever seeded under Project STORMFURY. The methods of generating and delivering the seeding agent were quite limited for the early experiments — namely on Esther in 1961 and Beulah in 1963. This technology had improved by the time of the Hurricane Debbie experiments in 1969. Also, more extensive monitoring of the low-level wind field was part of the plan, utilizing the Weather Bureau's DC-6 aircraft. On August 18, 1969, after five seedings at two-hour intervals, the maximum wind speed in Debbie was found to have decreased from about 50 m/s to 35 m/s or a reduction of about 30 percent (Figure 12). The hurricane was then left alone for the next 24 h, and by the time the aircraft returned to the storm on August 20th, the maximum wind speeds had returned to near 50 m/s. After repeating the seeding sequence of the 18th, the maximum wind speeds had reduced to about 42 m/s (Figure 13), or a reduction of about 15 percent. Although subsequent analyses [Hawkins, 1971; Sheets, 1973] indicated that a portion of the wind speed reduc-

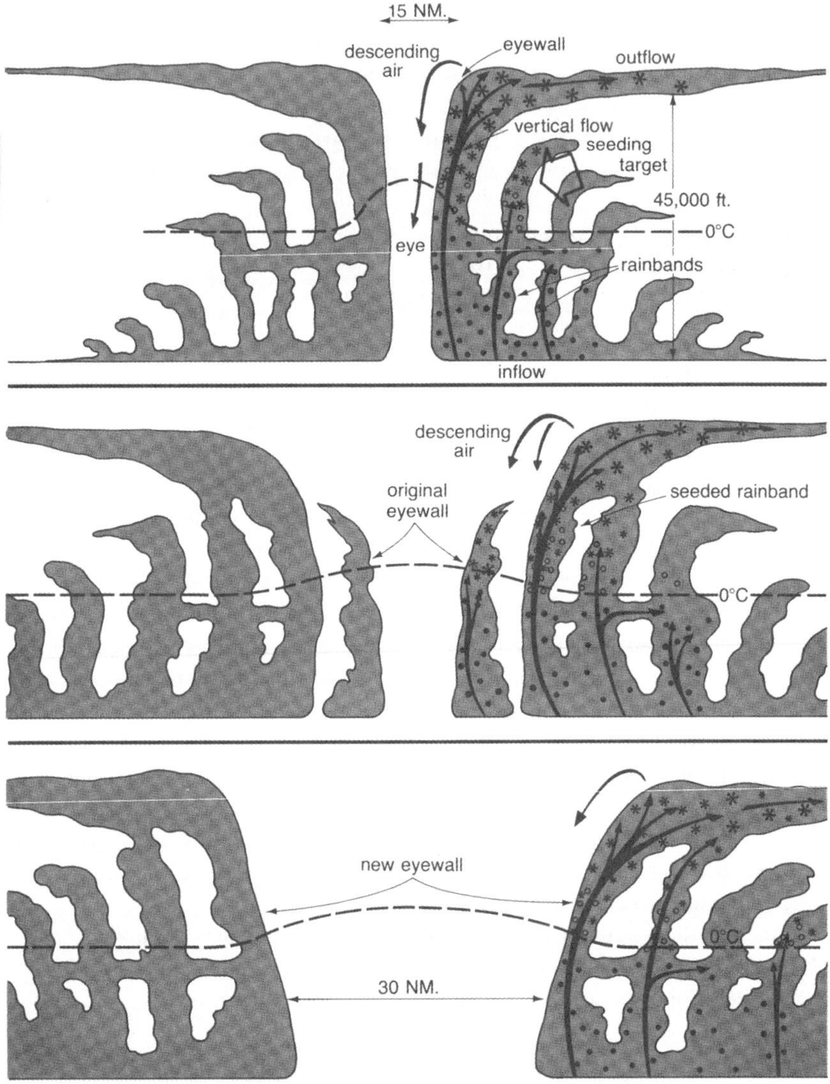

Figure 10. Schematic drawing of a vertical cross-section of a hurricane showing the hypothesized storm structure before (top), during (middle), and after (bottom) a Project STORMFURY seeding operation. Note that the vertical scale has been enlarged by about a factor of ten as compared to the horizontal scale.

tion on August 18 was probably caused by synoptic-scale influences, those same analyses strongly suggested that significant reductions resulted, on both days that could reasonably be attributed to the seeding events. In addition, although radar

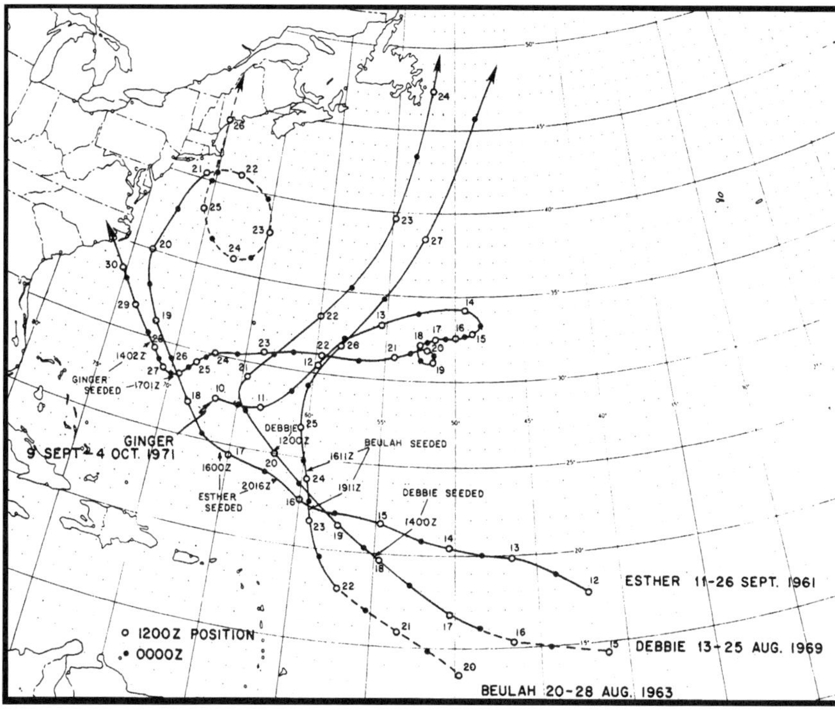

Figure 11. Tracks of all hurricanes seeded under Project STORMFURY from 1961 until the project was discontinued in 1980. Times and locations of seedings are indicated.

coverage was less than optimum, analyses of radar reflectivity sequences also supported the hypothesis showing an "eye" expansion after seeding [Gentry, 1970], consistent with the predicted results.

The results from the Hurricane Debbie experiments seemed so positive that many individuals believed the project should go operational, seeding major hurricanes that threatened land. A team of scientists, at Stanford Research Institute at Stanford University, did a decision analysis on all past seeding events (included the Esther and Beulah experiments). Dr. James Matheson of that group, reflecting their views, stated, "We claim they should consider seeding now, if a big hurricane comes straight for Miami". These scientists said "The Government may have to accept the responsibility for not seeding and thereby exposing the public to higher probabilities of severe storm damage and possible higher death tolls" [Pothier, 1972].

However, scientists involved with the project, although encouraged by the results, were much more cautious. Dr. Robert Cecil Gentry, Director of Project STORMFURY at the time, responded that:

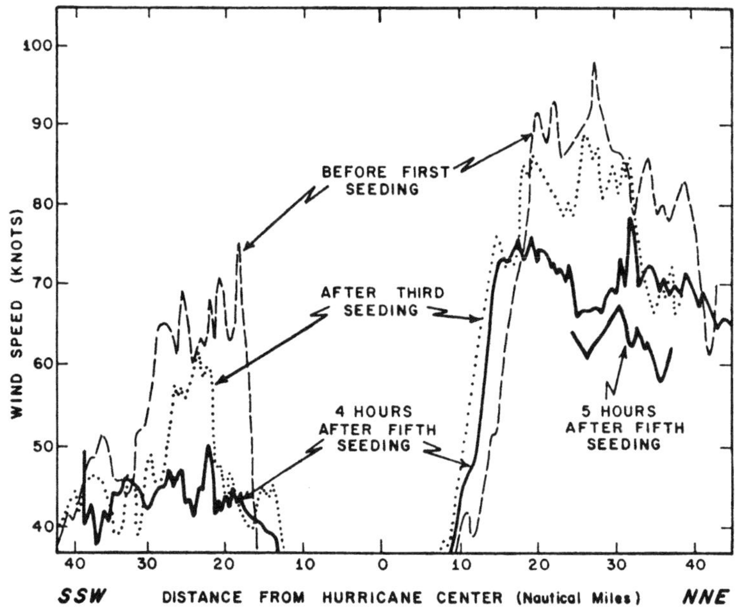

Figure 12. Hurricane Debbie wind-speed profiles recorded at 12,000 ft (3658 m) on August 18, 1969, before, during, and after seeding [after Gentry, 1970].

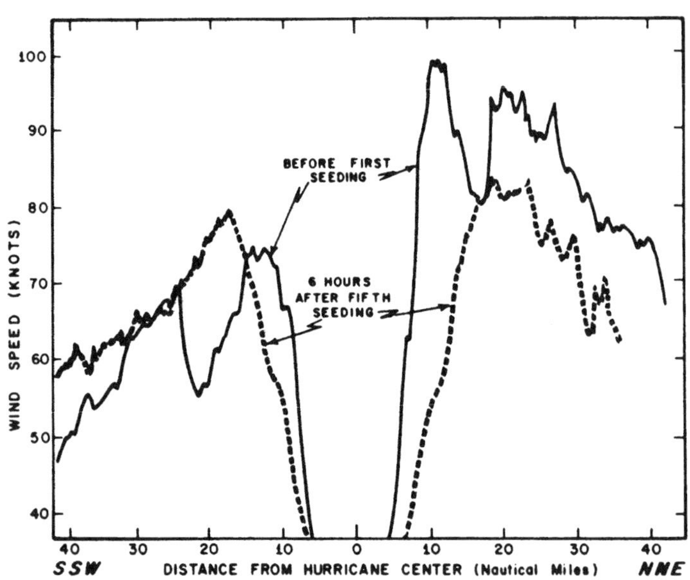

Figure 13. Hurricane Debbie wind-speed profiles recorded at 12,000 ft (3658 m) on August 20, 1969, before, during, and after seeding [after Gentry, 1970].

...first, a hurricane naturally undergoes sharp changes as it crosses a land area. STORMFURY scientists would not be able to tell whether changes in a storm were caused by seeding or the landfall. And second, it is still not a (proven) fact that STORMFURY (has developed a technology that) can weaken a hurricane...

Standard hurricane reconnaissance flight data over the years, although not as continuous and well documented as the Debbie research data, had shown cases of changes in the maximum intensity of hurricanes over time periods similar to those observed in the Debbie experiments.

A major lack in the Debbie experiments was the ability to monitor and document the entire hypothesized sequence of events. That is, the low-level wind field was reasonably well monitored, but the cloud microphysics and the three-dimensional structure of the convective elements were not. It was not possible for the DC-6s to get to the altitudes necessary to monitor the cloud microphysics for an extended period. The jet fighter aircraft delivering the pyrotechnics for dispensing of the seeding agent had no instrumentation to make such measurements.

The National Academy of Sciences' STORMFURY Advisory Panel (Professors Noel E. LaSeur, Florida State University (Chairman); Charles L. Hosler, Pennsylvania State University; Edward N. Lorenz, Massachusetts Institute of Technology; James E. McDonald, University of Arizona; and Jerome Spar, New York University) strongly agreed with the STORMFURY scientists that more information was needed before consideration should be given to making the program operational. They met in February 1970 to review the findings from the Hurricane Debbie experiments and to make recommendations for future activities under the program. The number one recommendation was:

...top priority...be given to the acquisition of aircraft and instrumentation necessary to obtain accurate and representative observations of liquid and solid water content of the eyewall and vicinity in the layer from approximately 20,000 to 35,000 ft. before, during, and after seeding.

A second recommendation was that:

...project aircraft continue to collect data in unseeded hurricanes to monitor the natural variability of those hurricanes for comparison with the behavior of 'seeded storms.'

As a result of the encouraging results from the Debbie experiments, the enormous potential benefits of a proven technology to reduce the destructive force of the hurricane, and the prestigious panel's recommendations, Congress appropriated $30 million to obtain the necessary aircraft and instrumentation to more thoroughly test the STORMFURY hypothesis. Eventually, a WC-130B aircraft was obtained from the Air Force for the project. More importantly, two new WP-

3D aircraft along with state-of-the-art instrumentation were obtained to support the program. Key instrumentation included Inertial Navigation Systems (INS), which greatly improved wind measurements, dropwindsonde capabilities, several cloud physics systems to measure liquid and solid water content and drop size spectra, digitized radar systems, including the first airborne Doppler weather radar system, and state-of-the-art recording systems. The result was the beginning of a whole new era in atmospheric and oceanographic data collection and associated research [Sheets, 1978].

Numerous research flights were conducted under the project utilizing these new airborne capabilities from 1977 through 1980. Unfortunately, no storms of suitable structure occurred within the target zone over the Atlantic during that period so no actual seeding experiments were carried out. Any hurricane selected for these experiments had to have a well-organized eyewall, not be forecast to possibly affect land within 24-h, and remain within the operational range of the project aircraft operating from the U.S. East coast, Puerto Rico, and/or Bermuda for at least a 24 h period for monitoring changes. These selection criteria along with a downturn in hurricane activity during this period limited possible experiments.

Also, cloud physics studies revealed mixed results with respect to the STORMFURY hypothesis requiring large areas of supercooled water. Numerous ice particles were frequently found in cloud areas that might be potential target areas for seeding, raising questions about the possible effectiveness of the artificial seeding agent for producing cloud growth in those regions. In addition, Shapiro and Willoughby [1982] obtained "long-term" monitoring data. In their study, aircraft followed each other sequentially at the same altitudes in the hurricane over a period of many hours. This was a different sampling plan from the simultaneous flights at multiple levels for a period of a few hours used in the "structure" studies, such as those collected in Hurricanes David (1979) and Allen (1980). Using these data, Shapiro and Willoughby documented natural changes (eye cycles) that were similar to those that had been observed during the Hurricane Debbie experiments. This was not all that surprising, as Gentry had stated that such changes took place naturally, but it did raise further questions about whether the changes in Debbie were natural or induced by the seeding.

One could interpret the fact that the proposed sequence of eyewall and associated wind speed changes occur naturally helps to support the hypothesis. That is, it demonstrates that IF convection can be enhanced in a secondary eye-wall or inner rain band, then the hypothesized reduction of wind speed would result. The double eyewall structure had been well known since Jordan observed it from radar reflectivities recorded in Hurricane Donna in 1960 [Jordan and Schatzle, 1961]. However, the long-term monitoring data had now provided the first opportunity to document the sequence of events associated with the evolution of this structure. The question however remains, "If and under what conditions might this convection be enhanced?"

Because of the scientific uncertainties, the dearth of opportunities to conduct actual seeding experiments, and the expense of keeping aircraft and crews operationally ready for such experiments, Project STORMFURY was discontinued in 1980. It was decided that more needed to be known about the internal processes of the hurricane before actual modification experiments were resumed, if ever.

Regardless of the direct results of Project STORMFURY, the resources and research generated by this project have, over the years, greatly contributed to the knowledge of the inner workings of the hurricane and its structure. This increased knowledge has led to improvements in operational forecasting of intensity and motion of hurricanes. For instance, the instrumentation of the NOAA WP-3 aircraft has permitted much better definition of the three-dimensional structure of the inner core of hurricanes.

SCIENCE AND PREDICTION PROGRESS RESULTING FROM THE TECHNOLOGY ADVANCES OF THE 1980s AND 1990s

New Results From More Sophisticated Radars Mounted on Research Aircraft

During the 1980s and 1990s, improved radars were mounted on both the fuselage and the tail of the NOAA P-3 aircraft, where they were used with better navigation equipment to derive the wind speeds and directions. Figure 14 (top) shows a horizontal cross-section of digitized radar reflectivity as recorded on a NOAA WP-3s lower fuselage radar in Hurricane Gloria on September 25, 1985. Superimposed is the flight track of the aircraft on a crossing pattern through the hurricane, with wind barbs plotted at one-minute intervals along the aircraft track. These flight level winds were derived utilizing the Inertial Navigation System (INS) capabilities, a significant improvement over the Doppler navigation system used in the DC-6s. Not only are the wind speed and direction measurements much more accurate than previously available, but the addition of quality digitized radar allowed those winds to be superimposed upon the precipitation field. Also, by assumption of "steady state" conditions for the one hour or so required to complete orthogonal passes through the center of the hurricane, additional information could be obtained. The single, tail mounted, 360° vertically scanning Doppler radar system measures the component of motion of particles, directly toward and away from the antenna. By combining these measurements along the flight track, the horizontal wind field, within a reasonable range of that antenna could be derived, not just the wind component along the flight path! Those Doppler winds are shown as wind barbs. The lower panel of Figure 14 shows a vertical cross section of the radar reflectivities recorded on the tail radar. Superimposed is an analyzed field of the horizontal wind speeds determined from the Doppler system.

The horizontal winds derived under the "steady-state" assumption represent the larger scale features of the hurricane, but not those of the convective scale.

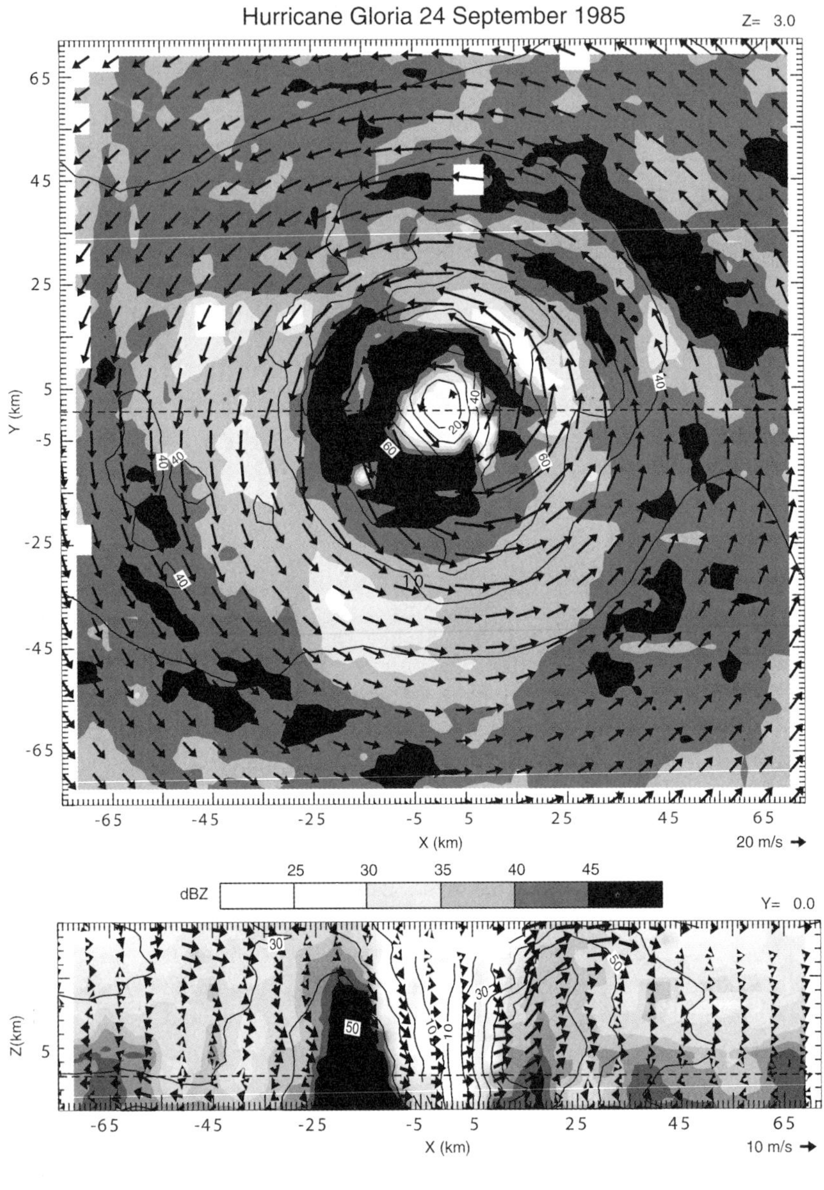

Figure 14. Flight level along-track and Doppler-derived horizontal winds at the 6.5 km level (upper panel) and a vertical cross-section of Doppler-derived horizontal wind speeds (lower panel) superimposed on radar reflectivities for Hurricane Gloria on September 25, 1985 [courtesy of Frank Marks].

The capability to represent those scales of motion, using single Doppler radar data from a moving airborne platform, awaited the development of new techniques. These were developed with amazing results [Lee et al., 1994]. Figure 15 shows a horizontal cross-section of the reflectivities for the eyewall region of Hurricane Olivia in the Eastern North Pacific on September 25, 1994 at an altitude of 6 km. Superimposed on these reflectivities are the horizontal winds at this level, represented by the arrows (scale for 50 m/s wind shown at bottom of figure), and an analysis of the vertical wind speeds (contours and numbers in white). Also shown are radial lines from the center outward along the azimuths of 326, 006, and 024 degrees. Figure 16 is a vertical, radial cross-section of reflectivities along the 326-degree azimuth of Figure 15, with superimposed two-dimensional wind flow along this vertical cross-section (arrows) and a contour analysis of the vertical wind speeds. Figures 17 and 18 show the same information at azimuths 006 and 024 degrees, respectively. Note the complexity of the flow in the eyewall region. Of particular interest is the inflow from upper layers of the eyewall with sinking occurring in the eye,[4] while at lower levels there was outflow from the eye to the eyewall. The strongest upward motion seems to be just upwind of the largest reflectivity values in the eyewall. This, of course, represents a major advancement in the ability to investigate and document the flow and energy processes in the inner core of a hurricane.

Results From Improved Capability to Measure Winds

One other major advancement during the past two to three decades was the implementation of dropwindsondes. Initially, these were based upon the Omega navigation system – Omega Drop Windsondes (ODWs). Burpee and others [1984] conducted studies on how these systems could be used to initialize numerical models for improved forecasting of hurricane motion. The first experiments designed to use ODWs to define the steering currents of mature tropical cyclones were conducted on the periphery of Hurricane Debby in 1982. The two NOAA WP-3 aircraft flew patterns in the environment of Hurricane Debby on September 15th and 16th (Figure 19) as the hurricane moved over the Atlantic between Hispaniola and Bermuda. Although several of the sondes failed partially or totally, records were sufficient to define a trough and a cut-off low (Figure 20) that were to affect the future motion of the hurricane. Burpee and his colleagues concluded that the sample was too limited "...to assess improvements in operational track forecasts that might result from the ODW observations". However, they also concluded that:

---

[4] A similar example of upper level inflow and sinking, inferred qualitatively from photographic evidence in Hurricane Edna, 1954, is illustrated in Chapter 2, Figure 6.

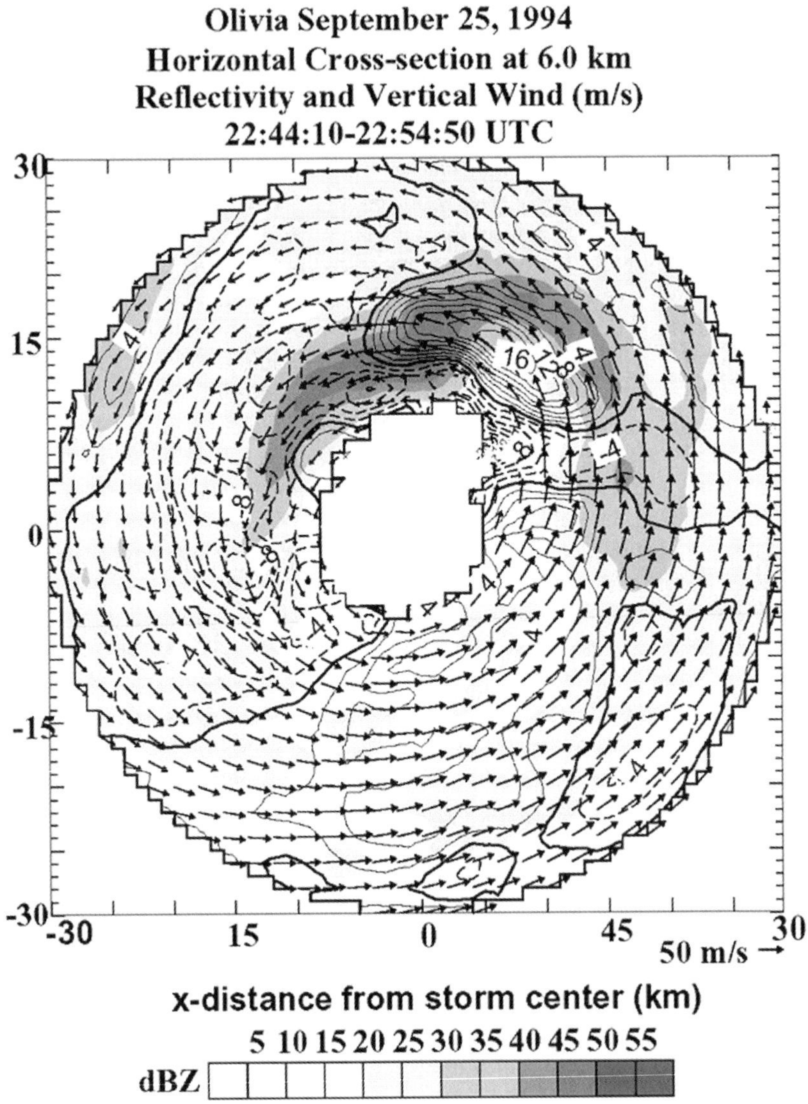

Figure 15. A horizontal cross-section of radar-reflectivities, Doppler-derived horizontal winds (arrows), and vertical wind speeds (analyzed contours and numbers) recorded at the 6 km level in Hurricane Olivia in the Eastern North Pacific on September 25, 1994 [courtesy of Frank Marks].

(1) ODWs are capable of defining the important features of a hurricane's environmental flow in the middle and lower troposphere at radii of 150-1000 km from the center of a tropical cyclone. It is feasible to transmit the ODW observations...to the NHC and NMC before forecast deadlines; and (2) the data can provide useful information for hurricane forecasters to use subjectively.

Over the next ten years, "synoptic flow experiments" were conducted in ten more hurricanes. The ODW performance, data transmission, and data utilization improved over this period. These data sets were used in initial analyses for three dynamical models to produce forecasts through at least 60 hours. These forecasts were then compared to "control" runs for those models where the ODW data were excluded from their initial analyses. The average error reductions in the consensus forecasts from the three dynamical models varied from 16% to 30% for these forecast periods. Burpee et al., [1996] state that

...The improvements are at least as large as the accumulated improvements in operational forecasts achieved over the 22-yr period from 1970 to 1991...and represent a major breakthrough in hurricane forecasting capability.

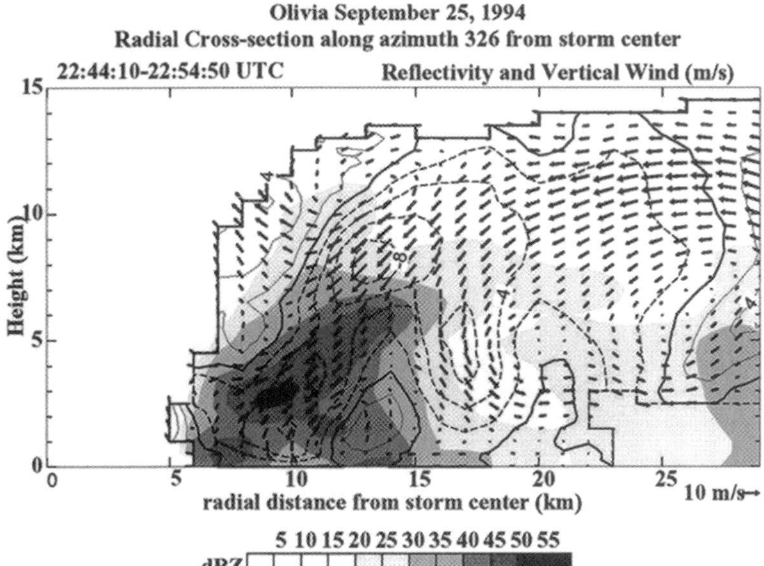

Figure 16. A vertical radial cross-section of radar reflectivities and Doppler-derived winds along the 326 degree azimuth of Figure 15 [courtesy of Frank Marks].

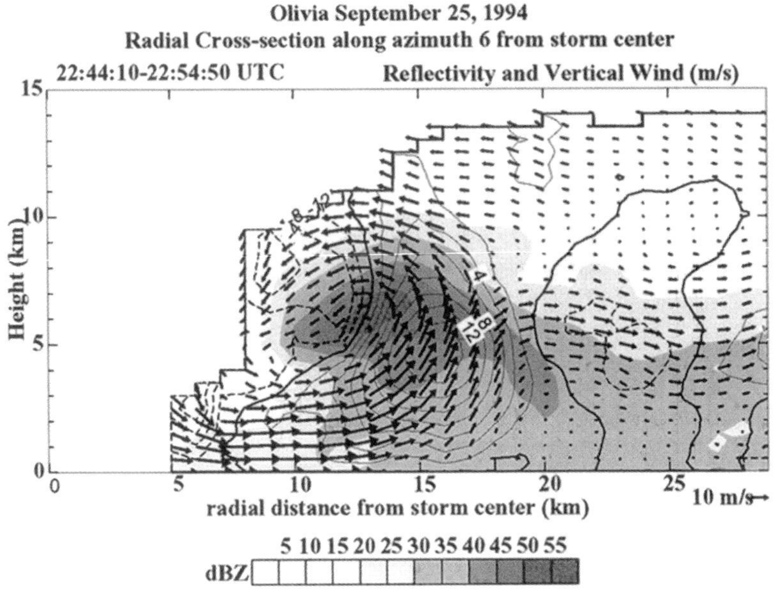

Figure 17. Same as Figure 16 except along the 006-degree azimuth of Figure 15 [courtesy of Frank Marks].

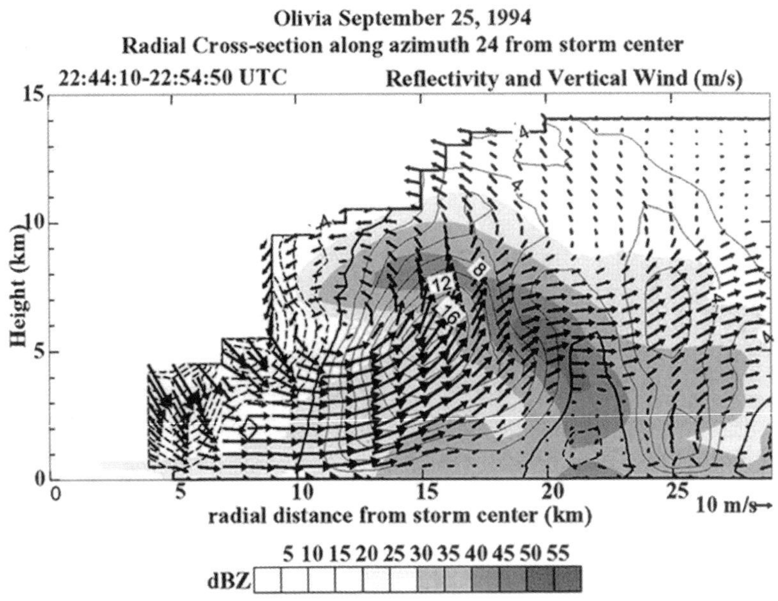

Figure 18. Same as Figure 16 except along the 024-degree azimuth of Figure 15 [courtesy of Frank Marks].

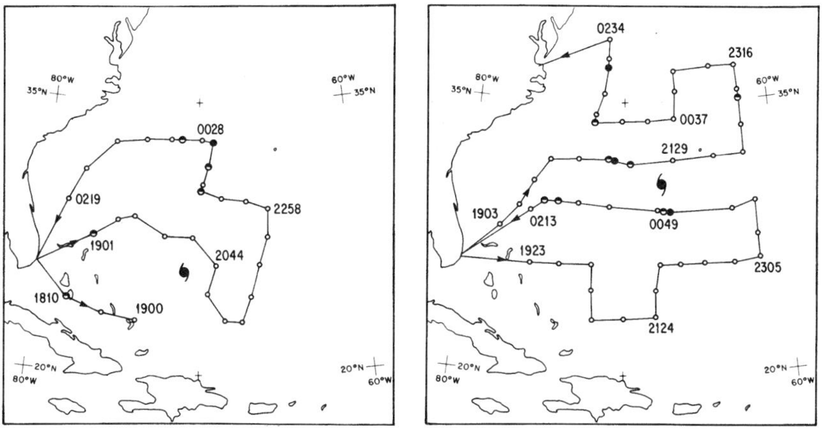

Figure 19. Flight tracks for dropwindsonde experiments conducted in the environment of Hurricane Debby (1982) corresponding to approximate map times of 0000 UTC on September 15 (left) and September 16 [after Burpee et al., 1984].

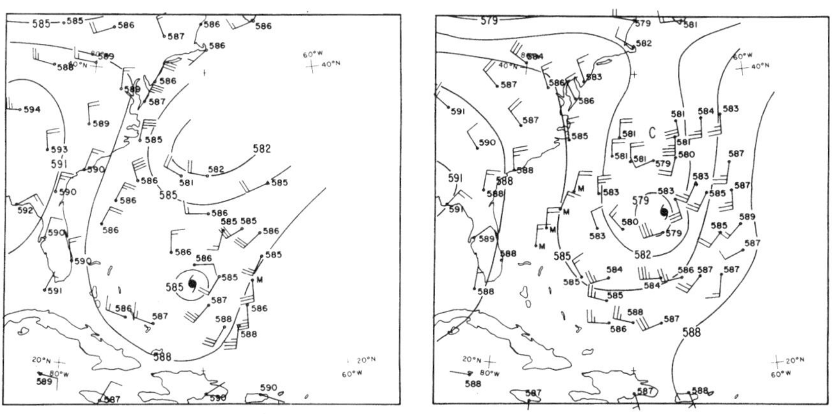

Figure 20. Five hundred (500) mb analyses at 0000 UTC on September 15 (left) and September 16 (right) of 1984. Geopotential heights in decameters are shown to the right of the location of each wind observation. Height contours are at 30-m intervals. M indicates missing heights. Wind speeds of 5 m/s are represented by a full barb and speeds of 2.5 m/s by a half barb with no barb for winds less or equal to 1 m/s. The hurricane symbol indicates the center of Hurricane Debby at map time [after Burpee et al., 1984].

These research data sets clearly demonstrated that the higher resolution dynamical models newly available for operational use, coupled with the rapidly improving computer capabilities, had the potential to significantly improve hurricane track forecasts. As one looked ahead, however, the current turbo-prop four engine aircraft (WP-3s and WC-130s) had significant operational limitations in altitude and speed for obtaining the desired data. That is, they could only sample the lower two-thirds of the troposphere. The models being developed by Kurihara and colleagues [1993] at NOAA's Geophysical Fluid Dynamics Laboratory (GFDL) also indicated that higher altitude data would not only be useful for improving track forecasts, but were also extremely important for intensity predictions, which were so far lagging in skill. Furthermore, the four-engine turbo-prop aircraft were generally more expensive to operate (larger crews and more maintenance) than a comparable twin-engine jet aircraft with significantly higher altitude capability. These speedier jet aircraft could also provide timelier and more comprehensive data sets for use in analyses and initializing the ever-improving dynamical models.

Introduction of Jet Aircraft

By the late 1980s, I had been advocating the acquisition of a modern mid-size jet aircraft to obtain the desired data sets [Sheets, 1990]. I also proposed that these data sets, together with land based upper-air soundings, could be used to "calibrate" nearby advanced satellite soundings. The satellite data would then be expected to degrade with distance from the "ground truth" site but would still provide much improved quantitative information in oceanic areas otherwise nearly void of accurate data. The results would significantly improve the initial fields for the dynamical models and could be expected to improve forecast guidance for hurricanes. However, this approach requires that satellite radiance information be used to obtain conventional state parameters such as temperature and humidity. This process has proven to be quite difficult. Most models that use satellite data now compare the observed radiances to the model's calculation of the radiances from the model-predicted temperature and humidity. Much work still needs to be done to fully realize the potential benefits of these combined data sets.

Budgetary priorities within NOAA prevented action on this proposal for several years. However, once again, hurricane disasters came along to change those priorities, at least within Congress, if not within NOAA. The 1988 and 1989 hurricane seasons produced more Saffir/Simpson Category 4 hurricanes, than had been observed in the Atlantic basin for two consecutive years since 1960-61. Hurricane Gilbert in 1988 was (and remains) the strongest hurricane on record in the Western Hemisphere. This massive hurricane caused catastrophic losses during its passage through the Caribbean (Jamaica, Yucatan Peninsula) and later in south Texas and the upper Gulf coastal regions of Mexico and inland. Also in 1988, Hurricane Joan caused large loss of life and property through the southern Caribbean (Venezuela) and into Central America (Nicaragua and Costa Rica). That same year, massive Hurricane Helene turned northward well east of the Caribbean. In 1989, large and powerful Hurricane Gabrielle followed a similar all-oceanic path.

However, less than one week following Gabrielle, islands in the northeast Caribbean, including the U.S. Virgin Islands and Puerto Rico, and the east coast of the United States were not so fortunate. Hurricane Hugo ravaged those island communities and made a direct strike on the southeast coast of the mainland U.S., with the center passing inland just northeast of Charleston, South Carolina. The result was several lives lost and damage in excess of $7 billion, the largest loss from a "natural disaster," to that date, in the history of the United States.

Meanwhile, Congressional hearings were held concerning the status of hurricane forecasting and warning systems and how they might be improved. A major topic of discussion was the requirement for the continuation of the aircraft reconnaissance program. As Director of NHC, I made presentations at those hearings and at a hearing in the wake of Hurricane Hugo. I was able to stress my belief that hurricane forecasting could be improved if a mid-size jet aircraft with appropriate instrumentation could be obtained to collect data in the near environment of the hurricane. Under the leadership of Senator Hollings of South Carolina and his staff, funds were allocated in FY 1994 for such an acquisition. The result was that NOAA purchased and instrumented a Gulfstream IV twin jet aircraft that became available for flying during the 1997 hurricane season.

The Global Positioning System – A Quantum Jump in Navigation

During this same period, the National Center for Atmospheric Research had been developing a much-improved sonde that used the Global Positioning System (GPS) to replace the Omega navigation system based ODWs. The Omega navigation system was being phased out around the world. The GPS-based sonde development was partially funded by the G-IV aircraft appropriation. Only limited data were collected during the 1997 season, but track forecasts, using the most sophisticated dynamical hurricane track prediction model (GFDL), showed marked improvements of track predictions with the GPS sonde data as compared to predictions without that data [Aberson and Franklin, 1999]. Those improvements were 8%, 31%, 32%, 12%, and 11% for the 12-, 24-, 36-, 48-, and 72-h position forecasts, respectively! Several flights have been made since 1997 for critical forecast situations with results generally continuing to indicate significant improvements in the forecasts, although generally not as large as the 1997 sample. Evaluations of those forecasts, however, also indicate the critical nature of the sampling strategy to optimize results. Work continues in that area today.

NEW TECHNOLOGY TO STUDY THE HURRICANE'S (LOWER) BOUNDARY LAYER

The GPS sondes not only provided more reliable and higher resolution data for the near environment of the hurricane for track and intensity model improvements, but the higher resolution data also have enabled detailed vertical profiles in the horizontal wind from flight levels to near the surface. Those profiles are causing some re-examination of operational procedures for extrapolating flight level winds to the near surface. They have given a better means to select the best wind profiles for high-rise building design, and which profiles we should use near the coast in strong hurricane situations.

Historically, one of the more difficult tasks faced by analysts and forecasters has been the determination of near surface (10-m) maximum sustained winds based upon the winds measured at reconnaissance aircraft's flight level. Those operational flight levels were generally at slightly above 3,000 m (~ 700 mb) in strong hurricanes. Based upon comparison of flight level to buoy data, Powell and Black [1990] recommended that a ratio of 63%-73% be used to reduce those flight level winds to the standard surface height. Many at the National Hurricane Center (NHC), including myself, believed that this ratio did not apply in the strong convective areas such as in the eyewall or vigorous rainbands. There we believed that convective systems transported high momentum air from aloft to near the surface. In fact, this transport could even result in an acceleration of the surface air, overcoming the frictional drag. For this reason, the operational practice at the National Hurricane Center for several years was to use a ratio of 80% to 90% of the flight level winds for estimating surface winds when strong convective elements were present in the storm. This practice would occasionally result in criticism of NHC intensity estimates, particularly by some members of the Wind Engineering Research Committee. Other members of that same committee were generally in agreement with the NHC's estimates.

Figure 21 shows the generally accepted vertical profiles of the horizontal wind near the surface (< 500m, see e.g., Chapter 7). The change in wind speed with height is basically a function of surface friction and the mechanical mixing introduced by that friction. The height of the undisturbed air would be much lower over ocean surfaces than in a highly urbanized area. If you were to consider such profiles to be valid for the core of a hurricane, as some design engineers apparently have, it is easy to see why they would be critical of the NHC surface wind estimates. For instance, the NHC estimated maximum sustained surface wind speed in Hurricane Andrew (1992) at landfall in South Florida was about 64 m/s. If you used the Figure 21 profiles with the 64 m/s surface wind, speeds would be one to two times that value at the upper levels of that building, with gusts much higher! Rather than considering the assumed wind changes with height to be wrong, some simply argued that the surface wind estimates were wrong.

Since 1997, well over 400 GPS sondes have been released in hurricanes and their environment. Figure 22 shows the release points and horizontal paths of three such sondes released in the eyewall region of 1997 Hurricane Guillermo in the Eastern North Pacific. Figure 23 shows the vertical profile of the horizontal wind for those three sondes. Sondes released on the inner edge of the NE eyewall (2348 UTC) and for the middle of the SW eyewall (2342 UTC) show that

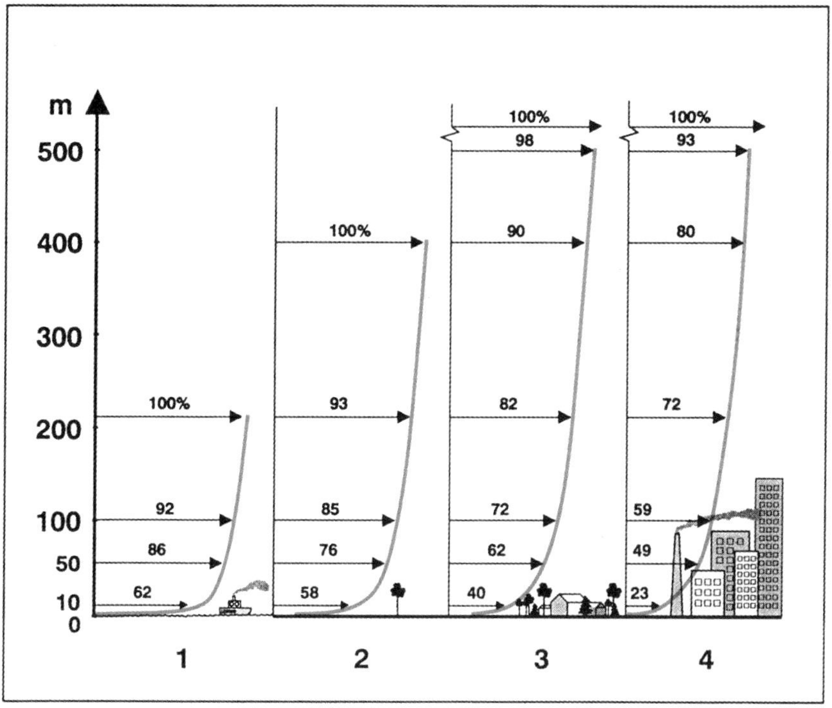

Figure 21. Wind velocity increases with height as a result of friction for: (1) at sea; (2) at the shoreline; (3) over land with trees or small buildings; and (4) over large cities with tall buildings. [Courtesy of Munich Reinsurance Co.]

the lowest winds were nearly the same as those at 3000 m above the surface. Also, these two profiles showed winds from 700 m down to about 300 m above the surface were substantially stronger than the near surface winds and those above 700 m. The sonde released outside of the convective zone of the north eyewall, generally shows smaller variations of wind with height, until the sonde fell to about the 400 m level, where wind speeds then decreased rather rapidly to the surface.

A similar set of dropwindsondes was released in the eyewall region of Hurricane Mitch in the Caribbean Sea, just north of Honduras, on October 27, 1998 (Figures 24 and 25). About 24 h earlier, Hurricane Mitch had been the strongest October hurricane of record in this hemisphere, with a minimum central surface pressure of 905 mb and an estimated maximum sustained surface wind speed of 80 m/s. By the sonde release time on the 27th, the pressure had risen to about 930 mb, with maximum sustained surface wind speeds estimated to be about 72 m/s. Also the convection in the eyewall had become quite asymmetrical. However, the east through northeast sector of the eye-wall remained

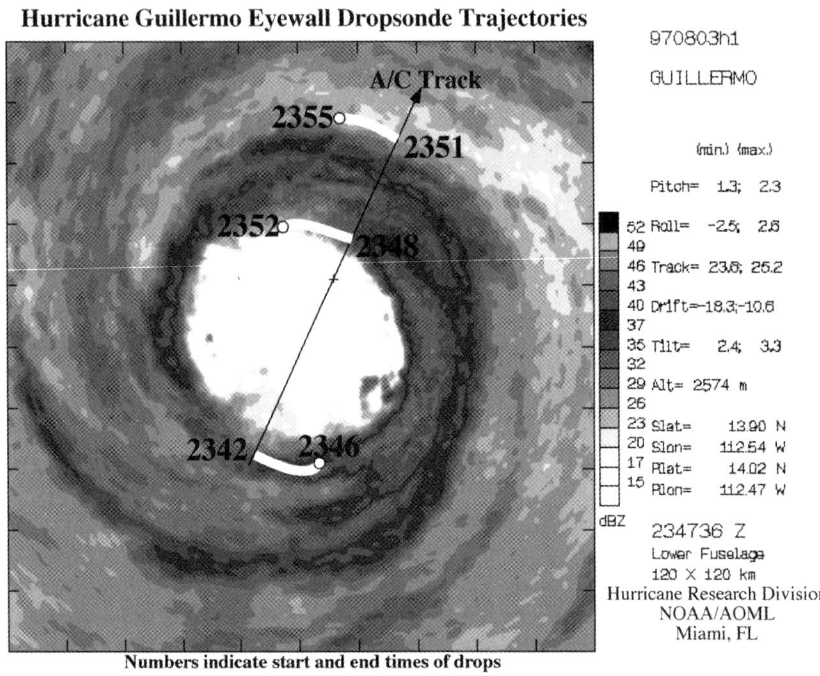

Figure 22. Radar-reflectivities recorded in Hurricane Guillermo in the Northeast Pacific on August 3, 1997. Superimposed are a portion of the flight track of the NOAA WP-3 research aircraft and the release and end points of three dropwindsondes released in or near the eyewall [courtesy of Michael Black].

quite convectively active. High radar reflectivities were recorded on the lower fuselage radar of the NOAA WP-3D research aircraft (Figure 24). It was in this eastern sector of the eyewall where the three sondes were released as the aircraft flew toward the center of the hurricane. The first sonde was released on the outer edge of the eyewall, but still near a convectively active area (2335 UTC surface splashdown). The second was released between two high-reflectivity zones, but was carried into the central eyewall high-reflectivity zone before splashdown at about 2336 UTC.

The third sonde was released in a high-reflectivity zone of the eyewall and, apparently, remained in such a zone through splashdown at about 2337 UTC, on the inner edge of the high-reflectivity zone of the eyewall. Note that only the outer sonde recorded a near-surface wind lower than the initial wind speeds measured just below the flight level of the aircraft. However, even this sonde, along with the other two sondes, measured wind speeds only a few hundred feet above the surface substantially stronger than those near the flight level did. The middle sonde measured the strongest wind speeds at the lower levels (1,000 ft.)

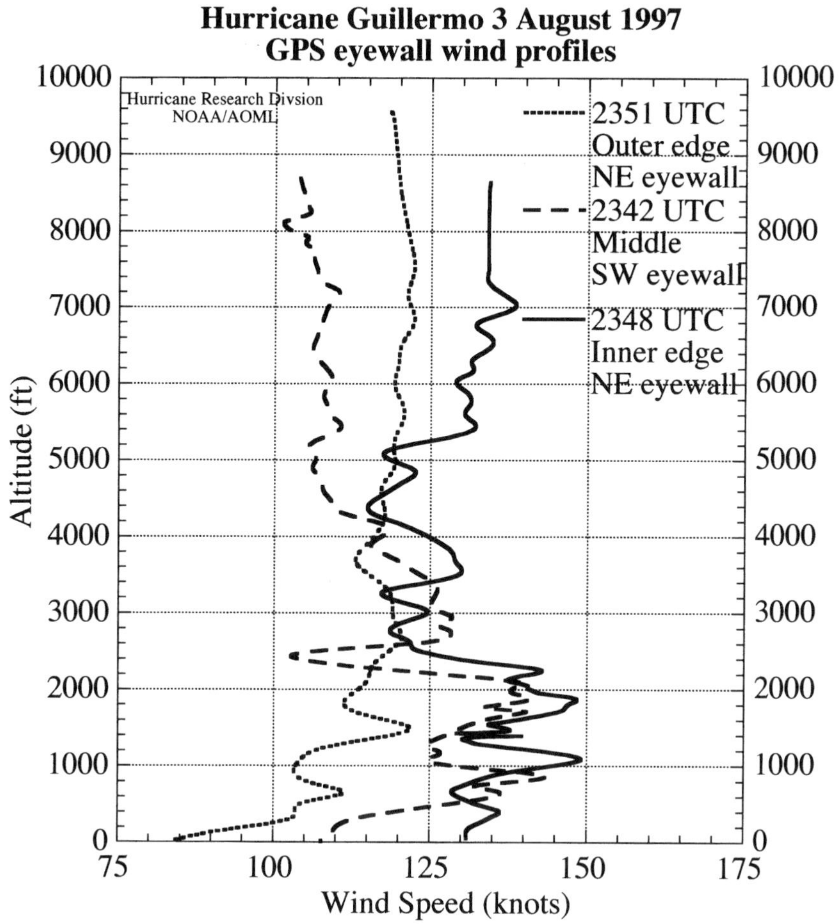

Figure 23. Eyewall and near eyewall vertical profiles of the horizontal wind speeds derived from dropwindsondes released in Hurricane Guillermo for the times and locations shown in Figure 22 [courtesy of Michael Black].

at about 167 kt, but failed shortly after that measurement. The sonde dropped on the inner edge of the eyewall measured a near-surface wind speed of about 145 kt. These measurements were substantially stronger than the 125 kt measured just below the flight level. Franklin et al. [2000] noted that the commonly used reduction factor of 0.9 for estimating surface winds from flight level winds measured at 10,000 ft. would have resulted in a substantial underestimate of the maximum surface wind speeds.

Although there are large variations from sonde to sonde, it is clear that the vertical profiles of the horizontal winds in the convective core region of a hurricane do not fit the form of the wind speed profiles shown in Figure 21 that are fre-

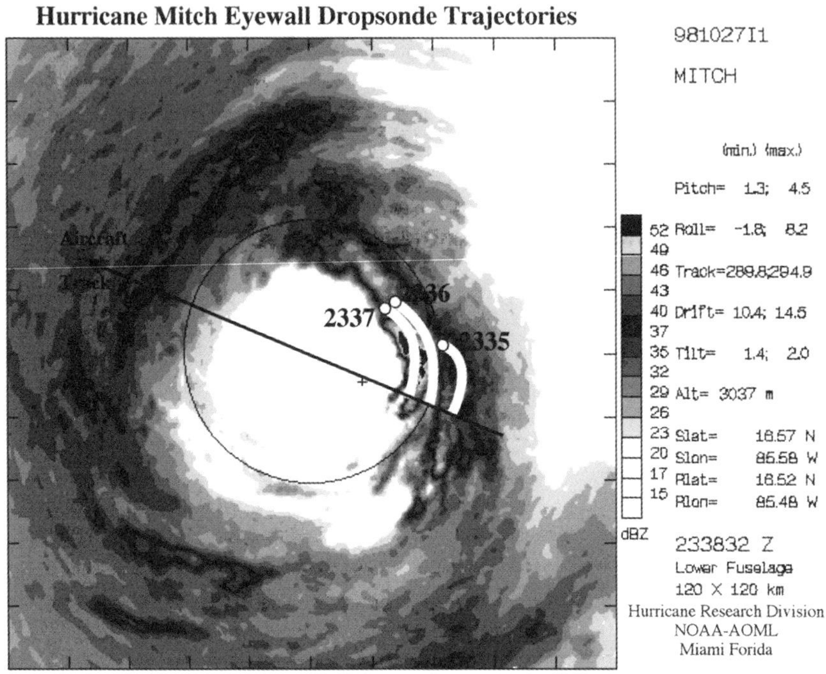

Figure 24. Radar-reflectivities recorded in Hurricane Mitch in the Caribbean Sea on October 27, 1998. Superimposed is a portion of the flight track of the NOAA WP-3 research aircraft and the release and end points of three dropwindsondes released in or near the eyewall [courtesy of Michael Black].

quently used for building design purposes. However, it should be noted that this does not mean that high-rise buildings, designed using the assumed dramatically reduced wind speeds at lower levels, are unsafe. Buildings are generally designed for the stronger wind speeds aloft. But, it may mean that lower buildings may be underdesigned for wind speeds that can occur in the highly convective region of the eyewall of a hurricane. This would have been the case with Hurricane Andrew of 1992 when it made landfall in south Florida.

Franklin and his colleagues [2000] analyzed and normalized (by the 700 mb level wind speed) some 183 sondes dropped in the eyewall and some 124 sondes dropped in the outer vortex to arrive at vertical profiles of the mean horizontal wind speeds in those areas (Figure 26). The results of their study show that, on average, from the surface to about 3000 m, the horizontal wind speeds are generally greatest at about 500 m and 800 m above the surface for the eyewall region. I believe that this profile would also apply to highly convective regions of rainbands as well as the eyewall. On average, surface winds would be about 91% and 78% of 3000-m level wind speeds for the eyewall region and outer vor-

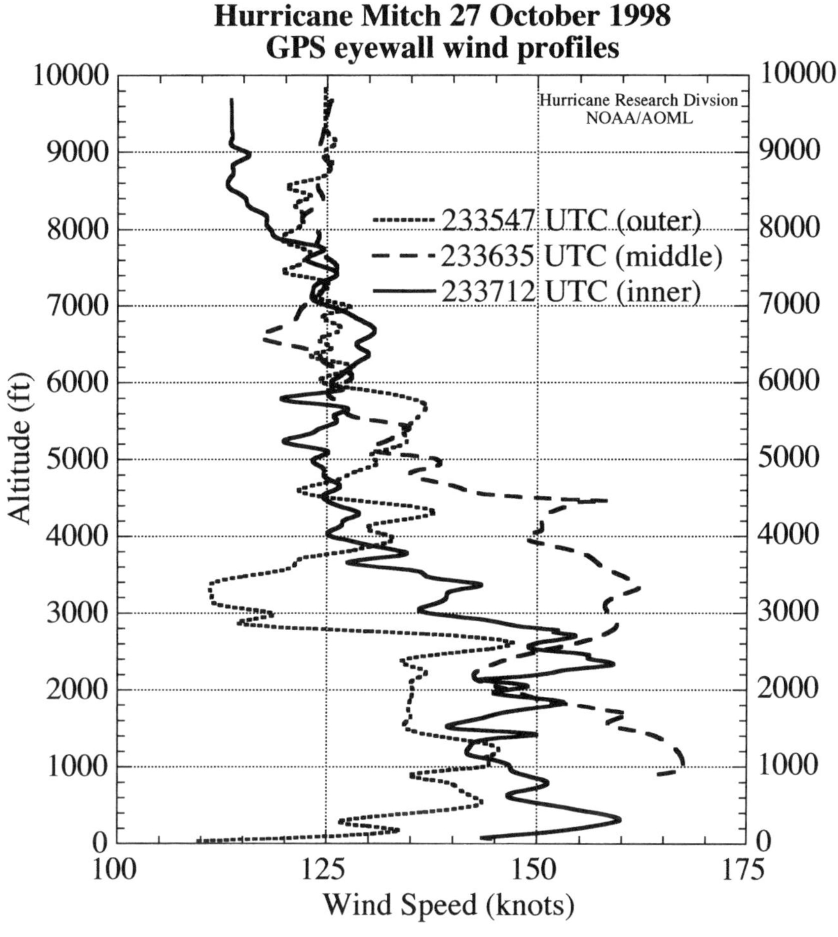

Figure 25. Eyewall and near eyewall vertical profiles of the horizontal wind speeds derived from dropwindsondes released in Hurricane Mitch for the times and locations shown in Figure 24 [courtesy of Michael Black].

tex region of the hurricane, respectively. All of this applies to winds measured over open water. These numbers for the outer vortex region are quite similar to what Powell and Black had estimated (73%) based upon buoy observations. However, the inner core/eyewall winds are underestimated using those buoy studies and are very close to what had been the general practice of the National Hurricane Center for the convective core region of the hurricane (80% to 90%).

One still must remember that these numbers are averages and differences will occur based upon the convective vigor of the hurricane and the stability of the boundary layer. Black and Franklin [2000] point to the convective factor for the large differences in low level wind speed profiles for rapidly intensifying

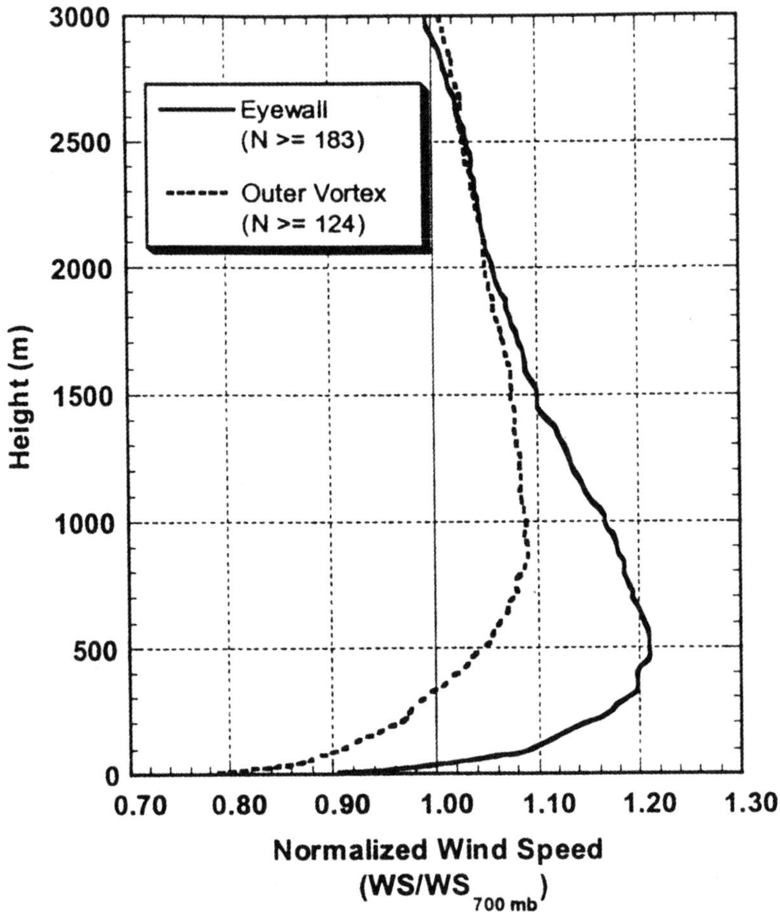

Figure 26. Vertical profiles of the "mean" horizontal wind speed normalized by the 700 mb wind speed, derived from some 183 GPS dropwindsondes released in or near the eyewall (solid line) and some 124 such sondes released in the outer vortex of several hurricanes [after Franklin et al., 2000].

Hurricane Bret as compared to the more quiescent Hurricane Bonnie (Figure 27). In the case of Bret, wind speeds in the lower 1000 meters were nearly twice as strong as those at 3000 meters with no dramatic dropping off of values in the lowest 200-300 meters. By contrast, wind speeds in the eyewall of Hurricane Bonnie were nearly constant from the 3000-m level down to about the 200 meters before decreasing rapidly near the surface.

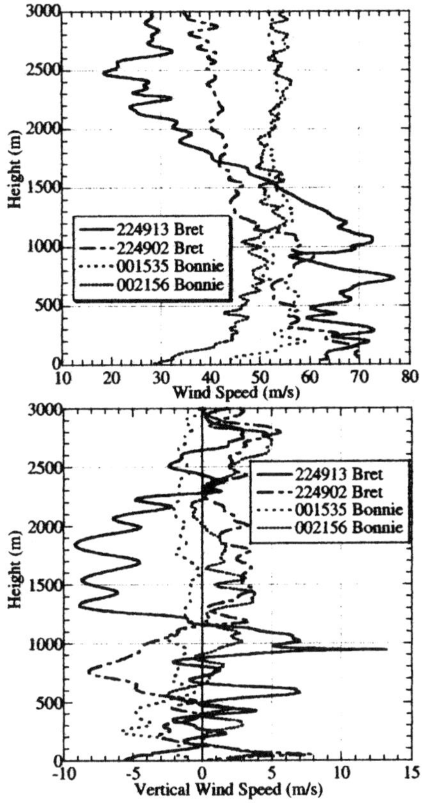

Figure 27. Vertical profiles of the horizontal wind speed derived from GPS dropwindsondes released in Hurricane Bret on August 21, 1999 and Hurricane Bonnie on August 24, 1998 [after Black and Franklin, 2000].

SUMMARY AND CONCLUSIONS

In summary, there have been great advancements over the past 40 to 50 years in measurement systems on surveillance and research aircraft for investigating hurricane structure. Those capabilities have provided greatly improved descriptions and understanding of the hurricane's energy processes and internal structure. This knowledge, coupled with satellite capabilities, advanced numerical models (using almost unbelievable computer technological advancements), and communications systems that allow use of high resolution data in those numerical models, have led to marked improvements in hurricane track forecasting.

However, even with all of these remarkable advancements, much of the original goals of the NHRP remain to be fulfilled. We do not yet understand why some systems will form strong tropical storms while others, which seem to have similar characteristics, do not. There remains limited skill in forecasting hurricane intensities and changes. Some are optimistic that such improvements will come soon with the use of upper level data in the near environment of the hurricane as well as sea-surface and sub-surface information. But this is not the only problem relevant to the general intensity of a hurricane. Doppler radar capabilities and GPS dropwindsondes are showing horizontal wind speed profiles, particularly in the lowest 1000 m of the atmosphere, that are much different from the figures used for construction of mid- to high-rise buildings. Again it should be noted that this does not necessarily mean that these buildings are not "safe" since they were generally designed for stronger winds at higher levels. It is clear that much remains to be learned about the low-level wind structure in hurricanes and how that varies in time and space, particularly as a strong hurricane moves on shore and inland. Such knowledge is not only of great importance for determining construction practices of the future, but also for determining appropriate responses before and after a hurricane strikes.

REFERENCES

Aberson, S. D. and J. L. Franklin, Impact on Hurricane Track and Intensity Forecasts of GPS Dropwindsonde Observations From the First Season Flights of the NOAA Gulfstream-IV Jet Aircraft. *Bull. Amer. Meteor. Soc.*, 80 (3), 421-428, 1999.

Black, M. L. and J. L. Franklin, GPS Dropsonde Observations of the Wind Structure in Convective and Non-Convective Regions of the Hurricane Eye-wall. 24th AMS Conf. Hurricanes and Tropical Meteorology Reprints, 14A.2, 2000.

Burpee, R. W., S. J. Lord and R. E. Tuleya, The Impact of Omega Dropwindsondes on Operational Hurricane Track Forecast Models. *Bull. Amer. Meteor. Soc.*, 77, 925-933, 1996.

Burpee, R. W., D. G. Marks and R. T. Merrill, An Assessment of Omega Dropwindsonde Data in Track Forecasts of Hurricane Debby (1982). *Bull. Amer. Meteor. Soc.*, 65, 1950-1958, 1984.

Franklin, J. L., M. L. Black and K. Valde, Eyewall Wind Profiles in Hurricanes Determined by GPS Dropwindsondes. 24th AMS Conf. Hurricanes and Tropical Meteorology Preprints, 14A.1, 2000.

Gentry, R. C., Hurricane Debbie Modification Experiments, August 1969. *Sci.,* 168 (3930), 473-475, 1970.

Hawkins, H. F., Comparison of Results of the Hurricane Debbie (1969) Modification Experiments With Those From Rosenthal's Numerical Model Simulation Experiments. *Mon. Wea. Rev.*, 99, 427-434, 1971.

Hawkins, H. F. and S. M. Imbembo, The Structure of a Small, Intense Hurricane – Inez 1966. *Mon. Wea. Rev.*, 104, 418-442, 1976.

Hawkins, H. F. and D. T. Rubsam, Hurricane Hilda, 1964: Structure and Budgets of the Hurricane on October 1, 1964. *Mon. Wea. Rev.*, 96, 617-636, 1968.

Hilleary, D. T. and F. E. Christensen, Instrumentation of National Hurricane Research Project Aircraft, NHRP No. 11, U.S. Weather Bureau, 71 pp, 1957.

Jelesnianski, C. P. and A. D. Taylor, A Preliminary View of Storm Surges Before and After Storm Modifications. NOAA Tech Memo ERL WMPO-3, 33 pp, 1973.

Jordan, C. L. and F. J. Schatzle, The "Double-Eye" of Hurricane Donna. *Mon. Wea. Rev.,* 89, 354-356, 1961.

Kurihara, Y., M. A. Bender, and R. J. Ross, Improvements in Tropical Cyclone Track and Intensity Forecasts Using the GFDL Initialization System. *Mon. Wea. Rev.*, 121, 2046-2061, 1993.

LaSeur, N. E. and H. F. Hawkins, An Analysis of Hurricane Cleo (1958) Based on Data From Research Reconnaissance Aircraft. *Mon. Wea. Rev.*, 91, 694-709, 1963.

Lee, W. C., F. D. Marks, and R. E. Carbone, Velocity Track Display – A Technique to Extract Real-Time Tropical Cyclone Circulations Using a Single Airborne Doppler Radar. *J. Atmos. Ocean. Tech.*, 11, 337-356, 1994.

Miller, B. I., Characteristics of Hurricanes, *Sci.*, 157(3795), 1389-1399, 1967.

Pothier, R., Scientists Face Moral Dilemma on Hurricane Seeding. *Miami Herald*, June 20,1972.

Powell, M. D. and P. G. Black, The Relationship of Hurricane Reconnaissance Flight-Level Wind Measurements to Winds Measured by NOAA's Oceanic Platforms. *J. Wind Eng. Indus. Aerodyn.*, 36, 381-392, 1990.

Rosenthal, S. L., A Circularly Symmetric Primitive-Equation Model of Tropical Cyclones and Its Response to Artificial Enhancement of the Convective Heating Functions. *Mon. Wea. Rev.*, 99, 414-426, 1971.

Shapiro, L. J. and H. E. Willoughby, Concentric Eye Walls, Secondary Wind Maxima, and the Evolution of the Hurricane Vortex. *J. Atmos. Sci.*, 41, 3189- 3211, 1982.

Shea, D. J. and W. M. Gray, The Hurricane's Inner Core Region: I. Symmetric and Asymmetric Structure; II. Thermal Stability and Dynamic Characteristics, *J. Atmos. Sci.,* 30, 1544-1576, 1973.

Sheets, R. C., Analysis of Hurricane Debbie Modification Results Using the Variational Optimization Approach. *Mon. Wea. Rev.*, 101, 663-684, 1973.

Sheets, R. C., Hurricane Anita – A New Era in Airborne Research. *Mariners Wea. Log,* 22 (1), 1-8, 1978.

Sheets, R. C., The National Hurricane Center – Past, Present, and Future. *Wea. and Forecast.*, 5 (2), 185-232, 1990.

Simpson, R. H. and J. S. Malkus, Experiments in Hurricane Modification. *Sci. Amer.*, 211, 27-37, 1964.

Staff, NHRP, Objectives and Basic Design of the National Hurricane Research Project, NHRP No. 1, U.S. Weather Bureau, 6 pp, 1956.

Chapter 4

Evolution of Prediction Models

Mark DeMaria and James M. Gross

INTRODUCTION

In order to issue coastal watches and warnings for tropical cyclones, it is necessary to estimate the future track of the storm center and the evolution of the surface wind field. In the first half of the 20th century, forecasts were primarily based on subjective judgments and extrapolation of current trends. In the second half of the century, scientists developed objective models for estimating future storm path and intensity. The early models were based on empirical relationships between storm properties and the storm's future motion. Track prediction models based upon basic physical principles were developed in parallel with empirical models. Later, these models were generalized to predict intensity as well as track. In this chapter, we review the evolution of objective forecast models from 1959-2000. We also evaluate official forecasts from the National Hurricane Center (NHC) for this period.

Two major technological advancements helped lead the way towards the development of objective forecast models. When aircraft reconnaissance was implemented in the mid-1940s, our ability to estimate current storm position and intensity was greatly enhanced. With more accurate estimates of storm position, there was a transition from qualitative estimates of future storm motion to quantitative track forecasts. At the Hurricane Forecast Center[1] in Miami in 1954, routine track forecasts (latitude and longitude of the storm center) out to 24 h began. These forecasts were extended to 48 h in 1961 and to 72 h in 1964.

[1]This Center became the National Hurricane Center (NHC) in 1956.

Hurricane! Coping with Disaster

10.1029/055SP05

The second major advance involved the development of computer technology leading to operational numerical weather prediction, which began on a routine, daily basis May 6, 1955 [Cressman, 1996]. Computer technology also greatly enhanced our ability to develop models for tropical cyclone prediction.

Although the implementation of operational tropical cyclone models began in the 1950s, there were several early studies that helped lay the foundation for this work. For example, in 1924, Charles Mitchell of the U.S. Weather Bureau estimated the average 24-h tropical cyclone motion in 2.5° latitude/longitude squares for each month of the Atlantic hurricane season, using data from 1887-1923 – an analysis that could be used for a 24-h track forecast, with initial storm location and month.

In 1947, in an unpublished "soliloquy," Grady Norton, the senior hurricane forecaster in the U.S. Weather Bureau from 1935-54, described another early, quasi-objective forecast method. This method used winds from upper air analyses near the top of the tropical cyclone in the area surrounding the storm to estimate storm motion for the next 24 to 36 h. The speed of the estimated wind vector is reduced by about 30% to account for what Norton termed "storm vorticity." Norton also stated that to apply this method, "just get your wind stream at the proper level, make a good rough estimate of what Coriolis force, friction and centrifugal force will do, and shoot!"

Some nine years later, in 1956, Herbert Riehl and collaborators at the University of Chicago developed the first operational objective Atlantic track forecast model: using statistical regression techniques to determine relationships between the 500 hPa[2] height gradient (proportional to the pressure gradient at midlevels) and storm displacement during the next 24 h. This model is an example of a technique that uses current synoptic information to estimate future storm motion. Empirical models that include synoptic information as predictors of future track or intensity are referred to as "statistical-synoptic." In later years, statistical models were developed that include parameters related to the storm itself, such as current and previous intensity and position and date, but they do not include any synoptic information. This type of model is referred to as "statistical."

As numerical weather prediction models improved, it became apparent that the statistical-synoptic track models could also be improved by including predictors from numerical forecasts of the storm environment. Empirical models that include predictors from other forecast models are referred to as "statistical-dynamical." The first statistical-dynamical model for the Atlantic basin was implemented in 1973.

Because of computer limitations, early numerical models for large-scale weather prediction were highly simplified barotropic models, which treated the

[2]One hPa (hectopascal) is equal to one mb (millibar) pressure unit.

atmosphere as a single vertical level, using equations that filtered all but the most fundamental aspects of atmospheric motion. After a brief flirtation with quasi-geostrophic models, which allowed for vertical variation of the atmospheric flow but maintained the filtered assumption, full primitive equation[3] atmospheric models became operational by the late 1960s. The development of numerical tropical cyclone models followed a similar path. Modified versions of the early barotropic models were developed for track forecasting. The first baroclinic model for hurricane prediction became operational in 1976. Because of the reduced influence of baroclinic processes in the tropics, barotropic hurricane track models continue to be run operationally, even though they have been long abandoned for general numerical weather prediction. In the following discussion, barotropic track models are referred to as "dynamical-barotropic," and primitive equation hurricane models are termed "dynamical-baroclinic."

The first dynamical-baroclinic hurricane model covered a limited area and used forecasts from global or other regional models with coarser horizontal resolution as lateral boundary conditions. As computer power increased, however, so did the resolution of regional hurricane and global forecast models. By the early 1990s, the resolution of global models had increased to a point where it became possible to partially resolve tropical cyclones as low-pressure systems. Thus track forecasts could be obtained directly from these models.

Modeling of tropical cyclones presents a challenging problem in which a fairly small-scale, circularly symmetric disturbance (the storm) is embedded in a larger-scale surrounding flow. Because of the lack of computer power, early models could not resolve the storm circulation and, in fact, several attempts were made to separate the prediction of the surrounding atmospheric flow from that of the symmetric storm circulation. Although these separation techniques were no longer necessary when computer power increased, a similar problem occurs in the determination of the model's initial condition. The observations are not adequate to fully describe the storm, so that the circulation in the model's initial condition can be misplaced from the actual storm position, and its magnitude is often too weak. Most models include special analysis methods within a few hundred kilometers of the storm to address these resolution problems.

The development of intensity forecast models has a shorter history than that of track models. Researchers at NHC developed a simple statistical model for intensity prediction called SHIFOR (Statistical Hurricane Intensity FORecast), which uses storm properties such as latitude, longitude, current and previous intensity, and date to forecast intensity changes out to 72 h. Scientists at the Hurricane Research Division in Miami developed a more general statistical-synoptic intensity model called SHIPS (Statistical Hurricane Intensity Prediction Scheme),

[3]Primitive equation models do not assume that the motion and wind fields are in balance, that is, the winds are not constrained to be geostrophic.

which has been run operationally since 1991. In 1997, it was converted to a statistical-dynamical model. In principle, the dynamical-baroclinic track models that were implemented beginning in 1976 could also have been used for intensity forecasting but, due to computer limitations, the grid spacing was considered too coarse to accurately represent the inner core of the storm. It was not until 1992, when the Geophysical Fluid Dynamics Laboratory (GFDL) hurricane model was run in real time, that intensity forecasts from a dynamical-baroclinic model became available.

VERIFICATION DATABASE

Realizing the importance of model verification, Charles Neumann of NHC developed a comprehensive database of track forecast models for the Atlantic tropical cyclone basin beginning in 1970. This database, which has been maintained and updated since its inception, consists of the initialization date and time and the predicted latitude, longitude, and intensity (if appropriate) at 12, 24, 36, 48, and 72 h from all of the guidance models that were available in real-time at NHC. Although individual model forecasts are not available before 1970, Tracy [1966] described average forecast errors for operational track models from 1959-1964. The errors from the NHC64 model from 1964-65, presented by Miller and Chase [1966], are also included in this study. The NHC official track forecasts are available from 1954 to the present. The model and the NHC official intensity forecasts are available since 1988.

Several parameters can be used to investigate long-term trends in model forecast errors (e.g., mean, median, root mean square error, and bias). For simplicity, the mean absolute error is calculated for each year of the sample and for each model that was available. The track is the great circle distance from the predicted storm location to the observed storm location. The intensity error is the absolute value of the difference between the predicted and observed maximum sustained one-minute surface wind (ten meters above the surface). The observed positions and intensities for the verification are determined from the NHC "best track," which provides values at 6-h intervals from a post-storm analysis of all available information. For consistency with operational forecasts and best track data, track errors are calculated in nautical miles, and intensity errors are in knots. For brevity, errors at 24, 48, and 72 h are emphasized.

Official NHC forecasts are issued three hours after synoptic time (03, 09, 15, and 21 UTC). Prior to 1989, forecasts were issued one hour later. Most dynamical models are initialized several hours after synoptic time and are not available until the official forecast has already been issued. These models will be referred to as "late" models. In contrast, most of the statistical-synoptic and statistical-dynamical models can be run using forecast fields from a previous run of a large-scale atmospheric model, and these are available in time for the official forecast. These models will be referred to as "early" models. In practice, late models are used for the next forecast cycle after the track and intensity predictions have been adjusted using current position and intensity. For simplicity, the NHC official forecasts and early and late models will be verified together, even though the late models use some information that is not available for the early models or official forecasts.

Under ideal circumstances, it would be advantageous to verify the models for a homogeneous sample for each year. However, many of the models were run on an experimental basis in real time before they became fully operational. For example, the GFDL model was run on some storms once or twice per day from 1992-94, before the model was transferred to the NOAA National Center for Environmental Prediction (NCEP) in 1995, when it was run on a more routine basis. If the requirement for a homogeneous sample was imposed, many of the forecasts from the other models would be eliminated from 1992-94. For this reason, the average errors in each year were calculated using all available samples for each model. In most cases, the samples are fairly close to being homogeneous. Notable exceptions are described in the next section. The average yearly sample size of the official forecasts is 138, 104, and 84, at 24, 48, and 72 h, respectively, although there is considerable year-to-year variability. For example, the years 1970, 1977, 1983, and 1991 had sample sizes less than 25% of average at some time periods. The model sample sizes are similar to or somewhat smaller than those of the official forecasts.

Several studies have documented the performance of the Atlantic objective guidance models for subsets of the sample included in this study. The track model forecast errors for 1959-64, 1973-79, 1983-88, and 1996-1997 were described by Tracy [1966], Neumann and Pelissier [1981], DeMaria et al. [1990], and WMO[4] [1999], respectively. WMO [1999] also documents the errors from one of the intensity models. In most of these studies, the verification sample was restricted to cases where the storm was of at least tropical storm intensity (maximum surface winds 35 kt or greater). In all of the results in this chapter, the samples were restricted to cases of tropical storm intensity or greater.

It is sometimes difficult to determine trends in tropical cyclone forecast errors because of year-to-year differences in forecast difficulty. For example, in 1988, most of the storms were at low latitudes and had fairly straight tracks. In contrast, in 1994, most of the cases were at higher latitudes, and some storms had unusual tracks, especially hurricane Gordon. McAdie and Lawrence [2000] described a method to correct for forecast difficulty in a study of the errors of the official NHC track forecast errors from 1970-1998. However, they also showed that it was possible to detect statistically significant trends in NHC forecast errors without the correction because the decrease in the error over the 28 years of the time series was relatively large. This result suggests that it should be possible to detect

[4]WMO stands for World Meteorological Organization.

trends in model forecast errors without correcting for forecast difficulty, since McAdie and Lawrence attributed most of the long-term reductions in the official NHC forecast errors to improvements in model forecasts.

TRACK FORECAST MODELS

In this section, the basic properties of track guidance models are briefly reviewed and verification results presented. Further details can be found in Tracy [1966], Neumann and Pelissier [1981], Anthes [1982], DeMaria et al. [1990], and WMO [1999] and the references contained within those papers. Table 1 lists the track models included in this verification.

TABLE 1. Operational Statistical (S), Statistical-Synoptic (SS), Statistical-Dynamical (SD), Dynamical-Barotropic (DBT), and Dynamical-Baroclinic (DBC) Track Forecast Models 1959-2000.

Model	Type	Timeliness	Maximum Forecast Length (h)	Years with Verification Data
HURRAN	S	Early	72	1970-1986
CLIPER	S	Early	72	1971-2000
RHS	SS	Early	24	1959-1964
MM	SS	Early	24	1959-1964
T-59	SS	Early	24	1959-1964
T-60	SS	Early	36	1959-1963
NHC64	SS	Early	48	1964-1965
NHC72	SS	Early	72	1972-1988
NHC73	SD	Early	72	1973-1987
NHC83	SD	Early	72	1987-1989
NHC90	SD	Early	72	1989-2000
NHC98	SD	Early	72	1998-2000
NWPBAR	BTD	Late	72	1959-1964
BAM	BTD	Early	72	1987-2000
SANBAR	BTD	Late	72	1971-1989
VICBAR	BTD	Late	72	1990-2000
LBAR	BTD	Early	72	1996-2000
MFM	BCD	Late	72(48 before 1984)	1976-1988
QLM	BCD	Late	72	1988-1994
GFDL	BCD	Late	72	1992-2000
AVN	BCD	Late	72	1992-2000
UKMET	BCD	Late	72	1995-2000
NOGAPS	BCD	Late	72	1995-2000

To solve the mathematical equations of the dynamical models on a computer, the basic fields are usually represented on a grid system. Models that use this method are referred to as grid-point or finite-difference models. The resolution of the grid affects the accuracy of the solution of the equations. The grid spacing of each of the grid-point dynamical models is included in the descriptions below. An alternative method for solving the dynamical model equations is to expand the basic fields (wind, temperature, pressure, water vapor) in terms of known functions and predict the amplitudes of the functions. Models that use this solution technique are referred to as spectral models. The nonlinear products in spectral models are evaluated on a "transform grid". For comparison, the resolution of the spectral models is described in terms of the spacing of the transform grid.

Statistical Models

HURRAN (HURRicane ANalog): Track forecasts made by finding an analog with a database of 782 historical cases from 1886 to 1979. The model also provides track probability information. It is not always possible to find an analog storm so the HURRAN sample was usually smaller than that of other models.

CLIPER (CLImatology and PERsistence): Regression equation model with current and previous storm motion, latitude, longitude, and time of year predictors. CLIPER is still run operationally because it is a benchmark for the evaluation of the skill of more general models.

Statistical-Synoptic Models

RHS (Riehl-Haggard-Sanborn): Regression model with 500 hPa geostrophic wind predictors. The RHS model is a modified version of what is considered the first objective Atlantic track forecast model.

MM (Miller Moore): Regression model with 700 hPa geostrophic winds, heights and height tendencies, and past storm motion predictors.

T-59 (Traveler's 1959): Regression model with surface pressure and past storm motion predictors.

T-60 (Traveler's-National Hurricane Research Laboratory 1960): Regression model with 500 hPa height and geostrophic wind, surface pressure, and past storm motion predictors.

NHC64: Regression model with 500 hPa height, height tendency, and geostrophic wind, surface pressure, 700 hPa heights, 1000-700 hPa, and 500-700 hPa thickness, and past storm motion predictors.

NHC67: Regression model that modifies a track forecast from climatology and persistence using current and 24 h old 1000, 700, and 500 hPa height predictors. This was the first statistical or statistical-synoptic model that provided forecasts out to 72 h.

NHC72: Regression model with output from the CLIPER model, and current and 24 h old 1000, 700, and 500 hPa height predictors.

Statistical-Dynamical Models

NHC73: Regression model with output from the CLIPER model, and current and 24-, 36-, and 48- h forecasts of 1000, 700, and 500 hPa geopotential heights. The geopotential height forecasts were from the National Meteorological Center's (NMC) primitive equation model.

NHC83: Regression model with deep-layer mean (1000-100 hPa) geopotential height predictors from the NMC/NCEP global spectral model. The predictors are determined in a storm-motion relative coordinate system, and separate regressions were performed for storms north and south of 25°N. The final prediction is statistically combined with a CLIPER forecast.

NHC90: Same as NHC83 but with an updated developmental data sample (1962-1981 replaced by 1975-1988).

NHC98: Same as NHC90 but includes a method to remove the tropical cyclone contribution from the geopotential height fields.

Barotropic-Dynamical Models

NWP-BAR: This is a generic name assigned to the series of barotropic hurricane track forecast models that were run at NMC from the mid-1950s to the late 1960s. The formulation and resolution of these models varied during this period. The very early models provided a track forecast by following a height minimum in a 500 hPa forecast. This technique was replaced by a method where the barotropic forecast provided a "steering flow" that was modified by including effects of the storm circulation.

SANBAR (SANders BARotropic): This model solves the barotropic vorticity equation using a finite-difference method with a grid spacing of 1.5°. Several modifications were made to the formulation of SANBAR, and in 1985 the grid spacing was reduced to 0.75° . The model was initialized with horizontal winds vertically averaged from 1000 to 100 hPa. Near the observed storm center, the analyzed fields were replaced by the sum of an axisymmetric vortex and the storm motion vector.

BAM (Beta and Advection): This model follows a trajectory using the wind forecast from the NCEP global model. The vertically averaged horizontal wind is used for the trajectory calculation, which includes a correction term to account for the drift of the storm due to the "beta-effect" caused by the variation with latitude of the Coriolis parameter. To remove the global model representation of the storm circulation, a spatial filter is applied to the wind fields. The version of the BAM model that was implemented in 1987 used winds that were vertically averaged from 850 to 200 hPa. Beginning in 1990, three versions of the BAM model were run, which included winds averaged from 850-200 hPa (BAM-Deep, BAMD), 850-400 hPa (BAM-Medium, BAMM), and 850-700 hPa (BAM-Shallow, BAMS). Strictly speaking, the BAM model is not purely barotropic, since the evolution of the large-scale flow is determined from a baroclinic model. However, since the prediction of the storm motion is determined from a single vertical layer (for each version of the model), the BAM model is considered a barotropic model. Also, the BAM model uses a 6-h old global model forecast so that it is an early model.

VICBAR (VIC Ooyama BARotropic): This model treats the atmosphere as a single shallow layer of fluid. The model equations (shallow water equations) are solved on a series of nested domains using a highly accurate spectral method developed by Vic Ooyama. The transform grid spacing on the inner mesh is 25 km, and the outer boundary conditions are from the NCEP global model. The model is initialized with heights and winds from the NCEP global model vertically averaged from 850 to 200 hPa. The fields near the storm center are replaced by an idealized vortex that represents the storm circulation. The size of the idealized vortex crudely matches that of the observed storm. Although VICBAR was never implemented operationally at NCEP, it has been run in real time on an experimental basis since 1990.

LBAR (Limited area sine-transform BARotropic): This model is a simplified version of VICBAR that was implemented at NCEP in 1996. The vortex initialization is the same as for VICBAR, but the shallow water equations are solved using a spectral method on a uniform resolution domain with a 75-km transform grid spacing. The lateral boundary conditions are from the NCEP global model. The domain size is smaller than that of VICBAR to help increase the influence of the boundary conditions. The primary advantage of LBAR is that the initial and boundary conditions are from the 6-h old global model forecast so that it is an early model.

Baroclinic-Dynamical Models

MFM (Moveable Fine Mesh): The MFM was the first operational dynamical-baroclinic tropical cyclone model for the Atlantic basin. It included 10 vertical levels with a 60 km horizontal grid spacing on a 3,000 km square domain. The horizontal domain moved to remain centered on the storm, and the lateral boundary conditions were obtained from the NMC limited-area Fine Mesh (LFM) model until 1984, when they were obtained from the NMC global forecast model. The storm circulation was initialized by adding fields from a version of the model with no environmental flow. Because of computer limitations, the MFM was only initiated for storms that threatened populated land and was only run to 48 h for the first several years after its implementation. The forecasts were extended to 72 h beginning in 1984. In the verification results below, the sample sizes for the MFM were considerably smaller than those of the other models, especially during the period from 1976 to 1983.

QLM (Quasi-Lagrangian Model): The QLM is similar to the MFM, but with increased horizontal (40 km grid spacing) and vertical (18 levels) resolution. The term “Quasi-Lagrangian” refers to an accurate mathematical method used to solve the model equations. The horizontal domain of the QLM was larger than that of the MFM (4,400 km square domain), but the domain did not move to follow the storm. The lateral boundary conditions were obtained from the NMC global forecast model. The initial storm circulation was represented by an idealized symmetric vortex that was added to a large-scale analysis.

GFDL (Geophysical Fluid Dynamics Laboratory): This model is an operational version of the GFDL research hurricane model. It has 18 vertical levels and three nested grids, where the horizontal grid spacing on the innermost nest is 1/6°. The inner two grids move to remain centered on the storm, and the boundary conditions on the outer mesh are determined from the NCEP global forecast model. The model includes a fairly sophisticated method for separating the initial flow from the global forecast model into vortex and environmental components. The vortex component is replaced by a vortex that is generated by a simplified version of the GFDL model. This model was run on an experimental basis in real time from 1992-1994 for selected storms and became fully operational at NCEP in 1995. In the verification results, the sample sizes for the GFDL model prior to 1995 are smaller than those of many of the other models.

AVN (Aviation run of the NCEP Medium Range Forecast Model): This is a track forecast from the NCEP global spectral model. From 1992-1999 storm circulation was represented by adding "synthetic" observations to the model data assimilation system. Beginning in 2000, a "vortex surgery" method is used, where storm circulation is determined from available data, and the resulting circulation is moved to the proper locations at the beginning of each forecast. The horizontal spectral truncation of the model was T126 when the model was first used for operational track forecasting. For a T126 truncation, the transform grid spacing is about 106 km. The spectral truncation has recently been increased to T170 (78 km transform grid spacing).

UKMET (United Kingdom METeorological office global model): This is a track forecast from the UK Met office global model. The symmetric storm circulation is initialized by adding synthetic observations to the data assimilation system. The synthetic observations also include an asymmetric component to represent the current motion of the storm. As of 1999, the horizontal grid spacing was 0.56° latitude and 0.83° longitude.

NOGAPS (Navy Operational Global Atmospheric Prediction System): This is a track forecast from the Navy's global spectral model. The storm circulation is represented by adding synthetic observations to the data assimilation system. The spectral truncation was initially T80 (166 km transform grid spacing) but was increased to T159 (84 km transform grid spacing).

Verification Results

Average errors for statistical, statistical-synoptic, and statistical-dynamical models are shown in Figure 1. Because forecast errors for individual years were

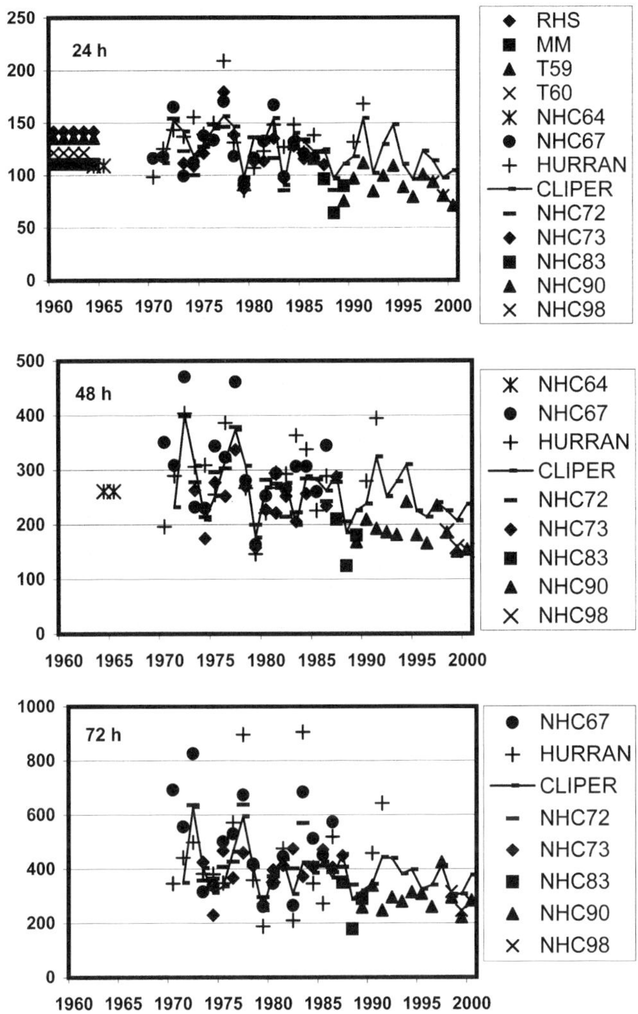

Figure 1. Track errors (nmi) from the statistical, statistical-synoptic, and statistical-dynamical models at 24, 48 and 72 h for each year of availability from 1960-2000.

not available for 1959-1965, the average error for the total period is plotted over this span of years. All of the diagrams begin in 1960, since there is no prior information. To make the plot less cluttered, only the CLIPER model errors for each year are connected with a line. The simple CLIPER model is often considered a benchmark for evaluating model skill. If the average error for a given model is below the CLIPER line in Figure 1, then it is considered skillful.

Although there is considerable year-to-year variability in the errors, a few trends are evident in the 24-h errors in Figure 1. There is, at best, only a slight downward trend in the model errors from 1970-1986, and most of the model errors are near or above the CLIPER line. In fact, most of the yearly average errors for this period are larger than those from the best models from 1959-1964 (MM and NHC64). A major improvement in the 24-h errors is evident beginning in 1987 with the NHC83 model, and continuing with NHC90 and NHC98. The average errors for these models were smaller than those of the CLIPER model for every year since 1987. The NHC83 model was run in real time beginning in 1983, but the forecasts were not available in the NHC data file until 1987. The results of DeMaria et al. [1990], which provided a validation of NHC83 for the period 1983-1988, suggest that the NHC83 model also had errors smaller than CLIPER prior to 1987.

The average errors at 48 h show trends similar to that at 24 h. There is only slight improvement from 1970-1986, but a significant improvement beginning in 1987 with the NHC83 model. There is a hint that the NHC73 model did provide some improvement from 1973-1987. This model had average errors that were smaller than CLIPER in 13 of 15 years during this period. As described above, NHC73 was the first statistical-dynamical track model.

The average errors at 72 h in Figure 1 are similar to the other time periods. The primary improvement occurred with the implementation of NHC83 and continued with NHC90 and NHC98. However, the good performance of NHC73 at 48 h did not continue through 72 h. For the period 1973-1987, the average errors from NHC73 were smaller than those of CLIPER for only six of 15 years.

For several years, the HURRAN errors (as shown in Figure 1) are the largest of any model and also appear to have the largest year-to-year variability. This result indicates that the analog technique as originally formulated does not provide competitive forecasts. Figure 1 also shows that there is little difference between the average errors from NHC90 and NHC98 during the three-year period of overlap. This result indicates that the removal of the tropical cyclone circulation from the global model forecast has only a slight influence on the predicted track using this statistical-dynamical approach.

Figure 2 shows the track errors from the dynamical-barotropic models. Only the deep version of the BAM model (BAMD) is shown, since the shallow version (BAMS) is only run for reference for highly sheared or dissipating systems and usually has much larger errors than BAMD. The errors for BAMM are similar to those from BAMD.

A trend of increasing accuracy is evident in Figure 2 at all three forecast periods. The SANBAR errors are smaller than those of the early barotropic forecasts (NWP-BAR) in most years, and the current set of barotropic models (BAM, LBAR, and VICBAR) generally have smaller errors than SANBAR. The improvements over this period are likely due to the improved numerical methods used to solve the forecast equations, better initialization techniques, and improvements in the large-scale models used for initial and boundary conditions.

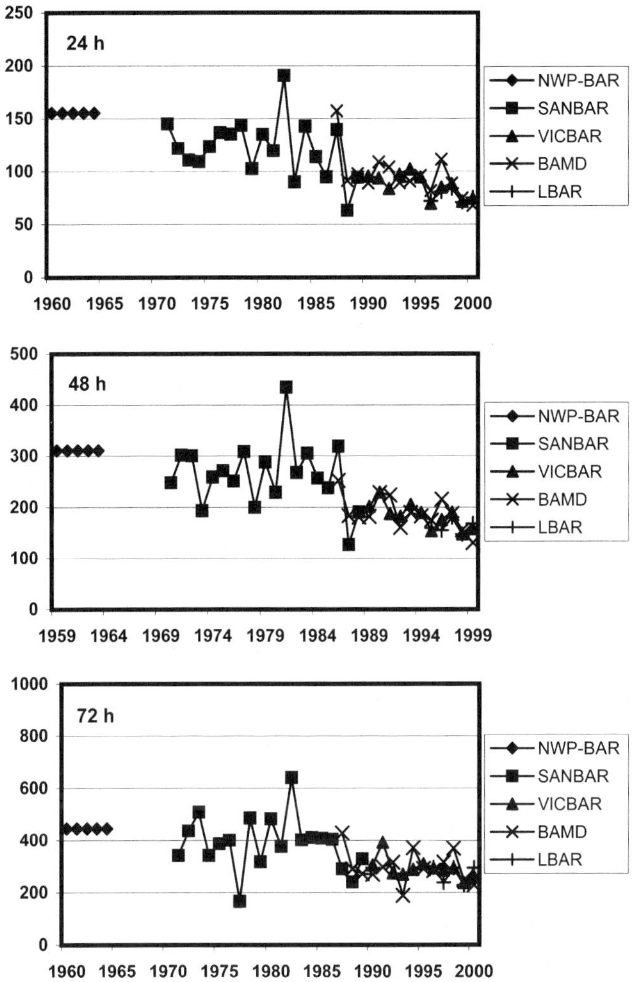

Figure 2. Track errors (nmi) from the dynamical barotropic models at 24, 48 and 72 h for each year of availability from 1960-2000.

Figure 3 shows the track errors from the dynamical-baroclinic models. At 24 h, the errors from the MFM model are comparable to those from the models at later times, except possibly the GFDL model. The performance of the MFM appears to degrade in the latter part of its sample (1987-89). During this period, the model was transferred to a new computer system at the National Meteorological Center (NMC)[5]. It is possible that errors were introduced during

[5]NMC became NCEP in 1995.

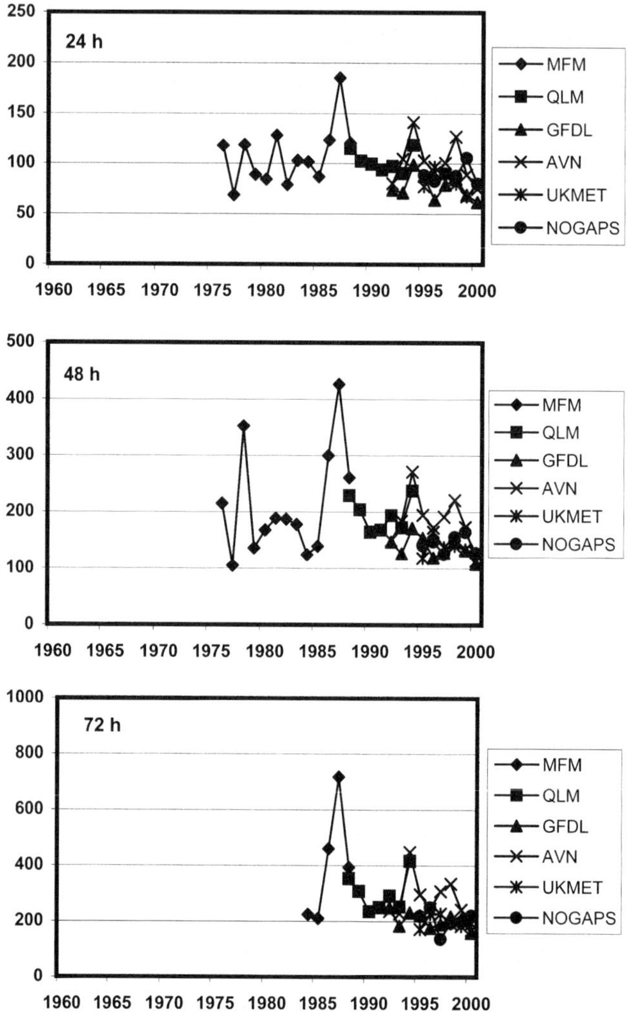

Figure 3. Track errors (nmi) from the dynamical baroclinic models at 24, 48 and 72 h for each year of availability from 1960-2000.

this transfer. The large-scale model used in the initialization of the MFM was also changed from a grid point to a spectral model during the late 1980s. This modification may also have had a negative influence on the MFM. This possibility is described in more detail when the model types are compared.

The AVN errors in Figure 3 are generally larger than those from the other global models (UKMET and NOGAPS). Several improvements were made to the tropical cyclone initialization and vortex tracking in the AVN beginning in 2000. In that year, the AVN errors at 24 h were comparable to those from the other global models.

The errors at 48 and 72 h in Figure 3 have similar trends to those at 24 h. The MFM errors are larger for the last three years of its sample, although there are also fairly large MFM errors at 48 h in 1977. The AVN errors are again larger than those of the other global models before 2000. The errors from the GFDL and global models appear to be smaller than those from the QLM.

The results in Figures 1-3 show that there have been improvements in all of the different types of forecast methods. The results also indicate that, in many cases, the improvements were incremental, rather than gradual. For example, the dynamical-barotropic models showed an improvement with the introduction of the SANBAR model around 1970 and a second improvement with the BAM and VICBAR models.

In order to compare the forecasts using the different methods, the errors from each of the model types were averaged for each year. Then, the yearly values for each model type were averaged over five-year periods. The statistical-synoptic and statistical-dynamical models were combined, since there was a transition from the first to the second type beginning in the early 1970s. The HURRAN model was not included since its performance was generally much worse than other available techniques. Thus, the statistical model average only includes CLIPER. The BAMS shallow model was excluded from the dynamical-barotropic sample, and the AVN was excluded from the dynamical-baroclinic sample, since the error characteristics of these models were not representative of other models of similar type available during the same time periods. The errors of the official NHC forecasts were also averaged over these same five-year periods.

Figure 4 shows the five-year average errors from each of four types of techniques and from the official NHC forecasts. The errors are plotted at the end of each five-year period. Since little information was available from 1966-1970 for the forecast models, the average errors from the 1961-1965 and 1971-1975 samples were used if both of these were available for that model type. As described previously, the simple CLIPER model is used as a benchmark for model skill.

Figure 4 shows that the early barotropic models had very large errors relative to the statistical models at 24 and 48 h. The forecasts improved with the implementation of SANBAR. However, it appears that little other improvement in the statistical or barotropic models occurred from 1960-1985, and the average performance of these models was only slightly better than CLIPER during this period.

The errors of the MFM, which became operational in 1976, were much smaller than those of the other models from 1976-1985. However, this difference might be explained by the fact that the sample size for the MFM was much smaller than that of the other models since it was only run for storms that had the potential to affect populated land areas during much of this time period. A homogeneous comparison of CLIPER and the MFM for the period 1976-1985 sug-

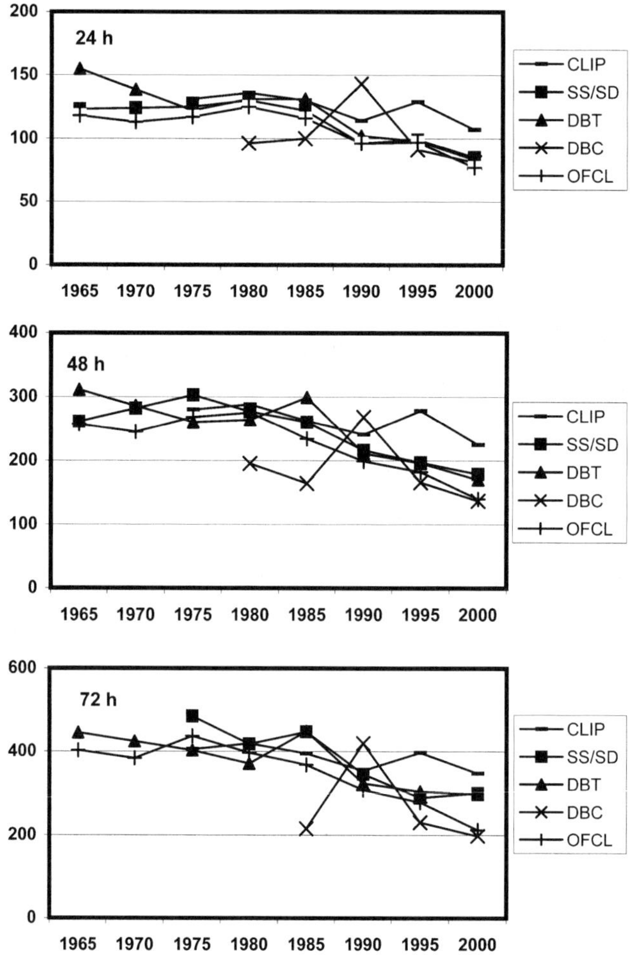

Figure 4. Five-year average 24, 48 and 72 h track errors (nmi) for the CLIPER model, the statistical-synoptic and statistical-dynamical models (SS/SD), the dynamical-barotropic models (DBT), the dynamical-baroclinic (DBC) models, and the official NHC forecasts (OFCL).

gested that the MFM errors were considerably smaller than those of CLIPER, although the difference was not quite as large as that shown in Figure 4.

During the five years ending in 1990, an interesting development occurred where the dynamical models appeared to go through a period of "dark ages". The performance of the dynamical models (MFM and QLM) degraded, and the barotropic and statistical-dynamical models improved when the BAM model, VICBAR, NHC83, and NHC90 were implemented. The degradation of the dynamical models might be related to the conversion of the MFM to the new computing system at

NCEP. The improvements in the global models may also have contributed to the increased errors in the MFM and the QLM. As described previously, it is difficult to accurately represent the tropical cyclone circulation in the initial condition of dynamical models. By the late 1980s, the tropical cyclone circulation in the global model was better resolved, but the storm position was often displaced from the observed position, sometimes by several hundred kilometers. Since the global model analysis is the starting point for the MFM and QLM, the displaced storm circulation probably had a negative influence on the initial conditions in these regional hurricane models, where the regional model vortex was steered by the displaced global model's tropical cyclone. The global model representation of the storm probably had a smaller influence on the statistical-dynamical models because the height predictors are very far from the storm center. (For example, in NHC90 the predictors are usually more than 600 km from the storm center.) The global model vortex probably did not have much effect on the dynamical-barotropic models because the BAM model applies a strong filter to the fields before calculating the trajectories, and the VICBAR and LBAR models replace the global model analyses with a symmetric vortex within several hundred kilometers of the storm center.

In the 1990s the "dark ages" of the dynamical-baroclinic models ended with the introduction of the improved GFDL model, as well as the continued improvements in the global models. The track errors during this period lived up to the early promise of the MFM and provided forecasts with smaller average errors than the other techniques, especially at 48 and 72 h.

The official forecast errors in Figure 4 are generally similar to those of the best available models. There was only a slight reduction in official forecast errors from 1960-1985, when the skill of the models was minimal. The official forecast showed significant increases in accuracy from 1985-2000, suggesting that the model improvements over this period played a major role in improving the official forecasts. A notable exception is the large difference between the official forecast errors and those of the dynamical-baroclinic model (the MFM) from 1976-1985. It is possible that, because this model was not routinely available for all forecasts, it had a much smaller impact on the official forecasts.

To illustrate the current track forecast skill, a verification of the CLIPER, NHC90, LBAR, GFDL, and UKMET models was performed with a homogenous sample from 1997-2000. This time period was chosen for consistency with the intensity forecast models, as will be described later. These five models were chosen as representative of the statistical, statistical-dynamical, dynamical-barotropic and dynamical-baroclinic (regional and global) techniques. To determine forecast skill, the errors from each model were normalized by the CLIPER model errors to give the relative error (RE) defined by

$$RE = 100(E_{model}-E_{CLIPER})/E_{CLIPER} \quad (1)$$

where E_{model} is the average error at a given forecast time for each model, and

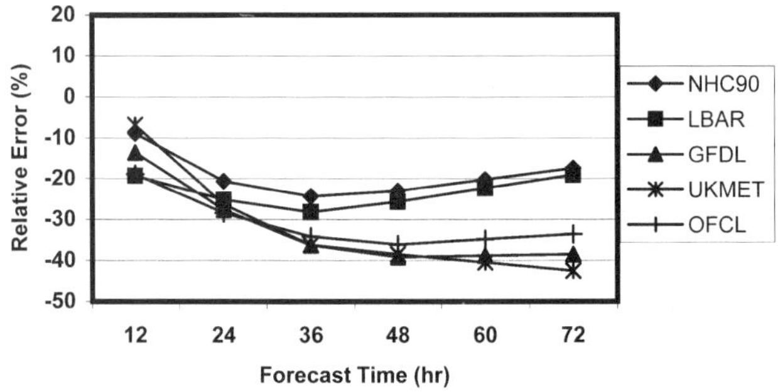

Figure 5. Relative track error (%) at 12-72 h of representative statistical-dynamical (NHC90), dynamical-barotropic (LBAR) and dynamical-baroclinic (GFDL and UKMET) models and the official NHC forecasts for a homogeneous sample from 1997-2000. The errors are relative to the corresponding CLIPER model errors.

E_{CLIPER} is the average CLIPER model error at the same forecast time. A negative RE represents the percent improvement over CLIPER, and thus, the model has forecast skill. When RE is positive, the model does not have skill.

Figure 5 shows the relative errors of the four track models. All of the models are skillful at all time periods from 12 to 72 h. The skill of LBAR and NHC90 are similar, but the GFDL and UMKET models have much more skill after 24 h. For the longer-range forecasts, the GFDL and UKMET models are able to improve upon climatology and persistence by about 40%. The official forecasts have about the same skill as the best forecast models.

INTENSITY FORECAST MODELS

In this section, we review intensity guidance models. It is important to note that the intensity models can be classified in a manner similar to the track models, save that there are no dynamical-barotropic intensity forecast models. The barotropic assumption precludes intensity changes.

Statistical Model

SHIFOR – (Statistical Hurricane Intensity FORecast): A statistical model that uses climatological and persistence predictors to forecast intensity change. Beginning in 1988, modifications to the model were made by Arthur Pike of NHC, which included a more recent developmental sample and a simplification of the model formulation. The developmental sample for SHIFOR is restricted to storms that remained over the water for the entire forecast period. The simple SHIFOR model is similar to the CLIPER track forecast model and is useful for providing a benchmark for the evaluation of more general intensity forecast models.

Statistical-Synoptic Model

SHIPS (Statistical Hurricane Intensity Prediction Scheme): A statistical-synoptic model that uses climatological, persistence, and synoptic predictors to forecast intensity changes. The primary predictors include the difference between the maximum possible intensity and the current intensity, the 850-200 hPa vertical shear of the horizontal wind, persistence (the previous 12-h intensity change), the 200-hPa eddy flux convergence of relative angular momentum, and the 200 hPa zonal wind and temperature. The maximum possible intensity is determined from an empirical relationship with sea surface temperature (SST). All of the predictors are determined from the NCEP global model analysis. The developmental sample for SHIPS includes only cases that were over the water, similar to the SHIFOR sample. This version of SHIPS was run from 1991-1996. Minor modifications were made after each year.

Statistical-Dynamical Model

SHIPS: Beginning in 1997, the SHIPS model was converted from a statistical-synoptic to a statistical-dynamical model. The primary atmospheric predictors out to 48 h were obtained from a numerical forecast model rather than from the analysis at the initial time. Another significant change to the model occurred in 2000, when the effects of land were included using a simple empirical decay model and the forecast fields out to 72 h.

Dynamical-Baroclinic Model

GFDL: The GFDL model described above in the track model section is also used for intensity prediction.

VERIFICATION RESULTS

Because the number of intensity prediction models is much smaller than the number of track models, it was possible to perform the verifications for a homogeneous sample without losing too many forecasts each year. All of the results in this section are for homogenous samples.

Figure 6 shows the average errors from the intensity forecast models and the official NHC forecasts for 1988-2000. From 1991-1996, the SHIPS errors were similar to or larger than the SHIFOR errors. However, when SHIPS was converted to a statistical-dynamical model in 1997, the errors from SHIPS became smaller than the SHIFOR errors during most years. The GFDL model errors are

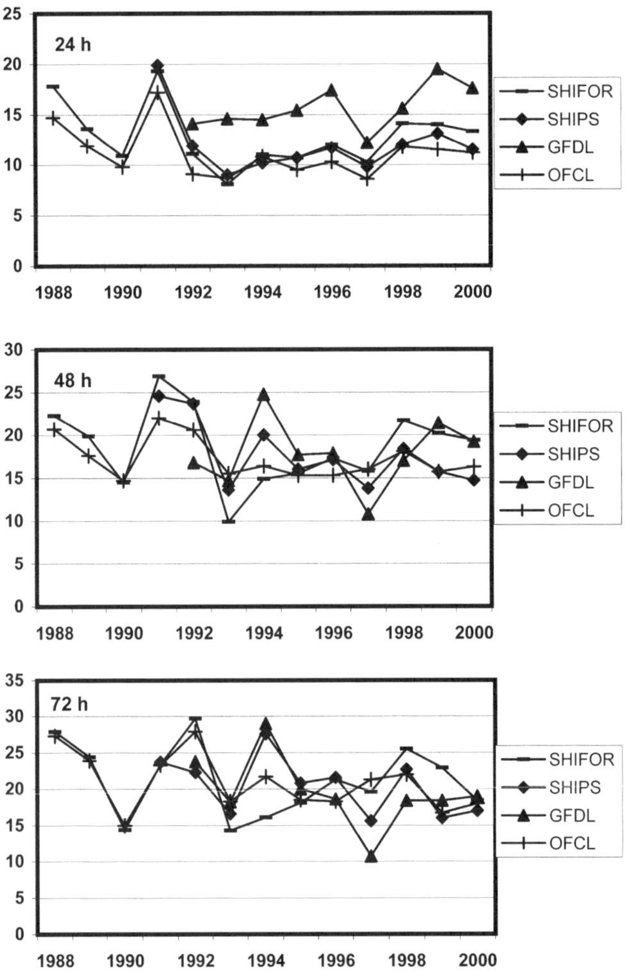

Figure 6. Average intensity errors (errors in predicting the maximum wind speed in kts) from the SHIFOR, SHIPS and GFDL models and the official NHC forecasts at 24, 48 and 72 h for each year of availability from 1988-2000.

quite large at 24 h relative to SHIFOR and SHIPS. At the later times, however, the errors become comparable to the statistical models, especially since 1997. The official forecast intensity errors are generally similar to the best available models, again indicating that model improvements translate directly into better official forecasts.

The skill of the intensity models can be evaluated in a manner similar to the track models by comparing the errors to those from a simple forecast based upon climatology and persistence. For this purpose, the errors for the SHIPS and GFDL models and the official forecasts were normalized by the SHIFOR errors using an equation analogous to Equation (1) for the track models. The relative errors were calculated for a homogenous sample of forecast cases from 1997-2000. This time period was chosen since the SHIPS model was converted to a statistical-dynamical model in 1997.

Figure 7 shows the relative errors from the SHIPS and GFDL models and the official forecasts. The relative errors for the GFDL model at 12 and 24 h are larger than 20% and are not shown in Figure 7. This figure shows that the SHIPS model and the official NHC forecasts have skill at all time intervals from 12 to 72 h and that the official forecast skill is greater than the models at 12 and 24 h. It is interesting that the relative errors from SHIPS are fairly similar to those from the statistical-dynamical track model (NHC90) in Figure 5. The GFDL model has no skill in the early forecast periods (12-36 h) but, by 72 h, has skill that is comparable to the SHIPS model. The lack of skill in the early part of the forecast period may result from the difficulty of initializing the more complex intensity forecast models with an assumed strong, small-scale hurricane circulation.

Comparing Figures 7 and 5 shows that the best intensity model at each forecast time reduces the error relative to climatology and persistence by only about half as much as the best track model at each forecast model. By this measure, the current level of skill for intensity forecasting is about half that of track forecasting.

SUMMARY AND DISCUSSION

This chapter reviews errors from Atlantic track forecast models and official NHC forecasts from 1959-2000. Errors from the first barotropic track forecasts are quite large relative to statistically based techniques. There is some improve-

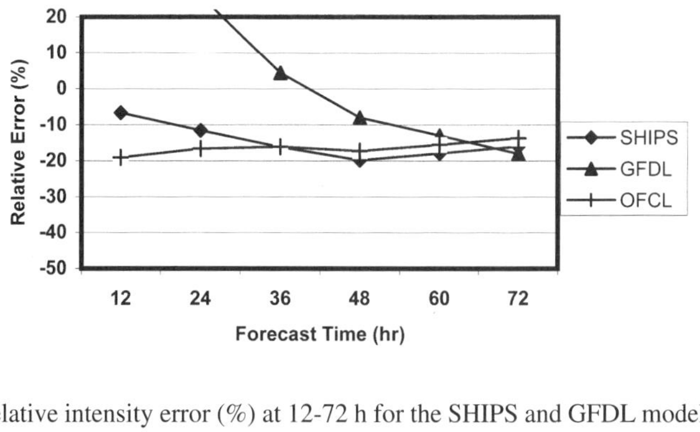

Figure 7. Relative intensity error (%) at 12-72 h for the SHIPS and GFDL models and the official NHC forecasts for homogeneous samples from 1997-2000. The errors are relative to the corresponding SHIFOR model errors.

ment in the barotropic model forecasts with the introduction of the SANBAR model in the late 1960s. However, progress in the statistical track forecasts was slow from 1959 to the early 1980s. During much of this period, statistical and barotropic models performed similarly. The first baroclinic model (the MFM) became operational in 1976, and had errors that were much smaller than the other techniques for the limited number of cases for which it was run. In the late 1980s and early 1990s, there was considerable improvement in the statistically based and barotropic models. The baroclinic model performance degraded during the same period, perhaps due to a computer conversion and to unexpected side effects from improvements in global models that provided initial and boundary conditions. During this period, statistical and barotropic models tended to outperform baroclinic models. Beginning in 1992, this situation reversed when the GFDL model was introduced, and global models progressed to the point where they could directly provide track forecasts. For a recent homogeneous sample of cases (1997-2000), the three basic techniques (statistically based, barotropic, and baroclinic) have comparable skill (relative to climatology and persistence forecasts) out to 24 h, but the baroclinic models are considerably more skillful in the longer time intervals.

Official NHC track forecasts show trends similar to those of the best models that were consistently available at the time. This result suggests that model improvements translate directly into improvements in the official forecasts. This demonstration of the rapid and successful applications of research is one of the most important conclusions of this chapter.

The performance of the Atlantic intensity forecast models from 1988 to 2000 is also reviewed. Other than the simple SHIFOR model, intensity forecast guidance models have only been available since 1991. For a homogenous sample from 1997 to 2000, the statistical-dynamical SHIPS model had some intensity forecast skill at 12-72 h, relative to climatology and persistence (SHIFOR). The baroclinic GFDL model had little or no skill at 12-36 h but was comparable to the SHIPS model at 72 h. The current skill of the intensity models and the official NHC forecasts is about half that of the track forecast models for forecasts of 24 h or longer.

Since 1959, the performance of track forecast models shows considerable improvement, especially during the last 20 years. It is likely that this improvement will sustain. Rapid advances in computer technology will allow for improved model resolution and the implementation of more advanced initialization and assimilation procedures. In addition, the number and accuracy of observations available for tropical cyclone analysis are increasing. A new dropsonde, which provides wind and thermodynamic measurements with much greater vertical resolution than was previously available [Hock and Franklin, 1999], has recently been developed. These dropsondes are routinely released during Air Force Reserve and NOAA aircraft reconnaissance missions. Important deductions from these sondes are discussed in Chapter 3. Continued development of unmanned aircraft (Chapter 13) also shows promise for increased *in situ* measurements in and around tropical cyclones. Remote sensing of tropical cyclones will also improve with the implementation of very high-resolution infrared sounders planned for the next generation of Geostationary Operational Environmental Satellites (GOES). Microwave remote sensing holds promise for tropical cyclone analysis, since these instruments can provide measurements below the cloud top. Finally, the radio occultation method, using the GPS constellation, can provide accurate and high-vertical resolution soundings that contribute independent information on the temperature and water vapor structure of the tropical storm and its environment (Chapter 11).

Improvements in intensity forecast models should also be possible, although this is a considerably more difficult task than improving track forecasts. The current version of the GFDL model is hydrostatic and has spatial resolution that is marginal for the representation of the storm's inner core. These limitations will be eliminated as computer power continues to increase. The lack of intensity skill of the GFDL model during early forecast periods indicates that improvements in initializing the model will be necessary. Nonetheless, the proper initialization of a non-hydrostatic model with microphysical and detailed boundary layer parameterizations will be a difficult task. There is also evidence that, in some cases, ocean response to the storm will need to be taken into account. Because of the complexity of the problem, it will likely be several years before baroclinic forecast models show the same increase in skill relative to statistical models as they currently do for track forecasts.

Although this review emphasizes track and intensity forecasts, there are other aspects of tropical cyclones that need to be considered. The areal extent of the storm-strength surface wind field has an impact on the storm surge, which determines the areas that require coastal evacuation. In addition, flooding rainfall rather than wind has caused the majority of the loss of life from U.S. landfalling tropical cyclones since 1970. Thus, in the coming years, additional emphasis should be placed on the prediction and verification of storm rainfall, and the areal distribution of the wind field.

REFERENCES

Anthes, R.A. Tropical Cyclones: Their Evolution, Structure and Effects. *Amer. Meteor. Soc. Monographs,* 41, 208 pp, 1982.

Cressman, G.P. The Origin and Rise of Numerical Weather Prediction. *Historical Essays on Meteorology, 1919-1995*, Boston: AMS, 21-42, 1996.

DeMaria, M., M.B. Lawrence, and J.T. Kroll. An Error Analysis of Atlantic Tropical Cyclone Track Guidance Models. *Wea. Forecasting,* 5, 47-61, 1990.

Hock, T.F. and J.L. Franklin. The NCAR GPS Dropwindsonde. *Bull. Amer. Meteor. Soc.,* 80, 407-420, 1999.

McAdie, C. and M.B. Lawrence. Improvements in Tropical Cyclone Track Forecasting in the Atlantic Basin, 1970-98. *Bull. Amer. Meteor. Soc.,* 81, 989-997, 2000.

Miller, B.I. and P.P. Chase. Prediction of Hurricane Motion by Statistical Methods. *Mon. Wea. Rev.,* 94, 399-406, 1966.

Neumann, C.J. and J.M. Pelissier. Models for the Prediction of Tropical Cyclone Motion Over the North Atlantic: An Operational Evaluation. *Mon. Wea. Rev.,* 109, 522-538, 1981.

Tracy, J.D. Accuracy of Atlantic Tropical Cyclone Forecasts. *Mon. Wea. Rev*., 94, 407-418, 1966.

WMO. *Tropical Cyclone-Related NWP Products and Their Guidance.* World Meteorological Organization Technical Document No. 966, Geneva, Switzerland, 1999.

PART II

Societal and Economic Challenges

Our ability to cope with the hurricane menace may begin, but does not end, with science and technology. It is a multi-faceted problem involving perceived needs for individual actions in response to proclaimed needs from forecasters and emergency management officials at all levels of government, including policy planning, to mitigate hurricane risks to life and property.

Chapter 5 serves as a bridge between the discussions of science and technology in Part I and the social and economic problems discussed in Part II. In recounting the events leading up to, during, and following the Galveston disaster of 1900, the entire spectrum of hurricane problems that must be faced individually and by a community are brought into focus here, then considered more specifically in the chapters on societal impacts which follow.

Chapter 6, in addressing longer range policy planning for mitigating losses from hurricanes, asks poignant economic questions as to where cost effectiveness must become a concern.

Chapter 7, after describing the purpose and applications of the Saffir/Simpson Damage Potential Scale, discusses the need for maintaining effective building standards and codes and reviews the lessons learned in the Miami area from Hurricane Andrew.

Chapter 8 presents a broader view of the prospects for reducing overall vulnerability, considering interactions between social, scientific, and technology concerns and capabilities.

Hurricane! Coping with Disaster

10.1029/055SP06

Sacred Heart Church in Galveston following the 1900 Hurricane. Photograph courtesy of the Rosenberg Library, Galveston, Texas

Chapter 5

The Great Galveston Hurricane of 1900

Neil L. Frank

The effectiveness of community efforts to cope with a hurricane depends upon an interactive process involving technical procedures for tracking and warning, societal issues including public response to warnings, and government-sponsored preparedness and safety measures.

In this first of four chapters in Part II, we turn to events leading up to and during the landmark tragedy at Galveston in 1900, as a bridge between Part I with its emphasis on technology and later chapters in Part II that address the socioeconomic impact of hurricanes. Here we display the limitations of technical procedures in 1900, as well as the societal impact, including public responses in the face of differing perceptions of imminent danger. In this saga the concern is primarily with the sudden lethal threat that confronted residents and responsible authorities alike in search of means for coping with the crisis.

In our view, this terrible event is a suitable benchmark for evaluating progress during the 20th century. Chapters 6 and 8 review and analyze other aspects of the societal and economic problems imposed by hurricanes, including policy planning. Chapter 7 stresses the urgency for proper engineering design and standards for new construction, and suggests appropriate guidelines.

INTRODUCTION

On September 8, 1900, a fifteen-foot storm surge associated with a great hurricane inundated Galveston Island killing at least 8000 people according to Rappaport and Partagas [1995]. The actual death toll will never be known

Hurricane! Coping with Disaster

10.1029/055SP07

because the magnitude of the disaster far exceeded the ability to accurately count and identify bodies. It is very likely that many of the dead were washed out to sea. Nevertheless, it was the deadliest natural disaster in this nation's history.

Improvements in tracking and forecasting hurricanes during the 20th Century, examined in Part I and later in Part III, have been very significant. Unfortunately, as discussed in Chapter 2, the rate of improvement in hurricane forecasting has not kept pace with the rate of increase in coastal population. Even today, a perfect hurricane warning could result in a disaster if people do not respond. Emergency managers and hurricane forecasters, faced with a rapid influx of new coastal residents, have been hard pressed to educate and properly prepare these residents for a direct hit by a major hurricane. And with these increases in populations-at-risk, the increases in warning time required for relocation of residents exceeds that which can be dependably provided by forecasters.

These and other problems that relate to the progress made, as well as our present ability to cope, may be instructive as we revisit the events before, during, and after the Great Galveston Hurricane of 1900. In considering overall progress during the 20th Century, together with what has happened on Galveston Island during that period, we will project an opinion as to whether Galvestonians are prepared for another Great Hurricane.

Weems [1957] and, more recently, Larson [1999] present excellent reviews of the Great Hurricane of 1900. I have used several of the eyewitness accounts from Larson's book in this chapter.

We are now reasonably confident that this hurricane had its origin in Africa. The logs from two different ships reveal the presence of a tropical weather disturbance well east of the Caribbean on Monday, August 27th, and again on Tuesday, August 28th.

About 100 weather disturbances are tracked each summer in the tropical Atlantic. Of these more than 50 have their beginnings in Africa. Every three to five days, May through November, an African disturbance enters the Tropical Atlantic and moves slowly westward under the influence of the trade winds. Many strengthen as they pass over the warm waters of the western Atlantic in late summer. Each year in the Atlantic and eastern Pacific over half of the named storms are spawned from African disturbances. With increases in ocean area observations since the two World Wars, it has been possible to trace nearly all major hurricanes that struck the United States to an African source. From the records available and the climatology of Atlantic hurricanes, we consider it likely that the Great Galveston Hurricane had similar roots.

From whatever source, this embryonic hurricane passed across the Leeward Islands into the eastern Caribbean late on Thursday, August 30th. A small drop in surface pressure was observed at Antigua on that date, suggesting that the disturbance might have been a depression at that time, with a closed wind circulation. It moved south of Puerto Rico and Hispaniola over the weekend and made landfall in extreme southeast Cuba early Monday morning, September 3rd. The strength at time of landfall in Cuba is unknown. However, the slow forward motion resulted in very heavy rains. One station in southeast Cuba reported 17 inches in 36 hours. The system finally moved into the Florida Straits near Havana early Wednesday morning September 5th. By afternoon the wind at Key West was blowing from the northeast, gusting to 45 mph. Late Wednesday evening, the pressure at Key West dropped to 996 mb, and the wind shifted to the southeast, suggesting the passage of a strong tropical storm or weak hurricane. The Weather Bureau in Washington, D.C. assumed the system was recurving to the north and east and decided to issue storm warnings for coastal areas in the southeastern United States. This was an unfortunate decision because the system continued moving between west-northwest and northwest toward the Texas coast, strengthening rapidly before it devastated Galveston late Saturday, September 8th.

Because no warnings had been issued for the Gulf of Mexico, the steamship *Louisiana* departed New Orleans Wednesday morning enroute to New York. During the night the skipper, Captain T. P. Halsey, noted a steady drop in pressure. Thursday morning, September 6th, one week after the system had entered the Eastern Caribbean, he observed the weather deteriorating rapidly with a sharp increase in wind. By afternoon, winds were gusting in excess of 100 mi/hr, and the barometer had dropped to 965 mb - near the threshold of a Category-3 hurricane on the Saffir/Simpson hurricane scale. Apparently, it had grown from a minimal hurricane to a Category 3 in approximately 24 hours. Nevertheless, the *Louisiana* survived the hurricane with minor damage and continued its journey to New York.

Again, because there were no warnings for the western Gulf of Mexico, another steamship, the *Pensacola*, left the Port of Galveston at 7 am, Friday morning, September 7th, some 36 hours before the arrival of the eye of the Great Hurricane. Captain Simmons was never aware that storm warnings were issued for Galveston Island thirty minutes later. Shortly after clearing the Bolivar Channel, the *Pensacola* encountered heavy seas and rough winds. Why Captain Simmons did not turn and run from the storm will never be known. He decided to stay on course for Pensacola. Faced with huge waves and winds near 100 mi/hr, Simmons ordered engines stopped Friday evening and dropped anchor.

The anchor chain parted Saturday morning at 10:30 AM and the *Pensacola* was at the mercy of the winds and the waves for the next 24 hours. Captain Simmons did not regain control of his ship until Sunday morning. He estimated they had drifted over 50 mi (80 km) during the night. Engines aboard the Pensacola were restarted and the ship steamed back into Galveston Sunday afternoon and docked amidst a mass of broken ships and dead bodies. Simmons had trouble navigating the harbor. Most of the familiar landmarks were missing or badly damaged.

The lowest pressure observed aboard the *Pensacola* was 965 mb on Friday night. The ship was never in the eye, where the pressure would have been much lower. The weather observations on the *Pensacola* and *Louisiana* show that the

Great Hurricane intensified very rapidly over the warm waters of the Gulf of Mexico from Wednesday until Friday. Unfortunately, forecasters did not realize this was happening.

It is unfortunate that Galveston remained unaware of what the *Pensacola* and the *Louisiana* were encountering. If they had known what the ships were experiencing, a train certainly would not have departed Beaumont early Saturday morning for Galveston. The plan called for the train to travel to the end of the Bolivar Peninsula, where it was to be transported to the Island by ferry. By the time the train reached the dock at midday, water was already up to the tracks and winds were near gale force. After numerous attempts to dock, the Captain of the ferry aborted the mission. The conductor ordered the train back to Beaumont. Ten of the 95 passengers ignored the order and waded in waist deep water seeking shelter in a nearby lighthouse. When the door to the lighthouse was finally opened, they found the spiral staircase jammed with almost 100 people. Amazingly, they all survived. All 85 passengers who stayed with the train perished when the train was washed off the track.

Furthermore, if there had been ship reports, a second train would not have departed Houston Saturday morning for Galveston. The Houston train made it across the bridge to the Island before being stranded by a washout. A second Houston train was dispatched on an adjacent track that was higher. By the time it arrived at midday, water was already over the tracks. Following a treacherous transfer of passengers from Train 1 to Train 2, the latter resumed a very slow, dangerous journey over flooded tracks to the downtown terminal. Train 2 lost all power about 100 yards from the terminal when rising water snuffed out the fire in the engine. Fortunately, all of the passengers were able to wade to the terminal. All bridges were destroyed, and the Island was cut off for several days.

The apathy on the Island at midday was appalling. Keep in mind this was only about eight hours before the arrival of the eye of a Category-4 hurricane. Most people felt the squalls would soon pass as they always had in the past.

A prime example of the apathy was the action of one of the passengers on Houston Train 2 who waded in waist deep water to keep a luncheon appointment. When he arrived at the office to meet his clients, he was told they had already departed for lunch. He then re-scheduled an appointment for 3 pm and waded back to the terminal to wait for the meeting.

Ironically, some of this apathy may have developed from public statements by Isaac Monroe Cline, including a local newspaper article he published in 1891. As meteorologist in charge of the Galveston Weather Bureau until 1901, they were statements he lived to regret when the Great Hurricane demolished Galveston in 1900. That they were conclusions of a broadly educated, erudite scientist, intellectual, and revered citizen of Galveston was sufficient for his constituents to consider his published accounts to be credible. Unfortunately, the statements were at best misleading. His statements and assertions, shared with various groups in Galveston were:

West Indies hurricanes are not a problem for Texas because they always recurve to the north before reaching the Western Gulf of Mexico.

Isaac was aware of the major hurricanes in 1875 and 1886 that destroyed Indianolia. He felt these were freak accidents. Indianolia was located on the south shore of Matagorda Bay and was the second biggest seaport in Texas in the mid-1800s before being destroyed.

Shallow water offshore from Galveston will protect the island from hurricane waves.

Just the opposite is true. Forecasters at the turn of the century did not understand the difference between wind waves generated by the hurricane and the storm surge. We now know the height of the storm surge is inversely related to the depth of the water and shallow water makes Galveston Island more vulnerable to the storm surge.

Buildings can be constructed to survive a hurricane.

Isaac had a chance to test this hypothesis when he rebuilt his home after it had been destroyed by fire in 1896. As we will see, the belief that his house was safe was, in part, responsible for the death of his wife.

These statements notwithstanding, I am not criticizing Dr. Cline's worthiness as a scientist. Isaac was a man of impeccable character and public credibility, holding advanced degrees in both physical and medical science. He was a notable pioneer in tropical meteorology with numerous authoritative publications on hurricanes. In 1901 he was appointed Meteorologist in charge of the Weather Bureau office at New Orleans and became one of the most highly regarded forecasters of his time. However, we will never know for sure whether his newspaper article in 1891 was at least partially responsible for the apathy.

The apathy in Galveston on September 8th came to an abrupt halt at Ritter's Café during the noon hour. Someone noticed there were thirteen men having lunch. They all had a big laugh. Shortly thereafter a gust of wind lifted the roof from the two-story building and the second floor came crashing down into the restaurant killing five immediately and seriously injuring another five. They sent one of the waiters for help, and he drowned in the flooded street.

News of this disaster quickly spread across the Island and people panicked. Hundreds converged on the posh Tremont Hotel. At the height of the storm, water covered the plush lobby and people retreated to the upper floors. It is estimated over a thousand people survived the hurricane in the Tremont.

On the east end of the Island, a tragedy was developing at the three-story Lucas Terrace Apartment. The apartment complex was surrounded by one-story, single-family homes. Before noon, rising water was entering the homes, and residents were forced to seek higher ground. Many sought refuge in the second and third floors of the Lucas Terrace Apartment. Early in the afternoon the water was so deep, it was impossible to leave the area. The people in the apartment were trapped. By mid-afternoon huge timbers from a government construction project on the east end of the Island began battering the Lucas Terrace. One by one the 64 rooms in the apartment were destroyed. Only one room survived with its 22 occupants. Several hundred people probably died in this one apartment building.

The Saint Mary Orphanage was located on the beach three miles to the west of town. Two buildings housed 93 children, ages two to 13, with a staff of ten nuns. Boys occupied the older building located to the east; the new building to the west was reserved for girls. Water crested the sand dunes before noon, and the boys were moved to the newer stronger building with the girls. The older building collapsed early in the afternoon, and when the newer building with all the children began breaking apart, the nuns took pieces of clothesline and tied six to eight children to each nun. This was a deadly decision because, when one or two of the children went down, it dragged everyone else down. The only survivors were three thirteen-year-old boys who were not tied to others. They spent the night clinging to a tree that had been uprooted. Ninety children and all ten of the nuns died.

Galveston lost all contact with the outside world shortly after 3:30 pm when the last communication line to the mainland was severed. Joseph Cline, Isaac's brother, worked for Isaac in the Weather Bureau office. Since there was no longer any way to forward weather reports to Weather Bureau headquarters in Washington, Joseph decided to leave the office and check on Isaac's family. He waded in chest deep water to reach Isaac's home. Joseph urged Isaac to evacuate and take the family to the weather office, but Isaac insisted they stay, confident that his house was safe. He was also concerned his sick and pregnant wife could not survive the trip. In addition to Isaac's family, 50 other people had sought refuge in the Cline house.

Shortly after 6 pm, rising water forced everyone in Isaac's home to the second floor. One by one houses around them collapsed until the Cline home was the only one standing. Between 7 and 8 pm the Cline house shuddered, lifted off its foundation and gradually turned over. Joseph grabbed two of Isaac's daughters and leaped through the front window. Isaac, his wife, and youngest daughter were trapped inside the house. Isaac was knocked unconscious and does not know how he got out of the house. When he regained consciousness, lightning flashes revealed a small child nearby. He rescued her from the water and was delighted to find that the child was his youngest daughter. Moments later they made contact with Joseph and the other two daughters. The Clines floated on debris for around eight hours before being deposited on a pile of broken homes four blocks from the original location of the Cline home.

Isaac's wife was not located until September 30th when a demolition gang uncovered a body in a pile of rubble that was believed to be part of the Cline home. The body was identified by the wedding ring.

Years later, Isaac said, "We probably would have weathered the storm, but for the trestle." For a nickel you could take a trolley ride from downtown to the beach. Several hundred feet of the trolley track was over the water. The hurricane dislodged the track and the supporting trestle. As the trestle moved inland, it gathered wreckage, and the mass of broken debris acted like a giant bulldozer knocking down everything in its path. It was the trestle that finally destroyed Isaac's home.

Words cannot begin to describe the horrendous conditions that the survivors faced. Initially, there were the corpses to contend with: both human and animal. Early Monday there was an attempt to identify and bury the dead humans, at least. But this effort was quickly abandoned when city officials realized the number of bodies greatly exceeded their capabilities.

A second plan, initiated Monday afternoon, called for bodies to be buried at sea. The bodies, with weights attached, were to be transported out into the Gulf of Mexico on barges. City officials found, however, that the only way they could entice men to work on this task was to offer them free whiskey. The offer drew men to the job, but they soon became drunk, and incapable of attaching the weights properly. Tuesday morning hundreds of bodies washed back ashore.

This left one alternative: the bodies had to be burned. For weeks, rising plumes of smoke could be spotted on the Island and the air was filled with the stench of burning flesh.

LESSONS LEARNED FROM THE GREAT GALVESTON HURRICANE

What Lessons did the Islanders Learn?

First, and most important, they learned Texas is not immune from West Indies hurricanes. All five of the Category-4 hurricanes that hit Texas in the 20th Century had their origin in Africa.

Second, they learned the storm surge is the killer part of the hurricane. Historically, nine out of ten people who have died in a hurricane have drowned in the storm surge. Although the greatest potential for loss of life remains the storm surge, a recent study by Rappaport [2000] indicates that, since 1970, freshwater floods have caused more than one-half of all deaths in the United States directly associated with tropical systems[1]. They also learned that storm surge is not a series of big waves. It is a dome of water more than 50 miles (80 km) wide that sweeps across the coast near the point where the eye makes landfall. Superimposed on the dome are wind-driven waves. The height of the storm surge is inversely correlated to the depth of the water. Accordingly, the shallow water offshore from Galveston makes the Island vulnerable to a high storm surge.

---

[1] Loss of life from hurricanes, as well as data on incidence of hurricanes and progress in prediction and warning skills, are based upon landfalls in the U.S. For example, they do not include loss from Hurricane Mitch in Central America in 2000.

Third, they learned that they did not understand the awesome forces associated with a hurricane; therefore, few people took any action to protect life and property until the water started to rise on Saturday, when it was too late.

Fourth, they learned if they were going to continue living on the Island, they must build some kind of a structure to protect the city. Almost immediately discussions focused on the need for a seawall. In September 1902, a contract for the seawall was signed. It was to be 17 ft high and more than 3 mi long. Before the project was completed in 1904, the U.S. Government paid to extend the seawall another mile to protect a fort. Down through the years the seawall was extended westward and is now over 10 mi long.

A second engineering project involved elevating portions of the city itself. The plan called for the Island to be filled to the level of the seawall on the beach with a gradual lowering of the grade to 10 feet at the center of the Island. To accomplish this task, a canal was dredged behind the seawall to the east end of the Island. Dredges were brought in from Europe to move fill from the bay to the Island. Two thousand one hundred fifty-six buildings had to be raised to accommodate the fill. In addition, streets, sidewalks, sewer lines, and the entire infrastructure had to be moved. The task of obtaining permits from over 2,000 landowners was an administrative nightmare. The contract for this fill project was signed in 1903, and the job was completed in 1911. Bixel and Turner [2000] devote a whole section in their book to these two remarkable engineering projects.

The wisdom of these two projects was verified in 1915, when another Category-4 hurricane hit the Island with a death toll of only 275. The seawall did exactly what it was designed to do; it protected the city.

Were These Lessons Applied in Adjacent Coastal Areas?

In the century that followed the Great Hurricane, have we been faithful in applying the lessons learned at Galveston in developing the Texas coast and other coastal areas?

The final westward extension of the seawall was completed in 1960 just before the arrival of Hurricane Carla in 1961 — the most recent Category-4 storm to strike Texas. Even though the eye crossed the coast near Port O'Connor, Carla was huge in size with tentacles reaching out to the Galveston/Houston area with a force comparable to the impact of a Category-3 hurricane. The water rose 9 ft along the beaches of Galveston, but the seawall protected the city. On the west end of the Island beyond the seawall, where ranching and fishing were the main activities, few buildings stood and thus damage was limited when Gulf waters spread across the Island.

By the time Alicia arrived in 1983, the west end of the Island had changed. The communities of Jamaica Beach, Pirates Beach, Terramar, Sea Isle, and Bay Harbor were all products of rapid development in the 1970s. With no consideration for hurricanes, homes in each of these villages were located on the beach with an average elevation of less than 5 ft.

Alicia, a Category-3 hurricane, produced weather conditions similar to what occurred in Carla. The seawall protected the city from the ten-foot storm surge. Beyond the seawall, however, there was extensive damage, particularly to homes near the beach. Of 207 homes on the beach, 99 were severely damaged or destroyed. Inland from the beach there was a remarkable decrease in damage. For example, on the second row (approximately 150 ft back from the beach) only 28 of the 169 homes suffered the same fate as those fronting the beach. More than 100 yds inland, homes were flooded but suffered very little damage.

This is an extremely important observation. If realistic setback lines for construction are established, beaches can be developed and potential damage from hurricanes can be reduced. This did not happen during reconstruction in the west end of Galveston Island after Alicia. Initially an attempt was made to enforce a Texas law that gives ownership to the State of all land seaward of the vegetation line. Because of erosion during Alicia, most of the homes on the first row were ahead of the vegetation line and were claimed by the State. However, after numerous lawsuits and intense pressure by the real-estate lobby, the State finally allowed the county to issue building permits for repairs to damaged homes.

The State and Galveston County missed a golden opportunity to do it right, and as a result, the west end of the Island is more exposed than before Alicia. Fifty-seven of the 99 destroyed beachfront homes and 23 of the 28 second-row homes were rebuilt. In addition, by 1994, 82 new beachfront homes had been constructed with another 83 on the second row.

In the decade prior to this publication, four new expensive beach projects have been started where beachfront lots have sold for about $200,000. As of 1999, there were 53 new beachfront homes and 36 new second-row homes in The Dunes, Kahala Beach, Indian Beach, and Acapulco Beach. In summary, before Alicia there were 207 beachfront homes on the west end of the Island. Today, there are over 300, and the number is growing every year.

The folly of rebuilding the west end of the Island with few restrictions was dramatically revealed during Tropical Storm Frances, which made landfall on the upper Texas coast in 1998. While Frances was only a tropical storm, the erosion was just as bad as had occurred during Alicia. The sand dunes that had been laboriously rebuilt during the 15 years that followed Alicia once again were swept away.

Again there was an attempt to enforce the Open Beaches Act. The Texas General Land Office identified 104 beach homes that were now on State land and were obstructing access to the beach. By law, the owners are required to remove these homes. Unfortunately, the Texas Attorney General refused to enforce the law and confusion reigned. This left the Land Office and local officials, who must issue building permits, in limbo. The result was that new developments on the west end were left just as exposed to hurricanes as Galveston was prior to the Great Hurricane.

POTENTIAL IMPACT OF AN EQUIVALENT "GREAT HURRICANE"

The following scenario is partially based on observations I have made of coastal damage in the aftermath of all major hurricanes that have hit the United States since Camille in 1969.

The height of the storm surge will range from 15 ft along the seawall to near 25 ft at the north end of Galveston Bay. Huge waves on top of the surge will break against the top of the seawall cascading tons of salt water onto the buildings on the north side of Seawall Boulevard, the street on top of the seawall. Water will be 5-10 ft deep in buildings in the center of town on the north side of the Island.

Out west beyond the seawall, water up to 10 ft deep will flow rapidly across the Island, causing catastrophic damage, particularly to beachfront homes. The damage will be much greater than what occurred during Alicia, and all 300 beachfront homes will be destroyed. Water will be deep enough to float the broken remains of beach homes, which will become battering rams — destroying homes on the second, third, and fourth rows. Over 1,000 homes on the west end of the Island will likely be destroyed.

Wave erosion will cut one or two channels through the Island, and road access to the west end might not be possible for weeks.

The two bridges connecting the Island to the mainland could also suffer damage, cutting off access to the Island for days or weeks.

The storm surge will cause water to rise 15-20 ft along the west side of Galveston Bay and in Clear Lake where the NASA complex is located. The beautiful waterfront recreational development in Kemah at the mouth of Clear Lake will probably be destroyed.

A 25-ft surge at the north end of Galveston Bay will cause serious flooding along the ship channel and close interstate highway I-10 East.

Winds up to 100 mi/hr will sweep inland more than fifty miles, causing extensive roof damage as far as Conroe to the north and Katy to the west. Building codes used in Southeast Texas are designed for winds up to around 75 mi/hr. When winds are in excess of 100 mi/hr, roofs start blowing off, weakening the building and causing walls to collapse. Winds over the inland areas could destroy over 100,000 homes.

The recovery process will be staggering. It could take weeks to restore water and power to many parts of Southeast Texas, and two to three years or more to rebuild all the destroyed homes.

It is a given that the west end of Galveston Island will be devastated during Category-4 or -5 hurricanes. What isn't known is how many people will die. This is a difficult question to answer because we do not know how soon people will respond to a call for action when a hurricane threatens. This is a major concern. Evacuation studies indicate it will take almost 36 hours to relocate residents-at-risk from Galveston Island and mainland coastal areas when a Category-4 or -5 hurricane threatens, even if all concerned act immediately when evacuation notices are issued.

Behavioral scientists tell us the first thing people do when faced with an official warning, particularly those who have little or no experience with a major hurricane, is to seek additional information to confirm the threat. Yet still another factor looms in this evaluation: when Carla hit Texas in 1961, the population of coastal counties from Matagorda Bay to Port Arthur was 1.6 million. Today it approaches 5 million.

As an additional concern, this explosive growth of coastal population occurred during a regional lull in hurricane occurrences. Out of the 36 hurricanes that hit Texas during the 20th Century, 21 have been in the Galveston/Houston area. But on closer examination, an interesting anomaly emerges. By dividing hurricane occurrences for the period 1896 through 2000 into three 35-year periods, we find the first third featured few but quite powerful hurricanes. The second was a very active hurricane period; the decade of the 1940s alone produced six hurricanes. During the final 35 years, there have been only five hurricanes, and with the exception of Alicia, they all have been weak. Hurricanes Chantal and Jerry in 1989 were the last hurricanes in Southeast Texas, and they were both Category 1 in strength. Clearly, the design of protective measures and plans of action to cope with the hurricane menace should not lean too heavily on hurricane climatology.

When the Great Hurricane threatened Galveston, neither residents nor local officials were prepared to respond appropriately. Today Island residents are well informed and better prepared. During the last 25 years, there has been a major effort to educate the people about hurricanes. In the 1970s, the Texas Marine Council started focusing on the hurricane problem and developed an excellent educational package. Every year newspapers and TV stations start the hurricane season with a hurricane special. Emergency managers, weather service personnel, and representatives of the media have given thousands of public talks. Free hurricane tracking charts are available at local supermarkets, with a map on the back showing what areas will be flooded by the storm surge. If there are people on the Island who are uninformed, they are in a state of denial and have their heads buried in the sands of ignorance.

However, the current lull in hurricanes enjoyed by Galvestonians may be coming to an end if the recent sharp increase in Atlantic "named storms" indicates a long-term future trend. Seventy-nine "named storms" occurred in the Atlantic basin from 1995 to 2000. This is the most active six-year period in the history of our hurricane records and represents a dramatic upturn from the 54 "named storms" that occurred in the previous six years (1989-1994).

Goldenberg and his collaborators [2001] suggested alternating patterns of quiet versus active hurricane periods related to decadal fluctuations in the temperature of the ocean. More hurricanes occur when the ocean water is warm. The tropical Atlantic was warm during the 1940s, 1950s, and 1960s when the number of hurricanes was above normal. The 1970s, 1980s and early 1990s featured cold water and fewer hurricanes. The recent increase in "named storms" appears to be in response to a warming of the ocean, and this could be the beginning of a long-term trend. If true, the early 21st Century may have some dramatic surprises for the United States.

The absence of major hurricanes, coupled with a large influx of new coastal residents in the years preceding this publication, could prove a deadly combination, if multitudes of new residents ended up wasting precious time trying to confirm the reality of a warning, rather than evacuating promptly. There remains, as well, the possibility of a meteorological surprise with inadequate warning time for people to respond. The nightmare hurricane forecasters fear most is explosive intensification of a weak hurricane as it approaches landfall. Either way, multitudes could be trapped on the Island, making the possibility of another Galveston disaster very real.

Psychologists tell us memories of bad events tend to dim with time. A wise philosopher once said those who ignore history are condemned to repeat it. Ignorance and fading memories are both evident today in planning and developing coastal areas.

So, what are we doing to prevent major lose of life in a future Great Hurricane? Very little, if the development of the west end of Galveston Island is an indication of what is occurring in other areas vulnerable to hurricanes. Our observations suggest that the Galveston Island development and vulnerability is not uncommon along the Gulf and Atlantic coasts. Despite the advances in understanding hurricanes, documented in other chapters, I consider the United States more vulnerable to hurricanes today than at any time in its history.

TABLE 1. Hurricanes Striking Southeast Texas

Years	Number of Hurricanes	Number of Major Hurricanes
1896-1930	5	3
1931-1965	11	4
1966-2000	5	1

REFERENCES

Bixel, P. and E. Turner, *Galveston and The 1900 Storm,* Texas University Press, 174 pp, 2000.

Goldenberg, S. B., C.W. Landsea, A. M. Mestas-Nunez and W. M. Gray, The Recent Increase in Atlantic Hurricane Activity: Causes and Implications, *Sci.*, 293, 474-479, 2001.

Larson, E., *Isaac's Storm,* Crown Publishers, 352 pp, 1999.

Rappaport, E. N., Loss of Life in the United States Associated with Recent Atlantic Tropical Cyclones. *Bull. Amer. Meteor. Soc.*, 2065-2073, 2000.

Rappaport, E. N., and J. F. Partagas, *The Deadliest Atlantic Tropical Cyclones, 1492-1994,* NOAA Technical Memorandum NWS NHC-47, 41 pp, 1995.

Weems, J. E., *A Weekend in September,* College Station, TX: Texas A&M University Press, 180 pp, 1957, 1980.

Chapter 6

Social Choice in Dealing with Hurricanes

Mary Frances Myers and Gilbert F. White

In contrast to Chapter 5, in which Neil Frank primarily addressed problems of coping with the lethal impact of hurricanes, we will examine available options to mitigate their social impact. We discuss public policies, both for long- and short-term planning, and consider how, in our democratic society, reducing damages from hurricanes can best be achieved within the bounds of social necessity and economic feasibility.

INTRODUCTION

The past 100 years have seen an expansion of population and development in coastal areas in the United States and subsequent losses due to hurricanes. Society has elected to deal with these extreme events in many ways, ranging from emergency response measures to a variety of mitigation techniques, such as land use planning, at all levels of government and within the private sector and non-governmental communities as well. This chapter first discusses issues surrounding the accuracy of reported losses due to hurricanes and the concept of beneficial uses of hurricane-prone areas. It next focuses on the wide range of mitigation measures used to adapt to hurricane threats including warning, emergency response, land use planning, building codes and standards, public education and outreach, insurance, improved predictive technology, and engineered solutions. We describe findings from pertinent social science research that provides insight into the evolution of knowledge regarding the effectiveness of societal choices for dealing with hurricanes. We conclude with some observations on the challenges that remain for dealing with hurricanes in the 21st Century.

Hurricane! Coping with Disaster

10.1029/055SP08

CREDIBILITY IN EVALUATING SOCIAL AND ECONOMIC LOSSES

Losses in human life and health, and the economic damage wrought by hurricanes, are difficult to estimate with a high degree of accuracy. There are a number of reasons for this difficulty, including the differences in methods followed by estimating agencies and the range of criteria used in assigning monetary losses.

Unfortunately, the task of calculating hurricane effects is divided among various agencies that do not use completely uniform criteria. As a result, while reports of lives lost as a consequence of a particular hurricane are highly likely to be comparable, reports of injuries, other health effects, and property damage are more likely to differ according to the criteria used by the reporting agency.

When it comes to economic and social losses attributed to a hurricane, the method of estimation may vary greatly from one agency to another. For example, a meteorological agency may use somewhat different criteria for crop loss than does an agricultural bureau, a crop insurance agency, or a social relief agency in calculating the basis for a farmer's claim for assistance. While there is some agreement on procedures among several agencies in many areas, there are no nationally uniform criteria for calculating losses.

Although there are several available estimates of the total losses caused by hurricanes during the past century, there are major reasons to question the comparability of these from one year to another and with losses from other natural causes. Overall, in terms of losses prevented, there is not yet a solid basis for evaluating the effectiveness of various public policies to deal with the societal effects of hurricanes.

In addition to inconsistencies among agency reports of hurricane impacts, a recent interdisciplinary review of coastal hazards, conducted by the John Heinz III Center [2000b] for the Federal Emergency Management Agency (FEMA), concluded that several major classes of economic and social costs typically were omitted in estimates made by the usual national, state, and local agencies after a hurricane. Some of the principal ones are:

1. The cost to the business community, particularly small firms, of such effects as business disruption, utility interruption, the indirect effects of temporary closure and delay in resuming normal business, and the indirect costs to complementary businesses and services.
2. The cost to individuals regarding social, health, and safety expenses that are not counted in the estimated cost regarding deaths and injuries; these may include long-term medical expenses, household preparations and evacuation cost, along with disruption of transportation facilities.
3. The cost to natural resources and the environment through destruction or disturbance of landscapes and ecosystems.

Methods of calculating monetary values for the categories above are generally lacking, but the absence of quantitative estimates does not mean that they are insignificant in influencing the course of mitigation or preventive activities in hazardous areas. Although difficult to calculate, they may be highly influential in private and public decisions to adopt preventive measures either before or after a major hurricane. A more precise set of estimates of actual and potential losses might influence both the character and magnitude of constructive measures by private property owners as well as public agencies. HAZUS, a loss estimation model developed by FEMA, provides a basis for local estimates of prospective flood and wind losses at the census tract level for every community in the nation by providing data along the more traditional lines of loss estimation. It is limited, however, in that it neglects part of the evidence that might be used to show the social feasibility of more comprehensive efforts to deal with hurricanes.

There are other problems in estimating losses from hurricanes. It is often particularly difficult to separate damages due to wind from damages due to flooding. This becomes problematic mainly for insurance purposes. If wind is the cause of damage to a structure during a hurricane, costs may be covered by a standard homeowner's insurance policy. If flooding causes the damage, damages are covered only if the property owner has a special flood insurance policy issued in accordance with the National Flood Insurance Program.

Another persistent problem is the conspicuous omission in most current estimates of actual and potential hurricane losses of the possible effects of long-term climate change and current sea-level rise upon the prospective losses. The Heinz Center study [2000b] calculated the progressive loss of land and consequent increase of hurricane damages along the Atlantic coast. For example, it projected an estimated average rise of sea level would cause the loss of 1,500 homes and the land they occupy each year along the South Atlantic coast. Yet, the loss of these homes from these effects is not considered in attempts to address losses due to hurricanes.

A further dilemma is that of estimating the effects of changes in sea level that result not only in increased flooding at a given location, but also in accelerated erosion of parts of the coast. The costs of land loss from erosion are not usually repetitive but are progressive as the coast retreats sporadically. Research coordinated by the Heinz Center [2000a] has sought to provide accurate estimates of the physical dimensions of this problem and of its significance for loss estimates and public policy with respect to national flood insurance and federal disaster assistance. While the erosion loss is not directly correlated with frequency of hurricanes, the two are related. A long-term cost of coastal development is the progressive loss of coastal ecosystems.

Finally, estimating the beneficial uses for coastal ecosystems, where they are not affected by human works to curb flooding or erosion, has been ventured in only a few areas. Such benefits might include, for example, the value of nature-based recreational experiences for tourists who visit untouched coastal waters, wetlands, reefs, beaches, and rocky shores. Other benefits of natural coastal ecosystems include the buffering effect from storms provided by dunes and beaches, flood water storage, and nutrient cycling, provided by coastal wetlands and streams as well as critical habitat provided for fish, birds, and other wildlife. All of these kinds of functions need to be considered in evaluating the net national costs or benefits of how society uses coastal areas. FEMA was directed by Congress to make a review of the value of natural and beneficial uses of floodplains in the National Flood Insurance Reform Act of 1994, but a report on that problem of estimation had not been completed in 2000.

All of these problems have long been recognized, as have ways to correct them. The work of the Heinz Center [2000, 2000b], previously noted, and a recent study by the National Research Council [1999] point to reasonable approaches to systematically collect and document losses due to extreme natural events such as a hurricane. We endorse the recommendations of those studies and suggest an additional step. In order to fully inform people about social choices to deal with hurricanes, not only must we have a good understanding of all the costs associated with the damages they cause, but we must understand the benefits of using the areas they impact as well. This includes both the beneficial uses of coastal ecosystems described above, and the benefits of other types of human occupancy (e.g., residential or commercial) of hurricane-prone areas.

APPROACHES TO DEALING WITH HURRICANES

Losses from hurricanes are a function of the relationship between extreme meteorological phenomena and society's built environment. Traditionally, society has dealt with hurricanes by trying to manage the events in order to limit losses from them. This management can be characterized by three different types of activities: (1) those which seek to modify the hazard, (2) those which modify the vulnerability of people and the built environment to damage from a hurricane, and (3) those which modify the impact of a disaster. These various strategies are frequently referred to as "mitigation tools" and are carried out before, during, and after hurricanes. In practice, they overlap, though each has its own aims and changes the effectiveness of the others.

Modify the Hazard

Modifying the hazard refers to human efforts to change the event – in this case, a hurricane. According to Simpson and Simpson [1966], the National Oceanic and Atmospheric Administrations (NOAA) initiated a program called Project STORMFURY in the 1960s to explore ways to modify hurricanes using silver iodide crystals to seed clouds in strategic areas of the hurricane. It was anticipated that the response to seeding would lead to an expansion of the storm core with consequent reduction in maximum wind speeds and a reduction in expected damage. From the half dozen experiments conducted in the course of a decade, it was not possible to demonstrate the postulated responses with statistical significance, and the program was abandoned. Further information on Project STORMFURY and hurricane modification can be found in Chapters 1 and 3. There is little expectation that any of the suggested methods can be proven successful or implemented in the foreseeable future.

Modify Susceptibility

Modifying susceptibility to hurricanes refers to both structural and nonstructural approaches that are designed to make society and the built environment it occupies less vulnerable to damage when a hurricane does strike.

Structural Measures. Structural measures, or protective shore works against ocean waves, involve two types of activities. One is the construction of sea walls, revetments, and bulkheads to protect shorefront property from wave attack. The second structural approach is to supplement or replenish natural features of the coastline, such as beaches and dunes, to replace eroded land and to provide a buffer between the ocean and vulnerable areas inland. The Army Corps of Engineers and others have spent hundreds of millions of dollars on such efforts, often designated as "coastal nourishment." An example is the project for nourishing the beach from Sea Bright to Asbury Park, New Jersey, at a cost of more than $1 billion [Heinz, 2000b].

For existing development, retrofitting vulnerable structures can make them stronger and less likely to suffer damage in the face of hurricane winds and storm surge, or flooding that occurs with many hurricanes. Activities such as adding hurricane clips to secure a roof to a building; installing storm shutters over windows to prevent flying debris from breaking the glass; and elevating utilities such as circuit breaker boxes, furnaces, and hot water heaters to higher levels in a house to prevent flood damage are inexpensive ways to reduce a building's susceptibility to damage from hurricanes.

FEMA has sponsored Building Performance Assessment Teams that survey the performance of primary structural, architectural, and workmanship features of buildings following disasters in order to recommend corrective and preventive action. For example, a team operating in southern Dade County, Florida, after Hurricane Andrew struck in August 1992, found that the loss of roof cladding was the most pervasive type of damage in that storm.

Nonstructural Measures. A nonstructural approach to reduce vulnerability to hurricanes is to do more intelligent planning for where and how new development occurs. A range of tools, including comprehensive land use planning, building codes, community facility planning, growth management, and capital improvement planning, are available to local governments so that when hurricanes do strike, damage to the built environment can be limited. State and federal government agencies are involved to the extent that they provide financial assistance for public buildings and infrastructure such as highways. Further, FEMA has fostered a variety of mitigation prototypes in selected areas. Examples of how these tools work include land use plans that call for setbacks or otherwise guide development away from hazardous areas (e.g., areas prone to storm surge). This allows communities to make decisions to avoid future damage. Building codes can require structures to be built to withstand certain wind speeds as well as to be elevated on piles, posts, or solid foundation walls to avoid flood damages. Both techniques may improve the structural integrity of a building and reduce its susceptibility to hurricane damage. Capital improvement plans that steer a community's growth away from hazardous areas can reduce vulnerability. Transportation plans that call for roadways to be elevated and curved as they approach the beach can reduce the extent and amount of overwash from storm surge, as opposed to roads perpendicular to the beach which act as conduits for flood waters and allow them to penetrate further inland.

A second non-structural measure concerns forecasts and warnings. Reliable forecasts and warnings of pending storms are critical to saving lives. This nation's scientific capability for predicting hurricanes improved greatly in the 20th Century. According to J.H. Sorensen [2000], hurricane warnings probably have improved more solidly in both prediction accuracy and warning integration than has the national warning system for any other natural hazard. The technology that now exists for detailed and continuous observations of hurricanes has contributed not only to better scientific understanding of them, but to the development of more prediction models (see, for example, Chapters 4, 8, 9, and 10). This sophisticated technology, in turn, poses challenges for implementing timely and effective systems for evacuating people in harm's way. Indeed, forecast and prediction are only one part of the picture. To use them effectively to reduce society's vulnerabilities, the information that scientists have must be relayed in a way so that their audience (be it an individual or a local emergency manager) is prepared to, and does, respond appropriately. A community's vulnerability to loss is very much affected by social factors —factors often not directly related to the hazard.

Hurricane warning systems involve complex processes. Providing effective warnings requires a chain of events, beginning with the technology to observe and predict a hurricane and ending with risk communication and human response. These components of warning systems are interdependent, making the system only as good as its weakest link. Good warnings will include information about the probability of a severe event, its location, and the timeframe involved, and also give people information on appropriate actions to take in response to the warning.

Disaster preparedness is a third important non-structural approach to mitigation. Disaster preparedness involves building an emergency management capability in advance of an event in order to facilitate an effective response. Critical to disaster preparedness is development of a vulnerability analysis that identifies the hazards of a particular place and the risk to that population. This must be followed by careful planning for the evacuation of threatened populations, recognizing that emergency evacuation plans may only be as effective as the capacity of the highways and transport used to move people, livestock, and goods at risk within certain time limits. In addition, plans for sheltering of and caring for victims; provision of emergency medical care, food, fresh water, and shelter; conduct of search and rescue operations; and provision of security and protection for property at risk must be made. Other specific functions also may be called for in the response phase. For example, hurricanes often cause loss of power facilities, which requires the provision of emergency power supplies.

A recent forum convened by the Atmospheric Policy Program of the American Meteorological Society [2000] called for a new assessment of the vulnerability of communities exposed to hurricanes and for improvements in preparedness, prediction, communications, and response strategies.

Modifying the Impact

The third class of approaches society uses to reduce vulnerability to hurricanes is to modify their impact, or to spread the loss from them. It is important to note that this strategy does not reduce damages; it just extends the cost of those damages to a larger or different population.

Insurance. One tool to relieve the financial burden on victims of disaster and taxpayers is insurance, spreading specified costs among a broad group of policyholders. Property damages caused by the high winds of a hurricane in the United States may generally be covered by an individual homeowner's property and casualty policy, though sometimes it is required as an extra "rider" in coastal areas. In the U.S. damages due to flooding caused by a hurricane are covered only if the property owner holds a special flood insurance policy under the National Flood Insurance Program.

Disaster Assistance. In addition to any insurance payments, disaster relief programs of governments and non-governmental organizations (NGOs) also serve to lessen the impact of disasters on victims. In the U.S. the availability of federal disaster relief is limited unless a presidential disaster declaration is made. If this is the case, the type of relief provided might include low-interest loans, small grants, temporary housing for individual victims, and support of 75% of the costs of damage to public facilities. In cases of the latter, state and/or local governments are required to pay the remaining 25%. For the more common, less catastrophic event, people must rely on insurance payments or funds from NGOs such as the Red Cross and other voluntary or church relief groups to recoup some of their losses.

Federal Approaches. The social choices for dealing with hurricanes have in many ways been guided by actions of the federal government. Rutherford Platt [1995] reports that during much of the 20th Century the nation's policy was geared primarily toward protecting uplands and existing buildings with engineered structures and beach nourishment. The first U.S. national program for land-use planning in hurricane-prone areas was initiated under the Coastal Zone Management Act of 1972. It provided the basis for a variety of federal and associated state and local measures dealing with a wide range of problems related to hurricane occurrence. In addition to this Act, the National Flood Insurance Program implemented in 1968, the 1982 Coastal Barrier Resources Act, and efforts of various public and quasi-public agencies to acquire and preserve coastlines in their natural state have contributed to the use of broader coastal management strategies.

In the late 1970s the federal government, initially through the Army Corps of Engineers and later in cooperation with FEMA and the National Oceanic and Atmospheric Administration (NOAA), began funding hurricane preparedness activities by developing a hurricane evacuation model and conducting studies up and down the coast. The studies provided maps of areas vulnerable to storm surge, evacuation clearance times, and other information needed to develop effective emergency response plans.

Summary of Approaches

This review of the various approaches used in response to hurricanes suggests that society has implemented a broad and wide-ranging management scheme to deal with these extreme events. However, as will be noted below, while much social research has been initiated on many of these approaches, there has not been a single comprehensive, integrated effort to assess the net effects of these approaches or national programs on land use and community welfare, including the health of coastal ecosystems. Therefore, it is not currently feasible to accurately assess their effectiveness.

WHAT RESEARCH TELLS US ABOUT THESE APPROACHES

An array of studies have, to varying degrees, informed practitioners and policymakers about how society's choices and methods in dealing with the hurricane hazard have influenced the losses suffered from those events. The National Science Foundation and other agency-funded research have addressed specific issues related to the hurricane hazard. These assessments include: (1) the environmental impacts of hurricanes, (2) risk perception by affected populations, (3) differential vulnerability of various groups, (4) building performance, and (5) analysis of disaster relief methods. In addition, there have been many ad hoc post-disaster reviews of various aspects of response and recovery from disasters, systematic internal agency audits of their hurricane response performance, and site-specific case studies of mitigation programs.

While it is impossible to review in detail this broad range of research, in general, it is fair to say that it reveals that the various approaches society has chosen to deal with the hurricane risk are interrelated. It is essential to acknowledge and understand the interrelationships in order to deal effectively with hurricanes and coastal storm problems. Further, the research reflects that there are many social, political, and economic factors that influence society's vulnerability to damage from hurricanes as well as the adoption and effectiveness of various strategies to mitigate those damages. A few examples follow:

1. Research shows that coastal management of the past 30 or so years has shifted away from "hard" structures. Their negative effects on coastlines have led some states to ban them entirely. On the other hand, beach nourishment is still used extensively, and the research findings are divided on whether or not this approach is an effective tool for reducing long-run hurricane vulnerability.
2. By and large, the improvement of forecast and warning technology over the past several decades has been credited with a reduced loss of life due to hurricanes, especially in the U.S. Whereas 100 years ago in Galveston, some 8,000 people died, no single hurricane has killed more than 60 people in the last 25 years in the U.S.
3. Along with technological advances, research demonstrated that the focus of efforts also needed to be on the warning delivery system so that an appropriate public response would occur. Various studies have concentrated on people's perceptions of and vulnerability to hurricane risk and the best ways to disseminate warning messages in terms of timing, content, and consistency. These study findings have improved the delivery of warnings as evidenced, for example, by the cooperative efforts of the Red Cross and government agencies to use the same language in their messages about hurricane preparedness.
4. At the same time, however, some observers caution that confidence in the improved warning technology has put society at more risk as population increases in coastal areas. New arrivals may have an unwarranted sense of security in hazardous locations and, in turn, a larger populace requires longer evacuation times. In addition, the forecast and response technology has grown in complexity. For example, with the advent of expanded communications technologies, people have multiple—and sometimes inconsistent—sources for warning information (e.g., various media outlets, the Internet, "official" government warnings). This can make the system more subject to catastrophic failure, such as when a moving hurricane takes an unexpected path and strikes an unprepared area.

5. Building codes have been shown to be an effective way of reducing the amount of damage a particular structure experiences in a hurricane. For example, Platt [1999] found that buildings constructed in compliance with National Flood Insurance Program elevation requirements in North Carolina faired quite well after Hurricane Fran, except for those right on the ocean where many were undermined by erosion. However, Mileti [1999] found that other field investigations of damaged buildings have revealed that often it is not the code that is the problem, but rather poor enforcement of the codes, poor workmanship, construction deficiencies, or the selection of building styles inappropriate for high winds.
6. Regulations, such as zoning ordinances and setback requirements, to keep development away from fragile coastal environments theoretically can reduce an area's vulnerability to hurricane damage. In practice, as with building codes, the effectiveness of these tools depends on how they are implemented at the local level. Ray Burby [1998] and his colleagues cited many factors that influence the effective implementation of regulatory tools. These ranged from public recognition of the risk, credible and persistent advocates, availability of staff resources, community wealth, political culture, mandates or assistance from state and federal government, previous hazard experience, and the availability of solutions that are technically and politically feasible.
7. Disaster relief policy in the country has evolved over the past half century. The trend has been a gradual, federal expansion in responding to events. This has been accompanied by an increasing sense of entitlement for reimbursement of losses on the part of state and local governments and individual disaster victims. This has resulted in an ever-growing cost to the nation for disaster relief. David Godschalk [1999] and his colleagues emphasized that, since 1988, federal policy has taken a significant turn with a new focus on mitigation. Since that time, federal funds have been available to not only rebuild and replace what was lost in a disaster, but to rebuild and replace in a better way or location, so a similar event in the future will not have the same serious consequences. Despite the good intentions, however, there is limited research available on the effectiveness of these disaster relief mitigation approaches. Indeed there are at least a few instances to show that disaster relief may be counterproductive in securing a safer future. Rutherford Platt [1999], for example, noted that North Carolina's coastal hazard mitigation program is one of the best in the country but found that after Hurricane Fran, "the abundance of available federal assistance ... can be blamed for undercutting the state in promoting hazard mitigation."
8. Insurance as an "adjustment" or social choice for dealing with the hurricane hazard had received little attention until 1992, when Hurricane

Andrew caused $15.5 billion in insured losses and caused nearly 40 companies to announce their intention to either cancel or not renew almost 850,000 policies in Florida. While a variety of state initiatives (e.g., moratoriums on insurance withdrawals and creation of the state-funded Florida Hurricane Catastrophe Fund) alleviated the situation in South Florida, the insurance industry itself has become more active in seeking ways to play a central role in hazard management programs. The basic findings of that research reported by Kunruether and Roth [1998] indicate that insurance can indeed be a player, but only through joint efforts with others and only by combining insurance with other strategies, including monetary incentives, fines, tax credits, well-enforced building codes, and land use regulations.

CONCLUSIONS

From our review of 100 years of social choice in dealing with hurricanes, we note that the reduction in loss of life (at least in the United States) in the latter half of the century reflects not only more accurate and effective hurricane forecasts and warnings, but more effective management of systems for evacuating coastal residents at risk. At the same time, from the research findings available, we offer a few other conclusions that are supported solidly in the studies currently completed:

1. We lack sufficient data on costs and benefits to individuals and municipalities and on the full experience of the range of private and public agencies to make good judgments of the net effects of various combinations of social choices in regard to the hurricane hazard.
2. Emergency response measures can be effective in reducing economic losses, but reliance on prediction and emergency technology may lead to more catastrophic losses in the future when more severe events occur.
3. We lack sufficiently comprehensive assessments of the impact of various programs and policies to be able to judge how effective they have been in reducing long-term damages from hurricanes.

In sum, the social choices the nation has made about public policy and land use in areas subject to hurricanes over the last 100 years have created a situation whereby almost half the population now lives along one of the U.S. coastlines. This puts more people and property at risk from hurricanes than ever before (see Chapter 5). Research about these choices suggests that we have extensive knowledge about the causes of hurricanes, the technical capability to, in most cases, make relatively accurate forecasts about when and where they will occur, and keen insights about how to mitigate damages from them. At the same time, economic losses from extreme events continue to rise at a rapid pace.

While losses are growing partly from the fact that the nation's capital stock is expanding, they also stem from the fact that our natural, social, and built environments, as well as their interactions are becoming more complex with each passing year. As such, even if the population was not expanding, losses due to hurricanes might well be on the rise because of the interdependence of these systems. For example, if the port facilities in New Orleans are demolished by a hurricane, the impacts can be felt thousands of miles away by wholesalers relying on supplies from that port or consumers who discover their energy costs are increasing because of damage to the petroleum facilities in the port.

CHALLENGES FOR THE FUTURE

The complex reasons behind the growing losses due to extreme events led to a major five-year study directed by Dennis Mileti [1999] at the University of Colorado's Natural Hazards Center. It involved 132 of the nation's experts on hazards and took stock of what is known and not known about natural disasters and programs to reduce damages from them. A primary conclusion of that work is that the nation's hazard management programs have been too narrowly focused on simple loss reduction, and they have been carried out in a closed framework that does not embrace the larger context of how society relates to its natural environment.

This traditional way of approaching hazard management programs focuses on short-term gains instead of long-term implications, and it places an artificial separation between hazard issues and other community issues. The Mileti study calls for a new paradigm for dealing with hazards—one that asks local citizens to look forward and consciously plan and design their future disasters, rather than simply respond to the results of a lack of planning. The approach is called sustainable hazard mitigation. It suggests that any hazard management program must embrace the notion of adjusting to the environment, incorporate a global systems perspective, and embody the concept of sustainability. Further, any such program must derive its authority from local consensus.

Specifically, to be "sustainable" over the long term, a locale would undertake actions to reduce losses from a hurricane only when those actions also: (1) maintain environmental quality, (2) maintain a certain quality of life for all residents, (3) promote disaster resiliency, (4) promote a vital local economy, (5) ensure intra- and inter-generational equity, and (6) are consensus based. The report suggests that hazard policies must be integrated into a broader context before any long-term progress in reducing losses can be made because extreme events are just one aspect of the natural environment within which they occur.

Promoting this new approach is not easy, as the challenge it poses requires a shift in the nation's attitude toward dealing with extreme events. In addition to the difficulty of switching gears to sustainable hazard mitigation, at least two additional challenges for research and public policy in the United States stand out for the hurricane community in the decades to come:

1. Assessing the effect that continuing climate fluctuations will have on sea level rise and coastal erosion and what this would mean for further mitigation programs.
2. Determining if changes in federal policies for offering flood insurance and disaster relief would have any effect upon the net losses and benefits of occupying coastal lands.

Before significant progress is made in reducing damages from future hurricanes, we believe the nation must turn its attention to meeting these challenges.

REFERENCES

American Meteorological Society, *Report of the Weather Channel Forum: Policy Issues in Hurricane Preparedness and Response*. Washington, D.C., 2000.

Burby, R. J. (ed.), *Cooperating with Nature Confronting Natural Hazards with Land-Use Planning for Sustainable Communities*. Washington, D.C.: Joseph Henry Press, 356 pp., 1998.

Godschalk, D. R., T. Beatley, P. Berke, D. J. Brower, and E. J. Kaiser, *Natural Hazard Mitigation Recasting Disaster Policy and Planning*. Washington, D.C.: Island Press, 575 pp., 1999.

The H. John Heinz III Center for Science, Economics and the Environment, *Evaluation of Erosion Hazards*. Prepared for Federal Emergency Management Agency, Washington, D.C., 203 pp., 2000a.

________, *The Hidden Costs of Coastal Hazards: Implications for Risk Assessment and Mitigation,* Washington, D.C. and Covelo, CA: Island Press, 220 pp., 2000b.

Kunreuther, H. K., and R. J. Roth, Sr. (eds.), *Paying the Price: The Status and Role of Insurance Against Natural Disasters in the United States,* Washington, D.C.: Joseph Henry Press, 300 pp., 1998.

Mileti, D. S., *Disasters by Design: A Reassessment of Natural Hazards in the United States,* Washington, D.C.: Joseph Henry Press, 351 pp., 1999.

National Research Council, *The Impacts of Natural Disasters: A Framework for Loss Estimation,* Washington, D.C.: National Academy Press, 80 pp., 1999.

Platt, R. H., The Evolution of U.S. National Policies on Coastal Hazards, in *Defense from Floods and Floodplain Management,* J. Gardiner, O. Starosolszky, and V. Yevjevich (eds.), Dordrecht, The Netherlands: Kluwer Academic Publishers, 568 pp., 1995.

Platt, R. H., *Did Public Regulations Matter: Rebuilding the North Carolina Coast After Hurricane Fran.* A report prepared under contract with the H. John Heinz Center for Science, Economics and the Environment. Washington, D.C., 37 pp., 1999.

Simpson, R. H. and J. Simpson, *Why Experiment on Tropical Hurricanes?* Transactions, New York Academy of Sciences, 28(8), 1045-1062, 1966.

Sorensen, J. H., Hazard Warning Systems: Review of 20 Years of Progress, *Natural Hazards Review,* 1, 119-125, 2000.

Chapter 7

Communicating Damage Potentials and Minimizing Hurricane Damage

Herbert S. Saffir

In contrast to Chapter 6, which discussed a wide range of societal problems encountered in coping with hurricanes, this chapter turns to the engineering aspects of hurricane wind loading on structures. After describing the origin and character of the Saffir/Simpson scale, with particular reference to wind damage potential, I examine design factors and building code requirements that must be met if structures are to cope with hurricane winds. From analyses of the ravages of Hurricane Andrew in South Florida, we will learn that the bulk of hurricane damage is related to correctable deficiencies in the design, application, and enforcement of practicable hurricane-resistant construction procedures.

CREATING THE HURRICANE SCALE

In 1969, I prepared a study for the United Nations on the technological aspects of design and construction to mitigate damage to low-cost housing caused by hurricanes worldwide. In the study, I proposed a hurricane scale to measure the storm's structural effects. At that time no scales based on structural damage existed for windstorms or hurricanes, although similar scales were used for earthquakes. The scale was proposed to provide comparisons of the effects of hurricanes of various intensities in diverse locations around the globe.

The scale was divided into five intensities of hurricanes based on the increasing amount of structural damage. The scale subjectively described the type and

Hurricane! Coping with Disaster

10.1029/055SP09

amount of structural damage that could occur for each category of storm, ranging from storms with winds of 74 mi/hr to those with winds of more than 155 mi/hr.

The scale has found broad use by hurricane forecasters and by local and federal agencies that are concerned with long-range disaster planning and preparedness as well as short-range evacuation of residents during hurricane alerts. During the early 1970s the scale was expanded by then-director of the National Hurricane Center, Robert Simpson, who added possible storm surge and surge disaster potential to the velocity and structural damage categories. Members of the National Hurricane Center (NHC) staff, including Simpson, made small adjustments of category ranges of wind speed, so that the wind speeds could be identified with central pressures. A storm surge along the coast during a hurricane will have considerable effect on structures on or along the coast. The factors involved in causing flooding from storm surge are:

1. The hydrostatic response of sea level to lowered surface pressure at the hurricane center.
2. The radial distance of the maximum wind from the hurricane center.
3. The bathymetry, or offshore water depth profiles in the landfall area.
4. The stage of astronomical tides at time of landfall.

Tables 1 and 2 define the scale and its values from 1 to 5. Table 1 describes the character of wind damage to be expected for each scale value. Table 2 defines the ranges of wind speeds, hurricane central pressure, and expected storm surge values (exclusive of astronomical tide stages). Wind speeds are sustained values of one-minute duration at lower elevations (10 m or 33 ft above the surface). To examine the increase of wind velocity with height, based on the roughness of the terrain, Figure 1 indicates wind speeds above ground at different heights, up to the gradient wind speed of 100 miles/hour for three different terrains.

The relationship between a hurricane's wind speed and its destructive force is shown in Figure 2. It shows that a Category-4 hurricane with 135 mi/h winds has about eight times the destructive force of a 45 mi/hr wind. A Category-5 storm with 155 mi/hr winds has more than 11.5 times the destructive force of a 45 mi/hr wind. Tropical storms reach hurricane force at 74 mi/hr.

The Saffir/Simpson scale has been of immeasurable assistance in helping emergency management organizations communicate the intensity of a hurricane threat and to advise of the necessity to take immediate protective action. Emergency management personnel have emphasized that they were helpless in communicating these directives to residents prior to the use of the hurricane scale. It has been used throughout the United States and was adapted for use in Guam and Australia.

Table 1. Hurricane wind damage potential scale

Scale No. 1–Winds of 74 to 95 miles per hour. Damage primarily to shrubbery, trees, foliage, and unanchored mobile homes. No real damage to other structures. Some damage to poorly constructed signs. And/or: storm surge 4 to 5 feet above normal. Low-lying coastal roads inundated, minor pier damage, some small craft in exposed anchorage torn from moorings.

Scale No. 2–Winds of 96 to 110 miles per hour. Considerable damage to shrubbery and tree foliage; some trees blown down. Major damage to exposed mobile homes. Extensive damage to poorly constructed signs. Some damage to roofing materials of buildings; some window and door damage. No major damage to buildings. And/or: storm surge 6 to 8 feet above normal. Coastal roads and low-lying escape routes inland cut by rising water 2 to 4 hours before arrival of hurricane center. Considerable damage to piers. Marinas flooded. Small craft in unprotected anchorages torn from moorings. Evacuation of some shoreline residences and low-lying island areas required.

Scale No. 3–Winds of 111 to 130 miles per hour. Foliage torn from trees; large trees blown down. Practically all poorly constructed signs blown down. Some damage to roofing materials of buildings; some window and door damage. Some structural damage to small buildings. Mobile homes destroyed. And/or: storm surge 9 to 12 feet above normal. Serious flooding at coast and many smaller structures near coast destroyed; larger structures near coast damaged by battering waves and floating debris. Low-lying escape routes inland cut by rising water 3 to 5 hours before hurricane center arrives. Flat terrain 5 feet or less above sea level flooded inland 8 miles or more. Evacuation of low-lying residences within several blocks of shoreline possibly required.

Scale No. 4–Winds of 131 to 155 miles per hour. Shrubs and trees blown down; all signs down. Extensive damage to roofing materials, windows, and doors. Complete failure of roofs on many small residences. Complete destruction of mobile homes. And/or: storm surge 13 to 18 feet above normal. Flat terrain 10 feet or less above sea level flooded inland as far as 6 miles. Major damage to lower floors of structures near shore due to flooding and battering by waves and floating debris. Low-lying escape routes inland cut by rising water 3 to 5 hours before hurricane center arrives. Major erosion of beaches. Massive evacuation of all residences within 500 yards of shore possibly required, and of single-story residences on low ground within 2 miles of shore.

Scale No. 5–Winds greater than 155 miles per hour. Shrubs and trees blown down; considerable damage to roofs of buildings; all signs down. Very severe and extensive damage to windows and doors. Complete failure of roofs on many residences and industrial buildings. Extensive shattering of glass in windows and doors. Some complete building failures. Small buildings overturned or blown away. Complete destruction of mobile homes. And/or: storm surge greater than 18 feet above normal. Major damage to lower floors of all structures less than 15 feet above sea level within 500 yards of shore. Low-lying escape routes inland cut by rising water 3 to 5 hours before hurricane center arrives. Massive evacuation of residential areas on low ground within 5 to 10 miles of shore possibly required.

Table 2. Saffir / Simpson Hurricane Intensity Categories

CATEGORY	CENTRAL PRESSURE Mb.(hPa)	CENTRAL PRESSURE Inches	WIND SPEED Mi /Hr	WIND SPEED Knots	STORM SURGE (Feet)
1	≥ 980	≥ 28.89	74-95	64-83	4-5
2	965-979	28.50-28.91	96-110	84-96	6-8
3	945-964	27.91-28.47	111-130	97-113	9-12
4	920-944	27.17-27.88	131-155	114-135	13-18
5	<920	<27.17	>155	>135	>18

WIND LOAD DESIGN AND BUILDING CODES

My original United Nations study explained how buildings react under the dynamic forces of hurricane strength winds. It also explored wind-load design on various building shapes and components and made design recommendations for building materials, such as steel, concrete, wood, and aluminum. Structural design considers the dead load of the building itself and live loads are added to it, along with Earth pressure, temperature, impact, wind loads, and earthquake loads.

Design professionals follow standard design methods that are set forth in building codes. They include standards for hurricane damage mitigation. The design professional must operate within these parameters. In the United States, the codes typically are regional in scope, but other countries generally adopt one

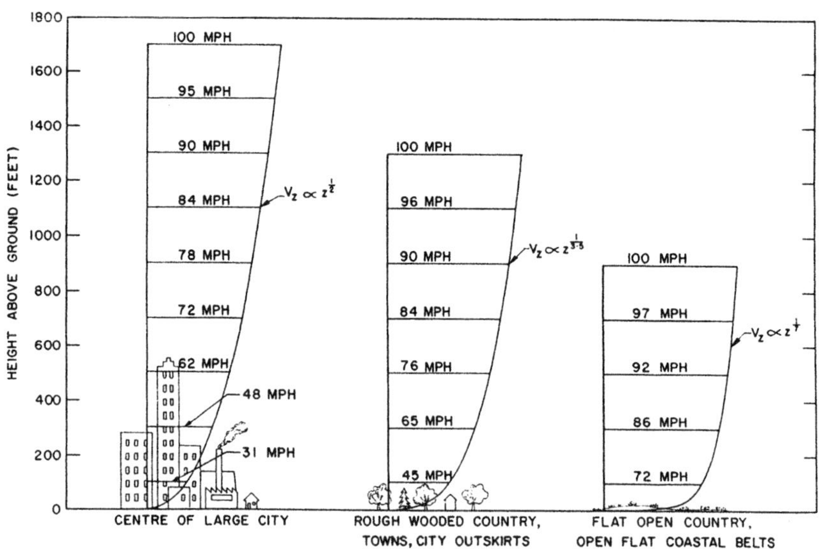

Figure 1. Wind Velocity Profiles Based on Terrain.

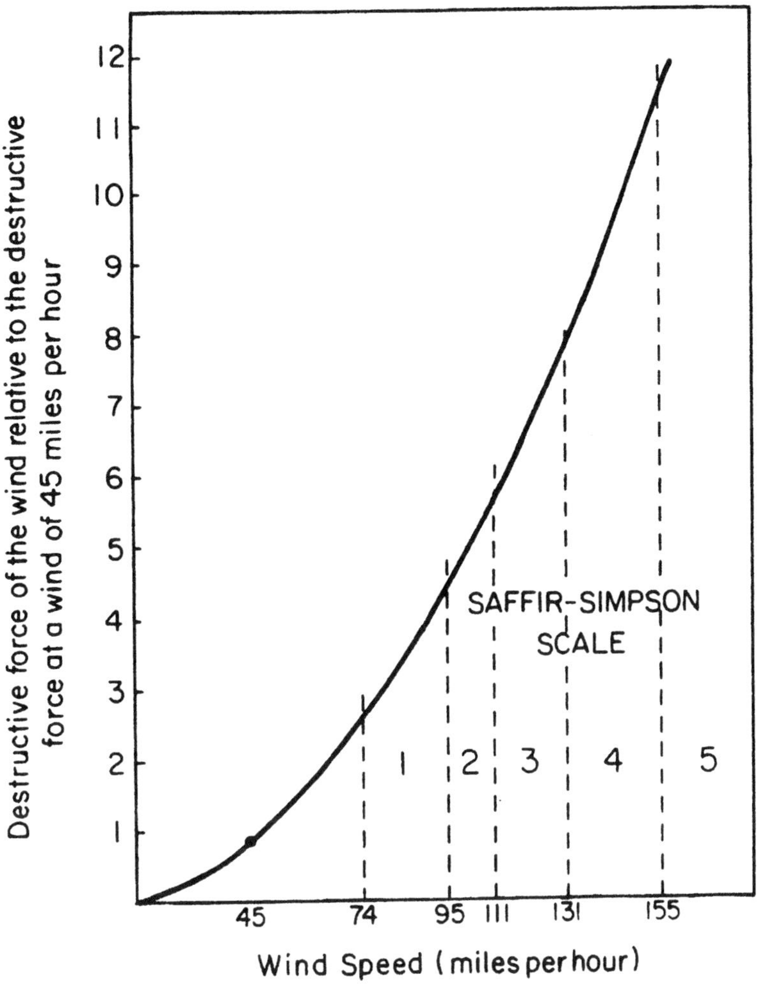

Figure 2. Wind Speed Related to Destructive Force [Pielke and Pielke, 1997]

code for the entire nation. Many improvements to the codes in hurricane-prone areas have resulted from investigations of the damage caused by hurricanes.

Building codes are legal documents that set forth basic requirements to protect the public's health, safety, and general welfare. They exist as a "check on fools and rascals" by providing a set of minimum standards to prevent errors in judgment from harming the public, according to the late Hardy Cross, a preeminent civil engineering professor at Yale University.

Hammurabi, King of Babylon, established the first building code around 1700 BC. It imposed a death penalty on builders who erected unsafe or faulty buildings. There are two types of building codes: specification and performance. A specification code describes in detail exactly what materials are to be used and how they are to be used, along with other specifics, such as sizes, spacings, and construction methods. A performance code prescribes the objective to be achieved but leaves broad leeway to the designers as to how to achieve the objective. In South Florida and many other coastal areas of the United States, local building codes include standards for hurricane damage mitigation. Many improvements to the codes in hurricane-prone areas have occurred after hurricanes struck.

SPECIFIC SOUTH FLORIDA BUILDING CODES

In Miami, officials developed the city's first extensive building code after a Category-4 hurricane[1] struck there in September 1926, causing extensive structural damage and an economic collapse. The code's wind load was rather meager. For example, the 1937 version of the code specified design wind loads for walls of buildings to meet the criteria in Table 3. These loads exceeded wind loads called for in other codes in the 1930s. Wind loadings were not considered in most cities, including New York City, which had no wind load requirement at all for buildings under 100 ft (30 m). Miami's 1937 building code established mandatory prescriptive requirements for one- and two-story buildings. It provided detailed drawings to show the continuous reinforced concrete footings and continuous tie beams required for all buildings regardless of the type of materials used. Subsequent hurricanes, such as Donna in 1960, forced various improvements to the uniform South Florida Building Code, which had been adopted in 1957 by Miami-Dade County and Broward County (Miami and Fort Lauderdale). Donna resulted in the first rational requirements for exterior windows, doors, and curtainwalls in a U.S. building code.

RECOMMENDED REQUIREMENTS FOR NEW CODES OR FOR UPGRADING HURRICANE-RESISTANT CODES

Disasters, such as Hurricane Andrew in 1992, often prompt building code changes. Building codes initially were developed over the years to prevent dis-

TABLE 3. Wind Pressure at Various Building Height

Height above ground	Wind pressure over entire building
25 ft.	25 pounds per sq. ft.
75 ft.	35 pounds per sq. ft.
> 5 feet	45 pounds per sq. ft.

[1] The minimum pressure of this hurricane was 934 mb. However, damage patterns in some parts of the city of Miami were consistent with those expected of Category-5 winds according to some analysts.

asters. The design to protect against wind forces has been an important part of the codes since wind caused the collapse of the famous Firth of Tay railway bridge in Scotland in 1879. The body of standards that has developed for the analysis, design, and construction to prevent windstorm and hurricane damage relies on a combination of meteorological and engineering sciences.

In considering potential damage from wind, the most important criterion is wind speed. The force exerted by the wind can be the damaging load that affects the strength and safety of the structure. The wind force increases with the square of the wind speed (Figure 2). Other factors also come into play, including the shape of the building, the location of the building component being considered, the topography of the site, and internal building pressures.

Some of the basic design recommendations for hurricane-resistant building codes that evolved after Hurricane Andrew are:

1. A determination of wind load requirements is one of the important steps to take in formulating this code. My opinion is that both the ASCE (American Society of Civil Engineers) 7-93 Standard (based on a sustained "fastest-mile" of wind) and the ASCE 7-98 Standard (based on a 3-second gust) are adequate. The basis for each of the standards is a 50-year frequency of occurrence, and neither standard provides for the Hurricane Andrew type storm. The owner or the designer must decide whether the importance of the structure being designed warrants an increase to 100-year frequency of occurrence. A 50-year frequency of occurrence may not be adequate; the building code is a minimum standard, not a maximum standard. FEMA, in its establishment of floor elevations for insured construction in case of flood, uses a 100-year storm to determine the required elevation; the structure itself, however, may only be designed for a 50-year wind event, which is inconsistent.
2. The subject of shop drawings needs to be clarified and enforced in the building code. In many cases shop drawings are design drawings and not just fabrication details for the shop. For instance, the standards pertaining to the residential roof truss design must be fully followed and enforced. Many failures that occurred during Hurricane Andrew were due to a lack of responsibility by the building architect-of-record, who refused responsibility for the truss design and installation and appurtenant bracing in the residence. Bracing items were omitted or were poor because the residential truss designer took no responsibility for this part of the building system. Sometimes this was left up to the builder by default. This also applies to other structural roof systems and curtain-wall installation.
3. Mandatory protection of building envelopes in new construction should be provided by requiring that the components of exterior walls, such as glazing, doors, and windows of enclosed buildings, be designed and

constructed to preserve the enclosed building envelope against wind pressure and impact loads from wind-borne debris or be protected by fixed, operable, or portable demountable storm shutters adequately designed for full loadings.

4. Mandatory testing of roof shingles and tiles as a system at 100 miles/hour must be required, and a testing standard for incorporation included in the building code. The use of wall systems other than concrete block in multistory residential and commercial buildings should be reexamined. Prefabricated systems should be banned if they obviously do not meet building code requirements.
5. The role of the structural engineer in the design of residential and commercial structures should be strengthened.
6. The architect- or the engineer-of-record must be required to participate in the structural inspection process. They must sign an affidavit prior to the issuance of the Certificate of Completion by the building official, certifying that, to the best of his/her knowledge, the building construction fully complies with the design intent and the building code. In addition, the approved permit plans reflect the structural as-built conditions.
7. As part of the permit set of plans, all drawings and specifications for the building component items, such as windows, doors, and roof trusses that have received product control approval, must be included in order that conforming inspections can be performed by building inspectors.
8. Design responsibility must be clarified for loads transmitted to the structure by mechanical equipment mounted on the roof. The engineer- (or architect-) of-record must approve such design.
9. For pre-engineered metal buildings, the following requirements should be part of the code:
   a. Roof and wall lateral and cross bracing, in addition to panel and deck diaphragm action for wind bracing.
   b. Doors anchored as part of the frame in the closed position, with all doors meeting the performance strength and impact requirements of the code.
   c. A heavier metal siding with a minimum gauge of 12 (0.06 cm).
   d. Limited or no use of self-tapping screws for fastening siding and roof panels; self-tapping screws can vibrate loose in a long duration wind.
   e. Continuous metal siding with decreased allowable deflection and siding subjected to the impact-load test consistent with the criteria established for the protection of building envelopes.
   f. As in the case of shop drawings, the engineer-of-record for the building shall take full responsibility for the entire process for pre-engineered metal buildings.
10. Standard specifications for one- and two-story "non-engineered" buildings must be included in the building code as a mandatory, not

advisory, feature. These detailed specifications should cover masonry residences and typical wood frame structures and include, but not be limited to, reinforced concrete continuous foundations, reinforced concrete tie (bond) beams, reinforced concrete tie columns, anchorage for roof trusses, and typical wood framing. Much of the "design" for "non-engineered" buildings will be performed in this section of the building code.

LESSONS FROM HURRICANE ANDREW

The South Florida Building Code was improved after Hurricane Andrew blasted the region in August 1992. Andrew made landfall 30 miles south of Miami. It caused about $30 billion in damage, making it likely the most costly natural disaster to hit the nation. Andrew had sustained one-minute winds of 145 mi/hr, with gusts reaching at least 180 mi/hr, probably higher in some locations.

Andrew's wind caused most of the tremendous structural damage, but other factors also contributed, including poor construction practices, lack of attention to the code, poor inspection procedures by the governmental building departments, and some deficiencies in the building code.

Even with its deficiencies, the building code prevented much worse structural damage from occurring, in my opinion. I served on the task force that revised the code after Andrew. In learning the lessons of Andrew, it is important to examine the task force's findings. The major portion of the observed property damage and building failures was attributed to:

1. The severity of Hurricane Andrew which exceeded the design storm used in the South Florida Building Code at that time, even though the design met existing code requirements.
2. Deficiencies in the building code (since rectified).
3. Design flaws in buildings damaged by Hurricane Andrew.
4. Penetration of building envelopes, unprotected and vulnerable to windborne debris (since rectified by adopting impact-load requirements).
5. Ineffective building code enforcement almost as "standard operating procedure."
6. An unclear division of responsibilities within the construction industry.
7. Contractors providing inadequate supervision of their field personnel.
8. Misapplication of approved products and materials.
9. Inadequate training at all levels in wind-resistant design, inspection, and construction.
10. A lack of public awareness about construction practices and regulations.

REFERENCES

American Society of Civil Engineers, *ASCE 7-93: Minimum Design Loads for Buildings and Other Structures,* American Society of Civil Engineers, 1993.

American Society of Civil Engineers, *ASCE 7-98: Minimum Design Loads for Buildings and Other Structures,* American Society of Civil Engineers, 1998.

American Society of Civil Engineers, *ASCE 24-93: Flood Resistant Design and Construction,* American Society of Civil Engineers, 1998.

Pielke, R. A., Sr. and R. A. Pielke, Jr., Hurricanes, New York: John Wiley & Sons, 1997.

Saffir, H. S., Florida's Approach to Hurricane-Resistant Design and Construction, *J. Wind Eng. Aerodyn.*, 1989.

Simiu, E. and R. H. Scanlan, *Wind Effects on Structures,* 3rd Edition, New York: John Wiley & Sons, 1996.

Simpson, R. H., and H. Riehl, *The Hurricane and its Impact*, LSU Press, 1981.

Simpson, R. H., The Hurricane Damage Potential Scale, *Weatherwise*, 1974.

Chapter 8

Reducing Vulnerability

Roger A. Pielke, Jr.

THE NATURE OF THE HURRICANE PROBLEM

Much has changed in the century since the Galveston Hurricane of 1900, but one aspect of hurricanes remains. When a hurricane strikes, it results in casualties, property damage, and disruption of communities. In recent decades, despite considerable variability in the frequency of hurricanes, their economic and societal impacts have been steadily rising. This increase is largely due to increasing population and development in exposed locations. There have been serious hurricane impacts since 1990, particularly Hurricane Andrew's (1992) large economic toll, and the human tragedy associated with Hurricane Mitch (1998) in Central America. These disasters have created the impression that growing losses are an indicator of increasing vulnerability to hurricanes. Consequently, many analysts believe that we face a serious policy problem.[1] As Neil Frank notes in Chapter 5, despite the advances made in understanding hurricanes, "I consider the United States more vulnerable to hurricanes today than at any time in history."

What then are the prospects for reducing hurricane risks to society? This chapter provides a context in which to consider the findings and assertions of the previous three chapters in the broad context of vulnerability to hurricanes. It argues

[1] There is a vast literature on vulnerability to hurricanes—and tropical cyclones generally—from scholars such as Cutter, Morrow, Peacock, Pulwarty, Diaz, Baker, Southern, Foley, and so on. For an extensive source of references to the broader literature see Pielke and Pielke [2000].

Hurricane! Coping with Disaster

10.1029/055SP10

that improving the use and application of knowledge already in hand can lead to even greater reductions in vulnerability in both the United States and other nations in the region.

In Chapter 6, Mary Frances Myers and Gilbert White present a three-part taxonomy of responses to hurricane vulnerability—modification of the hazard, susceptibility, and impact. As there is no clear evidence that society can intentionally control the behavior of storms (see Chapters 1 and 3 for discussion of deliberate hurricane modification), they argue that improved responses lie in the other two aspects of vulnerability. This chapter considers one example of each, susceptibility and impact, as illustrative of the obstacles blocking improved responses. The chapter then discusses prospects for improved *use* of hurricane forecasts and warning in the United States to mitigate the impact of storms. Finally, it discusses the profound vulnerability of some areas of Central America and the Caribbean as revealed by Hurricane Mitch in 1998.

CHALLENGES TO THE EFFECTIVE USE OF HURRICANE FORECASTS

Without a doubt, the overall development of hurricane forecasting and its use by decision makers in the United States is a triumph of science and technology. Never again will a hurricane strike the United States without advance warning, as occurred, for example, in the 1900 Galveston storm as documented by Neil Frank (Chapter 5).[2] To dream of unlimited opportunity for improved hurricane forecasts to greatly reduce vulnerability is untenable (see Chapter 4). At the beginning of the 21st Century, the path forward is not so clear-cut. The reason for this is that going from no or little warning capabilities to a mature warning capability (as occurred during the 20th Century) results in greater benefits than do marginal improvements to the generally mature warning system that the United States has in 2002. The situation is complicated even further by the fact that forecasts themselves can have profound *negative* impacts on society, or as Myers and White (Chapter 6) warn: "reliance on prediction and emergency technology alone can lead to more catastrophic losses in the future when large events occur." The simple reason for this is that the production and use of forecasts and technologies creates a new set of vulnerabilities. An example of this occurred in 1997 in response to record flooding on the Red River of the North, which flows from Minnesota to Lake Winnipeg, Manitoba. Decision makers and forecasters each misinterpreted the degree of uncertainty in what was (with hindsight) an accurate forecast. As a consequence, despite the long lead-time and skillful forecast, residents were ultimately unprepared for a flood that caused more that $2 billion in damages. Decisions made based on misinterpretations of the forecast were arguably inappropriate and enhanced rather than reduced risk.

[2] Of course, it is likely that a storm could rapidly intensify without advance warning, as occurred with Hurricane Opal in 1995. However, the storm itself would not go unnoticed.

A major challenge to the effective use of hurricane forecasts is "overwarning" —when a greater area is placed under a hurricane warning than actually experiences the effects of the storm. Overwarning occurs because forecasts of hurricane track and intensity are, to some degree, uncertain, and officials seek, for many reasons, to avoid having a storm strike in unwarned locations. Hence, hurricane forecasters typically post warnings for a greater stretch of coast than a storm actually affects. For example, the typical storm affects about a 125-mile length of coast, while warnings are typically posted for about a 400-mile length. This means that, on average, 275 miles of coast are warned but do not experience the direct effects of the hurricane. Of course, most storms do not make landfall perpendicular to the coast, and some storms parallel the coast for an extended period, increasing the challenge faced by forecasters and decision makers.

Experience in 1999 with Hurricane Floyd, a storm that moved parallel to the entire southeastern coast, dramatically highlights the challenges associated with overwarning. An anonymous reviewer notes that Floyd was "probably the most overwarned hurricane in the past 30 or more years" – illustrating the challenges of overwarning, at least in this atypical case. Hurricane Floyd, however, attained the dubious distinction of being the first billion-dollar storm in the United States in which the *costs of the forecasts* likely rivaled the costs of the storm's physical damage. Assuming for present purposes that the costs of evacuation are in the neighborhood of the oft-cited $1 million per coastal mile, then the evacuation of about 2,000 miles of coast related to Floyd's track along the U.S. East Coast totaled approximately $2 billion. According to the insurance industry, insured losses associated with Floyd totaled more than $1 billion, and uninsured flood costs added at least several billion to that.

In the case of Floyd, the comparison between forecast impacts and weather impacts reveals *overwarning* on track forecasts as one of the hidden costs of hurricanes.[3] Most storms, of course, do not lead to the same degree of overwarning seen with Floyd. Nonetheless, using the numbers cited above for the "typical" storm, the costs of overwarning are about $275 million per landfalling storm (i.e., 400 miles warned minus 125 miles affected). Overwarning is problematic because it is not free of cost and, thus, has significant economic consequences, as well as effects on how people perceive and respond to subsequent warnings (Chapters 5 and 6). In addition, if the degree of overwarning is large enough with respect to the track forecast uncertainty, it may make it difficult to incorporate improvements in forecasts into improvements in the warning process. Floyd was obviously an extreme example of overwarning, but perhaps as more and more people flock to the coast (Chapters 5 and 7), it may also be a cautionary tale of the sorts of problems to be expected more frequently in the future.

[3] Data for this section is based in large part on Jarrell and DeMaria [1999] and Powell and Aberson [2001], and the analysis draws on Pielke and Carbone [2002].

Unless hurricane predictions can be made nearly perfect and incentives to overwarn are eliminated, overwarning will continue to accompany hurricane forecasts. The size of the overwarned area, however, can be reduced in the forecast process depending on the amount of risk society chooses to bear. A larger warned area lessens the likelihood that a storm will "surprise" an unwarned community. But because overwarning is not free, issuing a larger warning involves a trade-off of overwarning costs with the reduction of risk of a surprise landfall. At the beginning of the 21st Century, these trade-offs are not being made quantitatively or even considered explicitly. For example, an anonymous reviewer notes, "The degree of overwarning has no objective basis. It results almost entirely from the personal views of the forecasters, particularly the Director of the National Hurricane Center." The reviewer further notes that "no hurricane made landfall outside the warned area in the history of the U.S. National Hurricane Center."

The following back-of-the-envelope calculation illustrates the dynamics of the risk versus benefit trade-off.[4] In 1997, the average error in hurricane track forecasts at 24 h was about 90 miles (the error is the distance between the point where a storm center is predicted to go and the point that it actually does go). And 95% of such forecast errors were less than 200 miles. By inference, this means that the average 400 miles of coastline warned per storm equates to a 95% confidence level (i.e., +/- 200 miles = 400 miles). Thus, at this level of error, for a particular storm there is a 1 in 20 chance that it will cross the coast over an area that was not warned at 24 h. If improved forecasts can reduce the magnitude of the 95% error in miles of coastline warned by, say, 20% from 200 miles to 160 miles, then, all else being equal, this would lead to a reduction in miles-of-the-coastline-warned from 400 miles to 320 miles. This equates roughly to a cost saving of $80 million per storm (80 miles times $1 million per mile), without changing the level of risk (i.e., the 95% confidence level).

Reducing the costs of overwarning depends upon two factors—improving forecasts and improving the use of forecasts. Improving forecasts occupies the attention of many scientists and research efforts in the public and private sectors. Improving the use of forecasts receives considerably less attention. Consequently, information is lacking that would allow officials involved in the hurricane warning process to evaluate trade-offs between the costs of hurricane track overwarning and the risks of unwarned impacts. This is a specific example of the general conclusion reported by Myers and White (Chapter 6) that "we lack sufficient data on costs and benefits ... of social choices." The same types of trade-offs occur with respect to hurricane intensity forecasts and indeed any prediction that is accompanied by uncertainty. Stewart [2000] provides an overview of the implications of the trade-offs.

[4] This example is presented for illustrative purposes only and is not meant to substitute for a rigorous analysis, the lack of which, as argued below, impedes more effective decision making with respect to hurricane warnings.

Some have suggested that instead of reducing the level of overwarning, it would be more important to increase the lead-time available to the emergency management community. This argument makes good sense, given how long it takes to evacuate many communities and how those times are increasing. But it is impossible to effectively compare the relative benefits of reducing miles-of-coastline-warned versus increased lead-time, because the value of improved lead-time has yet to be systematically studied. Such analyses are probably comparably important to the research on improving hurricane forecasts.

The hurricane case is probably representative of a more general situation throughout severe weather forecasts and warnings. Considerable benefits might result to public and private sector decision makers and the population at risk if they were to more effectively use the weather services that *they already have available* simply by understanding the impacts of the forecasts in the context of the decision environment.

VULNERABILITY IN THE CARIBBEAN AND CENTRAL AMERICA

In late October 1998, the remnants of Hurricane Mitch stalled over Honduras and Nicaragua, killing perhaps 10,000 people, affecting 6.7 million. As much as $8.5 billion in damages resulted.[5] With an estimated 10,000 deaths, Mitch was one of the deadliest Atlantic hurricanes in recorded history, ranking second to the 1780 "Great Hurricane" in the Lesser Antilles. Although the impacts of Hurricane Mitch were extreme in an absolute sense, they were not anomalous when placed into the context of increasing hurricane vulnerability in the region. Rather than being a freak climate event, "human intervention lies at the root of much of [Mitch's] damage [OCHA et al., 1999]." Some of the interrelated anthropogenic factors that increase the risk of natural disasters such as Mitch include rapidly increasing populations, widespread poverty, lack of access to adequate land, deforestation, and urbanization. The knowledge needed to address such issues goes well beyond science and technology, at least as conventionally understood. And while this book focuses on the United States, probably a greater unmet need for reducing vulnerability to hurricanes lies in other countries in the region. Because the U.S. has been relatively successful (so far), in responding to the hurricane threat, there would seem to be great potential and, perhaps, a responsibility to assist others in the region to achieve comparable successes. But to do so requires understanding the unique nature of vulnerability found in various locations.

Central America's population is growing rapidly with average annual growth rates over the past ten years ranging from 1.6% in Panama and Mexico to 2.6%

[5] This section is distilled from Pielke et al. [in review, *Natural Hazards Review*]. The extensive source material needed to document the numerous figures found here and support the analysis can be found in that paper.

in Costa Rica, Honduras, and Nicaragua. Population growth increases susceptibility because there are simply more people for a disaster to affect and because more people settle in disaster-prone areas. In addition, population growth can increase poverty. In some countries an inverse relationship exists between per capita Gross National Product (GNP) and total fertility rates. The countries having some of the highest fertility rates are the poorest.

Poverty is a critical factor underlying vulnerability. And poverty is endemic in the region affected by Mitch. For example, 48% of El Salvador's population was poor in 1994. Approximately 75% of Guatemala's population was living below the poverty line in 1995. Half of Nicaragua's population fell below the poverty line in 1995. Many Central Americans are below the international poverty line, defined as living on less than $1 per day at 1985 international prices, adjusted for purchasing power parity.

Poverty can increase the risk of natural disasters in several ways. Firstly, impoverished people often lack access to suitable land. When Hurricane Fifi hit Honduras in 1974, land ownership had become highly concentrated with 63% of Honduran farmers having access to only 6% of arable land. This was due to a variety of factors: large landowners driving out small farmers from their land to create large cotton estates; growth in land-intensive livestock production; ownership of large tracts of land by banana companies; and use of valley floors for large-scale irrigation. Peasants were forced onto steep hillsides where agricultural practices increased soil erosion and silting of rivers. By some estimates, Fifi killed as many as 10,000 Hondurans. In one town alone, 2,300 were killed when a temporary dam, created by landslides into a river, gave way. Yet after Fifi, trends continued, setting the stage for Mitch; the Honduran Central Bank estimated that by 1988, 48% of the country's valley lands were sown in cattle pastures.

Secondly, poor people migrate to urban areas in search of economically gainful activity if they are displaced from the land, for whatever reason. Because they frequently lack access to safe building sites, they build on steep hillsides or in flood-prone areas. Development of urban slopes and hilltops increases the risk of flooding lower-lying areas, where many Central American urban poor reside. Between 1980 and 1997, the percentages of Central Americans living in urban areas increased by anywhere from 3% in Guatemala to 10% in Nicaragua and Honduras.

Finally, poverty can be a factor in deforestation. Poor people may clear forests for agriculture, to build houses, and to gather wood for fuel. Deforestation increases soil erosion, which can enhance landslides, mudslides, and flooding. Between 1990 and 1995, Costa Rica, Nicaragua, Honduras, Panama, and Guatemala had some of the highest deforestation rates in the world.

Many of the factors that have increased exposure in Central America also exist in the Caribbean. While the rates of population growth are generally not as high as in Central America, countries such as Haiti, the Dominican Republic, and the

Bahamas experienced average annual growth rates of 1.6% and 1.7% in the last decade. The incidence of poverty in the Caribbean varies by country. In 1989, 48% of Dominican Republicans lived on less than $2 per day; in 1993 25% of Jamaicans lived at this level.

As in Central America, land is inequitably distributed in parts of the Caribbean. For example, 82% of Dominican Republic farmers have access to 12% of the agricultural land; 59% of farmers in Haiti occupy 22% of the land. This pattern is seen in Jamaica as well. Much of the land occupied by small farmers is either in mountainous areas or on hillsides. Some of the most substantial Caribbean losses from hurricanes in the last 50 years—Flora, 1963 (approximately 8,000 deaths) and Gordon, 1994 (1,145 deaths)—occurred because of floods, mudslides, and landslides on hilly terrain.

Deforestation is also occurring at a rapid rate in parts of the Caribbean. Between 1990 and 1995 Jamaica had an average annual deforestation rate of 7.2%; Haiti's was 3.4%. Between 1978 and 1988 Haiti's forested land decreased at a rate "bordering on desertification." Jamaica suffered unprecedented landslides from Hurricane Gilbert in 1988 due, in part, to the extensive removal of tree cover from steep slopes by several large-scale coffee plantations. And urbanization is rapidly increasing in many Caribbean countries due to high fertility rates, restrictions on cross-national migration, and internal migration of the rural poor. Between 1970 and 1995 the percentages of the population living in urban areas increased by as much as 33% in Trinidad/Tobago and Saint Vincent/the Grenadines. By 2000, over 64% of the Caribbean basin population is expected to be living in urban areas, up from 38% in 1960. With these kinds of demographic and economic issues, the societal benefit of improving forecasts of hurricanes pales in comparison to the benefits associated with reducing risk through other means.

It is perhaps seductive to think that there exists a "silver bullet" solution to reducing the future impacts of hurricanes in Latin America and the Caribbean. Ample evidence suggests this is wishful thinking. Scholars and practitioners have historically faced tremendous difficulties in addressing factors associated with hurricane losses, particularly in developing countries. Progress will be made incrementally, with careful attention paid to the role of scientific and technological capabilities in the context of the region's vulnerabilities. Much progress can result from the use of existing science and technology.

CONCLUSION

Although the examples in Central America and the Caribbean illustrated above originate in vastly different contexts, they do share one important characteristic. Reducing vulnerability depends upon more effective use of knowledge and technologies already in hand and addressing more fundamental societal problems such as poverty and other factors that contribute to it.

1. The case of forecast use in the United States suggests that while improving predictive abilities related to hurricanes may also improve societal responses, better responses can probably occur through more effective use of existing capabilities. To date, the needed data and analyses are lacking that would permit understanding forecast improvements in the context of risks and benefits. Efficient, effective use of advancements in the science and technology of hurricane forecasting will require attention to the *use* of products and services in the broad context of social and decision processes.
2. The risks posed by hurricanes in Central America and the Caribbean exemplify a profound vulnerability very different from that in the United States. Actions to reduce this risk may well depend upon a different approach to science and technology from that needed in the United States. As in the first example, it is probable that improved response can occur through a focus on implementing today's knowledge more effectively.
3. These examples suggest that, in the future, science and technology will continue to play an important role in reducing human vulnerability to hurricane impacts. But this role is much more subtle than it perhaps once may have been. In the United States the dramatic successes in minimizing the risks involved with hurricanes have been achieved by incorporating science and technology into hurricane policies and public response. In contrast, in many regions of Central America and parts of the Caribbean, there has been less incorporation of science and technology into the frail preparedness plans that may exist. That and other reasons cited above render these societies much more vulnerable. This is not to discount the need for continued advances in knowledge and technologies, but only to warn that the benefits to society of new developments will be highly limited if their effective *application* does not receive necessary attention.
4. The diversity of opportunities that stand before the science community share an important characteristic. Whether the context is the United States or developing countries, progress will be more effective when advances in science and technology proceed hand-in-hand with considerations of societal factors and alternative responses and policies for reduction of risk. In the U.S., there is a relative bounty of knowledge on the scientific, technological, and even societal aspects of hurricanes. The primary challenge for the future will be to integrate specialized knowledge into usable information that creates and applies new opportunities to minimize vulnerability. In my opinion, developing mechanisms to more effectively use existing knowledge is the greatest obstacle standing in the way of reducing vulnerability to hurricanes.

REFERENCES

Jarrell, J. D., and M. DeMaria. An Examination of Strategies to Reduce the Size of Hurricane Warning Areas. Preprints, *23rd Conf. on Hurricanes and Tropical Meteorology*, Vol. I, Dallas, TX, American Meteorological Society, 50-53, 1999.

Office for the Coordination of Humanitarian Affairs (OCHA), United Nations Development Programme (UNDP), United Nations Children's Fund (UNICEF), and Pan American Health Organization/World Health Organization (PAHO), *Joint Disaster Response and Recovery Mission to Central America as a Follow-up to Hurricane Mitch,* 1999. Internet: www.reliefweb.int/library/documents/mitch-jdrm/mitch222.htm

Pielke, R. A., Jr., and R.E. Carbone, Weather Impacts, Forecasts, and Policy: An Integrated Perspective. *Bull. Amer. Meteorol. Soc.,* Vol. 83, No. 3, 1233-1243, 2002.

Pielke, R. A., Jr., J. Rubiera, C. Landsea, M. Molina, and R. Klein. Hurricane Vulnerability in Latin America and the Caribbean. *Global Environmental Change, Part B: Natural Hazards.* (in review).

Pielke, R. A., Jr., and R.A. Pielke, Sr. (eds.), *Storms*, a volume in the nine-volume series of *Natural Hazards & Disasters Major Works* published as a contribution to the International Decade for Natural Disaster Reduction, London, Routledge Press, 2000.

Powell, M. D., and S. D. Aberson. Accuracy of United States Tropical Cyclone Landfall Forecasts in the Atlantic Basin (1976-2000). *Bull. Amer. Meteorol. Soc.*, 82, 2749-2768, 2001.

Stewart T. R. "Uncertainty, Judgment, and Error in Prediction", Chapter 3 in *Prediction: Science, Decision Making, and the Future of Nature*, pp. 41-57, Sarewitz et al. (eds.), Washington, DC, Island Press, 2000.

PART III

Potential Impact of Science and Technology

Present and Future

The assessment of progress over the span of a century depends not only on the units of measure one applies, but upon the viewpoint of the observer. It should not be surprising if ratings by an engineer, social scientist, hurricane forecaster, emergency manager, physical scientist, and a long-term coastal resident differ appreciably. Likewise the contributors to this volume, who come from differing backgrounds of experience and specialization, are not of one mind in evaluating 20th century progress. Part III speaks from the viewpoint of maturing physical and social sciences and advances in technology. It attempts to judge present abilities to cope with hurricanes as we enter the 21st Century and to project the extension and ultimate limits of progress we may encounter.

Chapter 9 analyzes the waypoints of scientific understanding, including a few excursions up blind alleys, and briefly what may lie ahead. Chapter 10 considers separately the progress in hurricane prediction and warning resulting from this scientific enlightenment. Chapter 11 demonstrates the increasing role of weather satellites both in research and in day-to-day monitoring and warning of hazardous weather. Chapter 12 follows with an example of on-going research using new science-driven technologies for explorations that were beyond reach less than two decades before this publication. Chapter 13 and 14 describe two new tools of observation that may help understand the interactive roles of ever-smaller scales of motion. Finally, Chapter 15 undertakes an overall assessment of how well a century of experience supported by new science and technology has equipped us to avoid a hurricane disaster rivaling that of Galveston in 1900.

Hurricane! Coping with Disaster

10.1029/055SP11

Chapter 9

A Century of Scientific Progress
An Evaluation

Kerry Emanuel

INTRODUCTION

In the early 20th Century, at the same time that the theories of relativity and quantum mechanics were developed, there was virtually no basic understanding of the physics of hurricanes. Neither the energy cycle that sustains such storms was known, nor the factors that controlled their movement. What was known about the structure and behavior of hurricanes was inferred largely from visual observations of clouds and damage patterns, and it was widely believed that the storm circulation extended upward only a few kilometers. By the end of the century, the structure and behavior of hurricanes had been thoroughly quantified, thanks to rapid technological progress. This progress made available such wonders as instrumented aircraft, weather radar, and satellites. The basic physics of the storms themselves, as well as their intricate substructure, had come to be well understood. This new understanding, coupled with the invention and rapid development of the computer, made it possible to forecast the motion of hurricanes with such accuracy that, given modern communications and transportation, loss of life from hurricanes has been virtually eliminated in the developed world.

This chapter chronicles the extraordinary progress in the scientific understanding of hurricanes through the 20th Century. The chapter begins with a brief review of progress through World War II, which marked an important turning point in the science of meteorology, and continues with an account of the rapid progress made in the first two decades after the war. In the 1960s, Charney and

Hurricane! Coping with Disaster

10.1029/055SP12

Eliassen [1964] proposed a very influential theory termed CISK (Conditional Instability of the Second Kind) to explain tropical cyclogenesis. The detailed synthesis of storm structure and behavior made possible by a suite of new observing systems is then described, followed by a review of the enormous progress in numerical modeling of hurricanes. Subsequent sections describe the ramp-up to our present understanding of various aspects of hurricanes including their genesis, intensification, movement, and transition to extratropical storms; their interaction with the ocean; and their control by and possible feedback on various oscillations in the climate system. An assessment of where we stand today, including outstanding challenges, concludes the chapter.

PROGRESS THROUGH THE SECOND WORLD WAR

By the early 19th Century, it had become well established that hurricanes are vortical storms. Arguably, the native inhabitants of the Caribbean region already understood this at the time of Christopher Columbus' voyages, given the detailed knowledge he obtained from them about the sequence of winds—enough to allow him to ride out a hurricane during his fourth voyage to the New World. There is also ample evidence that the early Chinese understood that typhoons are whirlwinds, and Francis Bacon described them as such at the end of the 16th Century. However, the fact that hurricanes have rotary wind fields was not firmly established in the modern world until the American scientist William Redfield published his observations in 1831.

An important step toward understanding the physics of hurricanes was the development of the first law of thermodynamics in the mid-19th century and the discovery of the latent heat of vaporization. Among the pioneers in applying these principles to meteorology was the American scientist James Pollard Espy. Observing that cyclones are invariably associated with clouds and precipitation, Espy came to believe that such storms are powered by the liberation of the latent heat of vaporization, and he undertook extensive laboratory experiments designed to determine what we now refer to as the moist adiabatic lapse rate. His work is particularly remarkable when we note that the formulation of the science of thermodynamics took place during the 1840s and 1850s, and the first law of thermodynamics was not introduced to meteorology until the 1860s. But already in 1841, Espy had concluded that the moist adiabatic lapse rate is "about five-eighths of a degree for one hundred yards of ascent, when the dew point is about 70 degrees." This is a remarkably good estimate considering the equipment and techniques at his disposal. Espy went on to deduce that if the environmental lapse is greater than the moist adiabatic lapse rate, convection may occur, and he believed that it was such convection that led to the formation of middle latitude cyclones.

It is a historical irony that the "thermal theory of cyclones", as the body of theory developed by Espy and his followers came to be known, was advanced in an attempt to explain extratropical cyclones when, in fact, it bears a remarkable resemblance to the idea of Conditional Instability of the Second Kind (CISK). Consider this exposition of the thermal theory of cyclones by Elias Loomis in 1846:

> *The heat liberated in the formation of this cloud raises the thermometer, causing a more decided tendency of the air inward toward the region of condensation....Relative elevation of temperature under the cloud gives increased velocity to the inward current of air. More cloud is thus formed, heat liberated....Thus the storm gains violence by its own action.*

Compare this to CISK, as described by Charney and Eliassen in 1964:

> *We should consider (the depression and the cumulus cell) as supporting one another—-the cumulus cell by supplying the heat for driving the depression, and the depression by producing the low-level convergence of moisture into the cumulus cell*

In fact, theoretical meteorology would be preoccupied with explaining extra-tropical cyclones for at least the next hundred years, with almost no serious theoretical work devoted to hurricanes until the 1940s. However, Father Benito Viñes, a Jesuit priest who came to Cuba in 1870 to serve as director of the meteorological observatory of the Royal College of Belen, in Havana, made a number of practical advances. Father Viñes worked tirelessly to establish a network of meteorological observing stations in Cuba and around the Caribbean and issued the first of many hurricane warnings in 1875. In 1877 he published his collected notes on hurricane behavior, "Apuntes Relativos a los Huracanes de las Antilles", which was soon published as "Practical Hints in Regard to West Indian Hurricanes" by the U.S. Army Signal Corps' national weather service.

Up through about the 1930s, several texts on hurricanes argued, mostly on the basis of the quick diminution of winds after landfall, that the storm circulation must be quite shallow, extending no higher than about 3 km. In 1935, Bernhard Haurwitz pointed out, on the basis of hydrostatics, that such a circumstance would require extremely high temperatures at low levels in the core. According to Tor Bergeron [1954], arguments that tropical cyclones must be shallow appeared as late as 1948.

WATERSHED YEARS: 1943-1964

As discussed in Chapter 2, on 27 July 1943, Army Air Corps Colonel Joseph B. Duckworth took an Air Force AT-6 trainer from an airfield in Texas and, together with Lieutenant Ralph O'Hair, became the first to penetrate the eye of a hurricane in an airplane. This marked the beginning of hurricane reconnaissance, which became routine by the late 1940s (see Chapter 3). At about the same time, the very first radar images of tropical cyclones revealed the structure of precipitation, showing for the first time an unambiguous depiction of the eye and spiral rainbands. Newly acquired measurements from aircraft and radar, as well as reports from ships and surface stations and occasional balloon soundings, allowed postwar meteorologists to construct reasonably accurate pictures of the geometry, circulation, and kinematic and thermodynamic structures of hurricanes, as well as a variety of tropical weather systems. These showed without a doubt that hurricanes are warm core systems that extend through the entire depth of the troposphere and perhaps into the lower stratosphere. The odd idea that hurricanes extend upward only a few kilometers from the surface was forever laid to rest.

Herbert Riehl, who had recently arrived at the University of Chicago from New York, having been sent there from Germany by his family in the 1930s, presented the first reasonably accurate description of the energy cycle of tropical cyclones in a 1950 paper. To the best of my knowledge, this is the first paper in which it is explicitly recognized that the energy source of hurricanes arises from the *in-situ* evaporation of ocean water[1]. By the next year, another German scientist, Ernst Kleinschmidt, could take it for granted that "the heat removed from the sea by the storm is the basic energy source of the typhoon". Kleinschmidt also showed that thermal wind balance in a hurricane-like vortex, coupled with assumed moist adiabatic lapse rates on angular momentum surfaces, implies a particular shape of such surfaces. He assumed that a specified fraction, q, of the rotational velocity is reduced by friction in the lower inflow layer. This is less than the rotational velocity that would be found if angular momentum were conserved in the inflow. He derived an expression for the maximum wind speed that depends strongly on the value of the arbitrary fraction q.

In his widely circulated textbook, now regarded as a classic, Riehl [1954] described hurricanes as heat engines and showed that, for air ascending in the eyewall to be appreciably warmer than that of the distant environment—a condition for conversion of potential to kinetic energy—the inflowing air had to acquire enthalpy from the underlying surface.

The work of Riehl and his colleagues, most notably Joanne Malkus, culminated in the publication of two important papers on hurricane energetics and thermodynamics in the early 1960s. The first of these once again emphasized that the horizontal temperature gradients that sustain tropical cyclones arise from heat transfer from the ocean. Making use of the fact that the horizontal pressure gradient is very weak at the top of the storm near the tropopause, that temperature lapse rates are very nearly moist adiabatic in the eyewall, and that the temperature of lifted parcels is a function of their boundary layer equivalent potential

[1] In 1944, Horace Byers recognized that the observation of nearly constant temperature following air flowing down the pressure gradient near the surface implies a sensible heat source from the ocean. The existence of isothermal inflow has been called into question by more recent observations.

temperature, Malkus and Riehl [1960] calculated a relationship between the surface pressure fall and entropy increase from the environment to the inner edge of the eyewall. In 1963, Riehl showed that this relationship is well verified in observations of actual storms and extended the Malkus and Riehl work in several ways. He derived a relationship similar to Kleinschmidt's but depending on different arbitrary parameters. Riehl came very close to formulating a correct energetic limit on hurricane intensity.

At about this time, Banner Miller developed a theory for the minimum central pressure in hurricanes. Miller also starts by assuming a moist adiabatic eyewall but explicitly ignored any increase in entropy from the outer region into the eyewall, opting instead to assume that the air in the eyewall starts out with the sea surface temperature and with a relative humidity of 85%. Miller then estimated a vertical profile of temperature in the eye itself by assuming dry adiabatic descent modified by mixing with the eyewall air. Once the eye temperature profile is constructed, the central surface pressure is calculated hydrostatically, assuming a level of zero horizontal pressure gradient at the standard pressure level nearest the level of neutral buoyancy for undiluted pseudo-adiabatic ascent in the environment. The calculated central pressures were in good agreement with the minimum pressures recorded in a limited sample of intense hurricanes.

It is important to note here that Miller's work departs in a significant way from the line of reasoning adopted by Riehl and Malkus. The latter had emphasized the crucial importance of enthalpy transfer from the ocean, while Miller regarded the hurricane as resulting from the release of conditional instability of the ambient atmosphere, requiring no enhanced air-sea enthalpy flux. He quotes Horace Byers' statement that compared the hurricane to "one huge parcel of ascending air" and states in his opening sentence that "the principal source of energy of the tropical storm is the release of the latent heat of condensation", a statement rather precisely analogous to a claim that elevators are driven upward by the downward force on the counterweights; both statements are simultaneously true and completely miss the point. In hindsight, Miller's estimate of the maximum intensity of hurricanes is energetically inconsistent. As the eyewall entropy is no larger than that of its environment, there can be no conversion of potential to kinetic energy in the secondary circulation of the storm. At the same time, the eye itself contains descending air with high temperature, a process that converts kinetic to potential energy. Miller's energy cycle in the net absorbs rather than produces kinetic energy and thus cannot maintain a system against dissipation. But in an important sense, Miller's analysis presaged the CISK thinking that became the dominant paradigm for tropical cyclone physics after 1964.

The decade of the 1950s witnessed rapid progress in understanding and predicting hurricane motion. The general idea that tropical cyclones move with the vertically averaged background flow was already evident in the work of Father Viñes and was well known at the time of World War II. By the end of the 1940s, however, Carl-Gustav Rossby had established that cyclonic vortices such as tropical cyclones alter the wind distribution in their environment so as to induce a poleward and westward drift of the vortices.

Perhaps the most important technological development of the 1950s was the advent of numerical modeling, beginning with the work of Jule Charney, Ragner Fjørtoft, and John von Neumann at the Advanced Study Institute in Princeton. They had run a simple barotropic model of the atmosphere to produce a series of 24 h forecasts of the 500-mb height field. Not long afterwards, in 1957, Akira Kasahara at the University of Chicago carried out the first numerical prediction of hurricane movement using a barotropic model. Kasahara forecast the evolution of the background flow with the hurricane vortex removed, supplemented with a routine that predicted the vortex motion relative to the background flow. Forecasts of two storms were made out to 48 h. This work inaugurated a long and productive period of research and development on the numerical prediction of hurricane motion (Chapter 4).

The network of surface and upper air stations that rapidly expanded during the war also led to the discovery of such tropical phenomena as easterly waves and to a new phase of research on tropical cyclogenesis. In 1948 two particularly important post-war papers were published, one by Herbert Riehl, the other by Erik Palmén. The latter demonstrated that hurricanes only develop over ocean water where the air temperature exceeds 27°C but not within five degrees latitude of the Equator. He also showed that these regions are conditionally unstable, whereas other parts of the Tropics (or during other seasons) are generally stable. Riehl emphasized that tropical cyclones always form from pre-existing disturbances and, in this and subsequent work, argued that the interaction of lower tropospheric disturbances with upper tropospheric troughs of tropical or extratropical origin is a key ingredient in tropical cyclogenesis. In 1955 Richard Pfeffer initiated another approach to examining the interaction between nascent or developed hurricanes with their environment. He argued that the initial spin-up of the circulation could be diagnosed in terms of eddy angular momentum fluxes into the storm.

These early investigators rejected the notion that tropical cyclones could begin spontaneously. In what is an enormously prescient work, Tor Bergeron [1954] tackled the problem of why tropical cyclogenesis requires a finite amplitude seed disturbance. Citing recent findings from the Thunderstorm Project, Bergeron pointed out that groups of convective storms are always accompanied by *cool, anticyclonic outflow* near the surface, which stabilizes the atmosphere and mitigates against the formation of a cyclone. He postulated that, were this to occur over a sufficiently warm ocean, the evaporatively cooled outflow would be reheated by contact with the sea. This would lead in some cases to what he termed the "inverting" of the cyclone, during which enough heat is added from the ocean to initiate convection and inflow of high enthalpy air. Following the earlier work of Riehl and Kleinschmidt, he again emphasized the need to increase the enthalpy of the inflowing air. Bergeron anticipated by more than three decades the finding that cold, low entropy convective downdrafts prevent most tropical disturbances from becoming tropical storms.

CONDITIONAL INSTABILITY OF THE SECOND KIND (CISK)

A review of the literature on hurricanes published during the first two thirds of the twentieth century reveals a striking separation between tropical meteorologists and the rest of the meteorological community. Bernhard Haurwitz, Horace Byers, Tor Bergeron and Carl-Gustav Rossby all made substantial but brief contributions to the science of hurricanes. Among the authors of major works on hurricanes published through 1963, only Herbert Riehl and Ernst Kleinschmidt published important work outside of tropical meteorology. Arguably, this separation of disciplines traces back to the failure of the thermal theory of cyclones, developed in the middle of the 19th Century, to explain the dynamics of extratropical cyclones. Already in 1877, Elias Loomis, a protégé of Espy and one of the original proponents of the thermal theory, expressed his doubts regarding the validity of that theory, stating that "It seems safe to conclude that rainfall is not essential to the formation of areas of low barometer, and is not the principal cause of their formation or their progressive motion." By the early 20th Century, the thermal theory of extratropical cyclones was dead; as expressed by the American scientist, Frank Bigelow, "The fact is that storms are produced by *horizontal convection* more than by *vertical convection.*" The great success of the Norwegian School, led by Vilhelm Bjerknes, and the subsequent linear theories of extratropical cyclogenesis by Charney and Erik Eady in the late 1940s showed that most of the observed properties of middle latitude storms could be explained without reference to condensation of water vapor. "As long as one is concerned with waves of small enough amplitude, the vertical motions will not be of sufficient magnitude to cause condensation, so that this factor may also be ignored," stated Jule Charney in his famous 1947 paper on baroclinic instability. Thus, Charney continued the tradition begun by Loomis of downplaying the effects of phase change of water on extratropical dynamics, a tradition that may be said to continue to this day.[2] The rapid progress and consequent popularity of extratropical meteorology in the decades after WWII led to a decline in the treatment of moist processes, and especially of moist thermodynamics, in university curricula. By the 1970s it was quite possible to obtain a doctorate degree in meteorology without having had any serious exposure to thermodynamics involving phase change of water.

---

[2] Erik Eady, whose 1949 paper is cited together with Charney's as the foundation of the theory of baroclinic instability, worried about the effects of condensation, noting that baroclinic growth rates calculated by ignoring condensation are too small. In a widely ignored portion of his otherwise well-known paper, Eady included the effects of a pre-existing, zonally oriented band of saturated air in his linear analysis of baroclinic growth and noted the strong tendency toward frontogenesis at the boundaries of the cloudy zone as well as an increase in the rate of growth.

But it was Charney and his Norwegian colleague, Arnt Eliassen, who in 1964 brought extratropical dynamicists back into tropical meteorology. During the previous year, Vic Ooyama of New York University spent some time at MIT with Charney, discussing his ideas about cumulus convection and hurricanes. They published separate papers in 1964, evincing more or less the same idea, which became known as CISK (Conditional Instability of the Second Kind), a term introduced in the Charney and Eliassen papers.

As reviewed by Michio Yanai [1964], CISK almost certainly had its roots in the failure of early attempts to simulate tropical cyclone intensification with numerical models. In each of these early attempts, a conditionally unstable atmosphere was used in the initial condition; invariably, the model developed grid-scale motions that represented its attempt to release the instability. In no case did the simulations proceed long enough for a vortex to develop.

With hindsight, we can say that these first attempts to numerically simulate tropical cyclone intensification were grounded in the view that both convection and tropical cyclones result from the release of potential energy stored in the tropical atmosphere, as evidenced by conditionally unstable soundings. But it was known since at least the 1938 paper by Jacob Bjerknes that convectively unstable updrafts seek very small scales, whereas a tropical cyclone's inner core is appreciably wider than a typical convective cloud.

Charney, Eliassen, and Ooyama attempted to resolve this dilemma by arguing that tropical cyclones represent an organized mode to release conditional instability, with the convergence of the cyclone's Ekman boundary layer serving as the organizing agent. They maintained that the rate-limiting factor for the growth of the cyclone-scale circulation was the advective supply of latent heat. From Charney and Eliassen [1964]:

> *We should look upon the pre-hurricane depression and the cumulus cell not as competing for the same energy, for in this competition the cumulus cell must win; rather, we should consider the two as supporting one another—the cumulus cell by supplying the heat energy for driving the depression, and the depression by producing the low-level convergence of moisture into the cumulus cell.*

A key concept that emerges in this work is the idea that organized latent heat release drives a circulation, which in turn feeds moisture into the system; the key feedback is between latent heating and moisture supply. Charney and Eliassen demonstrated this concept using a two-layer model in which the magnitude of the heating was assumed to be proportional to the Ekman-induced vertical velocity at the top of the boundary layer. The heating itself was applied at two model layers. Thus, while the vertically integrated heating was proportional to a measure of the circulation, its vertical distribution is determined by other means or simply specified.

Later work by other scientists showed that the vertical distribution of heating is crucial in CISK models. Broadly, if the vertical heating function is sufficiently weighted toward high altitudes, the most rapidly growing disturbances have comparatively large scales. In Charney and Eliassen's two-layer model, the heating was assumed to be the same at both levels, which produced the linear growth rate peaks at the smallest scale, albeit with a broad plateau out to scales of a few hundred kilometers.

Almost absent from Charney and Eliassen's paper was any mention of the supply of enthalpy from the sea surface, which had been regarded as so crucial by the Riehl-Malkus-Kleinschmidt school; passing mention is made of the role of the ocean in keeping the boundary layer at ambient humidity. Ironically, the major role of the surface in CISK is as a momentum sink, supplying the drag necessary to produce Ekman convergence. The linear growth rates in CISK are actually greater over land than over ocean, thanks to the larger drag coefficient.

The CISK formulation draws attention away from the supply of energy to tropical storms and replaces it with a concern for the supply of moisture. In effect, it takes for granted the existence of a large reservoir of available potential energy in the tropical atmosphere and puts the release of that energy into a mathematical straight jacket, forcing it to occur on large scales. The difference between the Riehl-Malkus-Kleinschmidt view and CISK is illustrated in Plate 1. In CISK, buoyancy is generated by large-scale convective overturning, releasing the available potential energy of the ambient atmosphere; in the Riehl-Malkus-Kleinschmidt view, the warmth of the inner region is directly related to the elevation of boundary layer entropy by surface fluxes. The critical unstable feedback in CISK is between the large-scale circulation and the convection; in the older view it is between the circulation and the surface fluxes.

It would be difficult to overstate the enormous influence of the CISK paper. Now, for the first time, there was a simple mathematical formulation for representing the collective effects of cumulus clouds on larger scale circulations—so simple that a first year graduate student could easily formulate linear models of tropical disturbances. Before long, extratropical meteorologists[3], who would not formerly have been caught dead with moisture on their hands, were enthusiastically describing all manner of circulation systems using the CISK framework. But as the number of papers on CISK multiplied, so too did its conceptual interpretations. Some stuck to the idea that organized convection is "caused" by moisture convergence, others saw it as a way of accounting for the need for vertical motion at the top of the boundary layer to overcome convective inhibition. By the early 1980s, the different flavors and interpretations of CISK had become so profuse as to lead Vic Ooyama, one of the theory's developers, to state that the term had lost any definite meaning.

---

[3] The present author not excepted.

In many ways, the advent of CISK was a setback for tropical meteorology and for the study of hurricanes in particular. It took the background tropical atmosphere, as many did before and have done since, as a clear, conditionally unstable atmosphere in which large-scale circulations, like tropical cyclones, are particular mechanisms that release instability. It ignored the fact that a tropical atmosphere, with no large-scale circulation, comes into a state of radiative-convective equilibrium, in which clouds are continuously releasing available potential energy at the rate it is generated by radiative cooling of the atmosphere and surface heat fluxes. Such a state should be the natural basic state for the study of tropical circulation systems, just as a balanced zonal flow serves as a natural basic state for the study of baroclinic waves in middle latitudes. This state has very little available potential energy. Indeed, the evidence available to date suggests that the interaction between large-scale circulations and convection *per se* is a stable one; growth of large-scale circulations appears to require variable surface fluxes, ocean surface temperature gradients, or variations in longwave radiation owing to variations in clouds and water vapor. This is entirely consistent with Bergeron's 1954 findings that convective showers, whether individually or in clusters, produce low entropy, anticyclonic outflow near the surface.

Although papers using the CISK framework continue to be published, the weight of evidence suggests that it is a false hypothesis. The numerical modeling work beginning in the late 1960s should have laid CISK to rest, but for reasons explored later, it did not do so entirely.

SYNTHESIS OF NEW OBSERVATIONS

The second half of the 20th Century saw a rapid increase in the number and quality of observations of hurricanes. Although aircraft reconnaissance was introduced during World War II, the early measurements of meteorological quantities were fairly crude. For example, without inertial navigation or Doppler radar to determine ground speed, surface wind speeds were estimated visually through the late 1950s. But by the mid-1960s, both kinematic and thermodynamic quantities were being measured with reasonable accuracy by a combination of reconnaissance aircraft and dropsondes. One great advantage of aircraft reconnaissance during this time was the use of high-altitude aircraft such as the B-57; this allowed an investigation into the properties of the outflow layers of hurricanes. Such capability disappeared in the late 1960s and was not replaced until NOAA purchased a Gulfstream IV aircraft in the late 1990s.

A great step forward came in 1960, when the first image of a hurricane was transmitted from a polar orbiting satellite. By the 1970s, virtually all tropical cyclones (with the possible exception of some of the elusive, so-called "midget" storms) that formed anywhere over the Earth were recorded by satellite observations. This provided an almost immediate improvement in warnings since, before satellites, quite a few storms went undetected through at least part of their life,

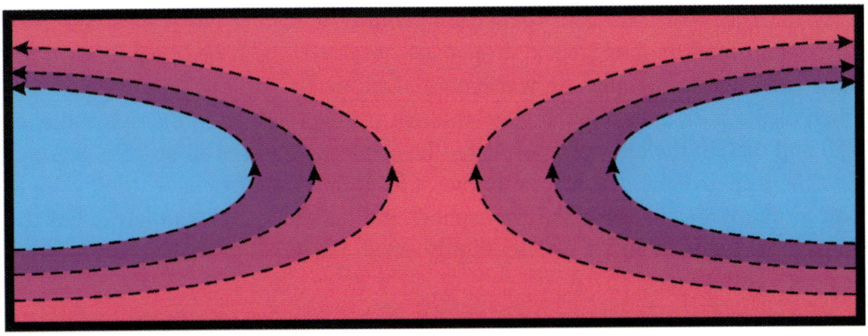

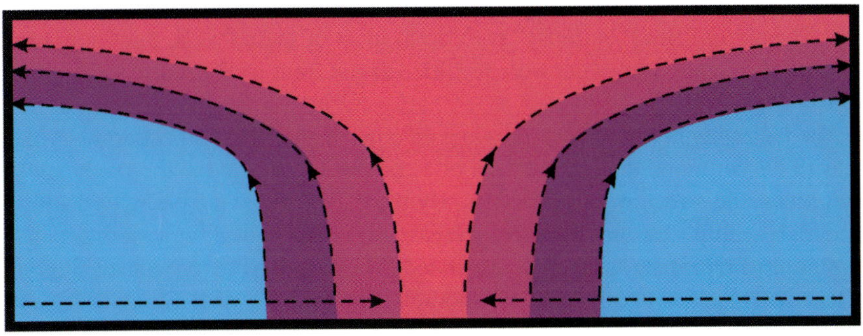

Plate 1. Distributions of saturation moist entropy above the boundary layer and moist entropy in the boundary layer of a hurricane. In the CISK view (a) the warmth of the core is owing to the upward advection of high entropy air from the boundary layer, while in the Riehl-Malkus-Kleinschmidt view (b) the warmth of the core results from enhanced enthalpy transfer from the ocean, giving an increase in entropy in the inflowing boundary layer air. In (a) the ambient atmosphere must be conditionally unstable; (b) can work within a conditionally neutral environment.

owing to the paucity of non-satellite observations over much of the tropical oceans.

It is not my purpose here to provide a review of the history of observational development, for that is done within Chapters 2, 3, and 11. Rather, I discuss the synthesis of hurricane structure and behavior that the new observations made possible.

By the late 1960s, the axisymmetric structure of mature hurricanes had been well determined by aircraft and dropsonde observations. It had been known for some time that hurricanes are warm core vortices; the aircraft data showed that much of the horizontal temperature gradient is concentrated in the eye and eyewall and that, in the upper troposphere, the eye temperature can be 15 K warmer than their environment at the same pressure. In strong hurricanes, the cyclonic flow in the lower and middle troposphere can be nearly axisymmetric, but the upper level anticyclonic is usually highly asymmetric, often concentrated in one or more outflow jets.

An example of the detailed picture of tropical cyclones developed by the late 1960s is given in Plate 2, which shows the distribution with radius and pressure of the (pseudo-adiabatic) equivalent potential temperature, constructed from aircraft measurements at five altitudes: 500 m, 750 mb, 650 mb, 500 mb and 180 mb. The aircraft flew "butterfly" patterns at most of these levels; the figure represents an average over what were regarded as the front (NW) and rear (SE) sectors of the storm.

The large number of aircraft missions into hurricanes that had occurred by the late 1970s permitted the construction of a composite view of the thermodynamic and kinematic structure of hurricanes. By the early 1980s, increasingly detailed and high quality measurements from research aircraft allowed for a more detailed synthesis of hurricane structure and also led to descriptions of the statistical properties of convective updrafts and downdrafts in hurricanes. With the advent of airborne Doppler radar, the science of hurricanes took a big step forward, yielding unprecedented views of the detailed kinematic structures of the storm and accompanying mesoscale features. The invention of lightning detection networks allowed for a detailed mapping of lightning in hurricanes near land, showing interesting relationships to the structure and evolution of the storms.

The increasing abundance of aircraft data, together with better and better analyses of the large-scale environments of tropical cyclones, led to greatly enhanced understanding of the control of hurricane structure, intensity, and movement by the large-scale environment, thanks largely to the work of William Gray and his colleagues at Colorado State University. The movement of hurricanes was related observationally to environmental flows, and the environments of developing systems were compared to those that did not develop. Progress was also made toward understanding environmental influences on tropical cyclone size, structure, and intensity. This work showed, among other things, the important influence of environmental wind shear on tropical cyclone intensity change.

Observations of the effects of hurricanes on the upper ocean also came to light in the 1960s. Beginning with the 1967 work of Dale Leipper, it became obvious that hurricanes have a profound effect on the uppermost 200-300 m of the ocean, deepening the mixed layer by many tens of meters, cooling the surface temperature by as much as 5°C, and causing near-inertial surface currents of 1-2 m/s, detectable at depths up to at least 500 m. By the early 1980s, James Price had established that most of the cooling is caused by entrainment, which is created by turbulence generated from the strong shear of the near-inertial currents across the base of the mixed layer. This nicely explains the observation that much of the cooling is offset to the right of the storm's track (in the northern hemisphere), where the turning of the wind stress as the storm passes resonates with inertial oscillations.

THE ERA OF NUMERICAL MODELING

In the late 1950s, two tracks of numerical modeling efforts were begun. The first, a 1957 paper by Akira Kasahara, sought to understand and predict the motion of existing storms using barotropic models. The second by way of a 1961 paper by Kasahara, sought to understand the development and intensification of tropical cyclones using axisymmetric models with thermodynamics.

The track model development begun by Kasahara continued through the 1960s, culminating in the first operational forecast model by Frederick Sanders and Robert Burpee in 1968. While these models were increasingly successful, they met stiff competition from statistical forecasting methods, which were also improving rapidly (see Chapter 4 in this volume). But by 1995, the U.S. National Weather Service had adopted the Geophysical Fluid Dynamics Laboratory hurricane prediction system developed by Kurihara and his colleagues. This is a triply nested, high-resolution primitive equation model, which consistently out-performed statistical methods of track forecasting. As the forecast models rapidly improved, lack of observations and inadequate knowledge of how to assimilate them started to become a significant factor in limiting the quality of the forecasts, and the models themselves became tools for exploring what kind of observations could best improve forecasts. For example, in 1996 Robert Burpee and colleagues showed that data collected from dropwindsondes deployed from reconnaissance aircraft significantly improved track forecasts.

Modeling the intensity and intensification of hurricanes progressed somewhat more slowly. After many unsuccessful attempts to model the intensification of hurricanes, a breakthrough came with Ooyama's 1969 publication. Ooyama used an axisymmetric model whose vertical structure consists of a boundary layer of fixed thickness surmounted by two layers of variable thickness. Surface heat fluxes and drag are calculated using bulk aerodynamic formulae, assuming a fixed ocean temperature. Cumulus convection is represented by a scheme that in some respects resembles that of H. L. Kuo, in that the rate of convective heating depends on boundary layer moisture convergence, but it also depends on the degree of condi-

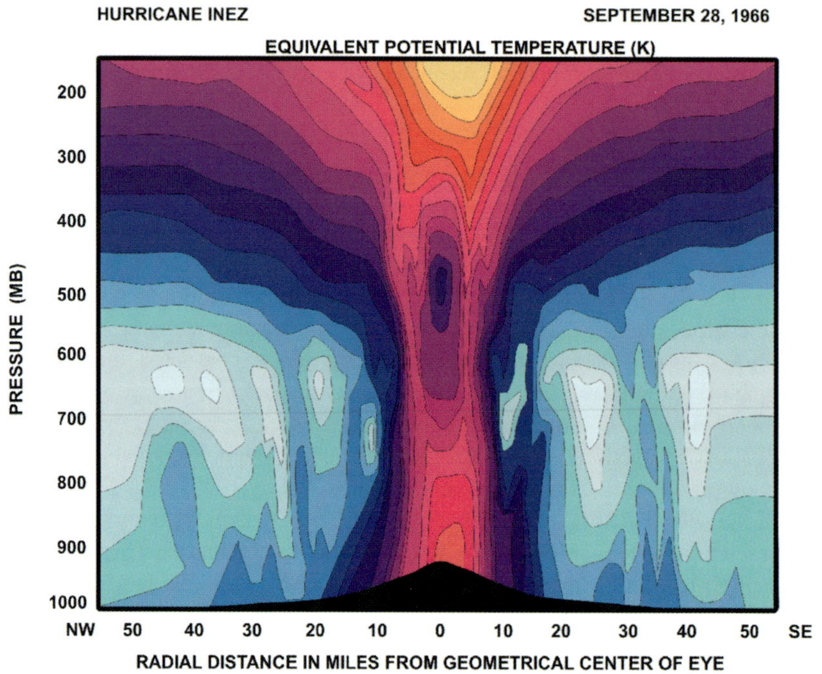

Plate 2. Equivalent potential temperature (K) as a function of pressure and radius from the center of Hurricane Inez on 28 September, 1966, based on aircraft data at 500 m, 750 mb, 650 mb, 500 mb and 180 mb. Contours are at intervals of 2 K with a minimum value of 336 K (light blue) and a maximum value of 376 K (yellow). After Hawkins and Imbembo (1976).

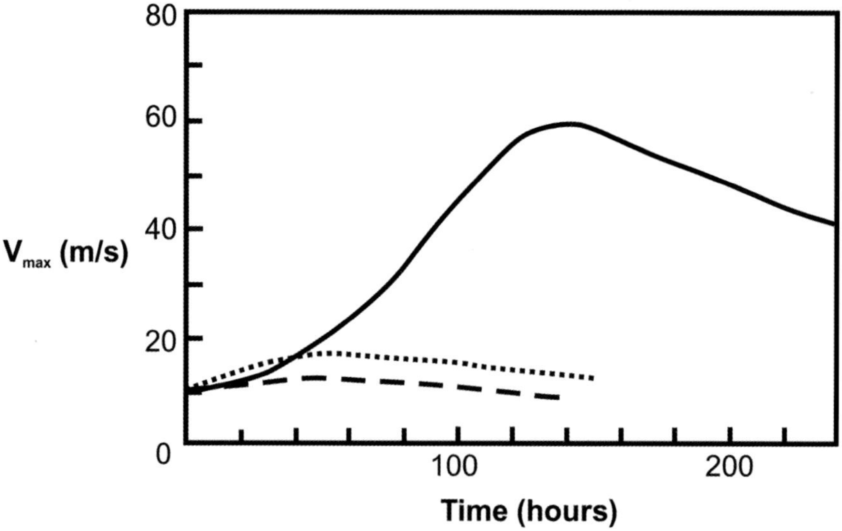

Figure 1. Evolution of the maximum wind speed in the lower layer of Ooyama's model, for three experiments: The control run (solid) and two runs with no surface enthalpy flux. The second of these (dotted) has enhanced conditional instability of the initial state. After Ooyama [1969].

tional instability and the relative humidity above the boundary layer. Ooyama did not include a representation of radiative cooling of the atmosphere.

Figure 1 shows the evolution of the maximum azimuthal (rotational) wind speed in Ooyama's control simulation, together with another run in which surface heat flux was eliminated. Both the structure and temporal evolution of the storm, simulated by the control experiment, strongly resemble those found in nature, though the rapid decline following peak intensity is not always observed. In this model, the decline is likely due to the lack of radiative cooling. No steady state can be achieved when a heat source is active in the absence of a heat sink.

Ooyama [1969] presented a series of experiments that showed (consistent with the Riehl-Malkus-Kleinschmidt view) that the intensity of the model storm varies directly with the coefficient of heat transfer and inversely with the drag coefficient. Subsequent work confirmed this dependence of maximum wind speed on the exchange coefficients, and in 1996 George Craig and Suzanne Gray demonstrated that the rate of intensification exhibits a similar dependence on these coefficients, with more rapid intensification for *smaller* drag coefficient, in contradiction with the predictions of CISK.

And yet Ooyama [1969] left an opening for adherents of CISK theory. The initial condition in the numerical experiments was strongly conditionally unstable—an ambient boundary layer parcel would be about 10 K warmer than its environment when lifted to the top layer. This is why there is some slight intensification of the initial vortex, even when the surface enthalpy exchange is omitted (see Figure 1). Ooyama himself proved that the initial state was linearly unstable to tropical cyclone-like perturbations, and there is every indication that the very early development of the vortices in his simulations was indeed a result of CISK. Axisymmetric models with no surface heat fluxes whatsoever and enough conditional instability do indeed produce intense vortices, as demonstrated by Yamasaki in 1977. However, it is not clear whether Yamasaki's result was at least partially a consequence of forcing convection into an axisymmetric straight jacket[4]. (The fact that we do not observe tropical cyclone-like disturbances in demonstrably unstable air over land is telling.) As recently as 1997, Ooyama's 1969 simulations were held up as supporting the idea of CISK.

The use of a linearly unstable initial state must with hindsight be regarded as a weakness of the Ooyama work. It has been known at least as far back as Riehl's 1948 work that tropical cyclones always develop from pre-existing disturbances, such as easterly waves or frontal disturbances, suggesting that the tropical atmosphere is indeed linearly stable to tropical cyclones. In 1997 Dengler and Reeder showed that had Ooyama taken his initial state to be conditionally neutral, he would have found that development would only occur if a vortex of sufficient amplitude is used in the initial condition, an aspect more in keeping with observations. (In 1987 Richard Rotunno and I demonstrated that finite amplitude vortices can intensify to full hurricane strength in an environment in statistical equilibrium.) Nonetheless, there can be no question that Ooyama's work was an important development in understanding and simulating hurricane development.

Following the publication of Ooyama's 1969 paper, work on numerical simulation of hurricane development and structure progressed rapidly. By the late 1970s, computer speed had improved to the point that cumulus convection could be explicitly (albeit crudely) simulated in an axisymmetric model, as shown by Stanley Rosenthal in 1978[5], and Richard Anthes published the first three-dimensional simulations in 1971. These latter showed, for the first time, the development of spiral rainbands. By the 1980s, simulations were being performed by Hugh Willoughby, Steven Lord, and colleagues using nonhydrostatic models incorporating complex microphysics including ice physics, and by the end of the

---

[4] Yamasaki's simulated vortex was of very small dimensions, with a radius of maximum winds of only about 5 km.

[5] As a curious historical footnote, Rosenthal discovered that he did not need to parameterize convection in his tropical cyclone simulations because in one such simulation he inadvertently omitted the cumulus scheme and found that the model developed a reasonable storm anyway.

century, three-dimensional, nonhydrostatic models were simulating a very realistic structure of actual storms. An example of such a simulation is shown in Plate 3, using the nonhydrostatic MM5[6] model run at 1.67 km horizontal grid spacing.

Although there was rapid progress in modeling hurricanes from 1969 onward, it took until 1995 for the skill of track forecasts from a numerical model to overtake that of statistical methods (see Chapter 4 for a complete review of these developments). Much of this delay is attributable to the fact that the quality of global analyses was insufficient for track forecasting until the late 1980s. This was, in turn, due both to the paucity of observations over the oceans and to the relatively poor quality of data assimilation techniques. The slow pace of advance in these areas can be attributed in part to the structure of research and operational meteorology in the U.S., which leaves neither a clear path nor a system of incentives to bring research products into operational use. The U.S. National Academy of Sciences, in their year 2000 report, recently condemned this circumstance by appropriately subtitling it, "Crossing the Valley of Death." A notable exception is the development and operational implementation of the Geophysical Fluid Dynamics Laboratory (GFDL) model, the first numerical model to beat statistical models in track forecast skill. But, at the time of this writing, the skill of hurricane intensity forecasts made by numerical models has yet to overtake that of statistical schemes except at longer forecast times (72 h), as discussed in Chapter 4. Thus, it is clear that there is a long road ahead.

BACK TO THE OCEAN

By the late 1980s, the failure of CISK to explain or predict the major features of tropical cyclones, including their scale and dependence of intensity and intensification on the environment, led to a renewed interest in the Riehl-Malkus-Kleinschmidt view of local surface enthalpy flux as the primary energy source for tropical cyclones. Observations of the entropy distribution (e.g., Plate 2) made it clear that there is a strong surface entropy source under the eyewall; indeed, Plate 2 more strongly resembles Plate 1b than it does 1a. In 1986, Douglas Lilly and I, independently, developed analytic models of the nonlinear, axisymmetric steady state tropical cyclone. Lilly's approach was based on the conservation of energy and angular momentum along streamlines, whereas mine assumed, as had Kleinschmidt's, that the vortex is neutral to slantwise moist convection. We both derived essentially the same expression for the maximum wind speed, which depends on the surface exchange coefficients, a thermodynamic efficiency, and a measure of the thermodynamic imbalance between the tropical ocean and atmosphere. In many respects, these expressions are similar to one

[6] Mesoscale Model version 5, developed by Pennsylvania State University and the National Center for Atmospheric Research.

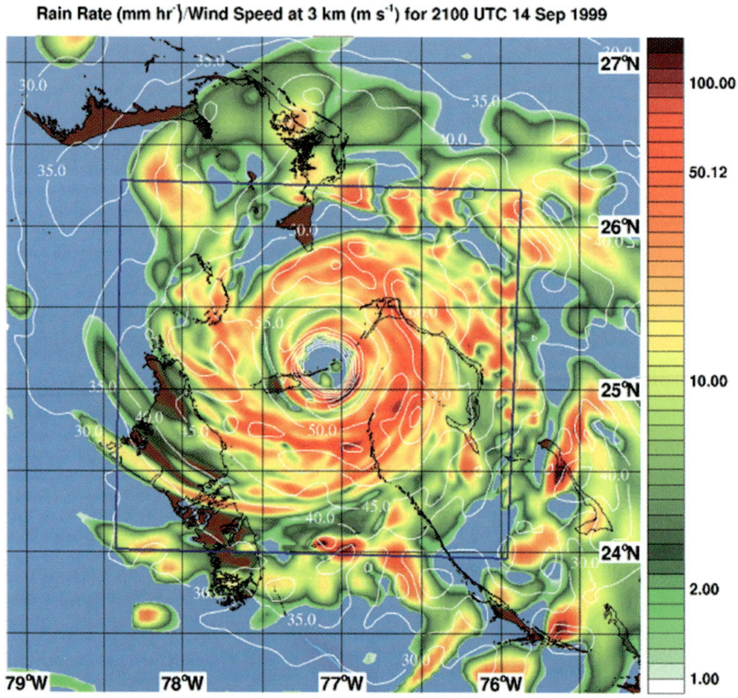

Plate 3. Rain rate (colors) and wind speeds above 30 ms^{-1} (white contours) from a simulation of Hurricane Floyd, 1999, using the nonhydrostatic MM5 model run at 1.67 km horizontal grid spacing. Note the spiral rainbands and the outer and inner eyewalls. Figure courtesy of Dr. Shuyi Chen.

derived in 1963 by Herbert Riehl; but unlike Riehl's, they did not depend on details of the outer structure of the storm and included an explicit dependence on the depth of the circulation. They suggest a strong sensitivity of hurricane intensity to those boundary layer processes that determine the exchange of heat and momentum with the ocean and ocean temperature near the eyewall.

Unfortunately, little is known about how the surface exchange coefficients behave at high wind speeds in nature. As is apparent in Plate 2, most of the entropy increase in the inflow occurs very near the eyewall; it is here that hurricanes are sensitive to the exchange coefficients. (Note that, for this reason, the centers of hurricanes can approach very near to land before their intensity begins to diminish). Measurements at low to moderate wind speeds suggest that the drag coefficient increases with wind speed, due to increased surface roughness, but the heat exchange coefficient remains approximately constant. When extrapolated to hurricane wind speeds, this would yield too much drag to explain the observed intensity of hurricanes. This suggests that other physical processes must come into play to enhance the enthalpy exchange and/or diminish drag. In 1999 Edward Andreas and I suggested that the relevant mechanism is re-entrant sea spray, which transfers significant amounts of enthalpy to the air. As of this writing, there is much on-going research on air-sea exchange at extreme wind speeds.

The expressions derived by Lilly and I suggest that the observed ocean cooling of order 1°C under the storm core could have a significant feedback on hurricane intensity. But the first simulation of a hurricane using a coupled ocean-atmosphere model, by Simon Chang and Richard Anthes in 1971, showed ocean feedback having little effect on storm intensity. This led to a period of roughly two decades during which ocean feedback was regarded as unimportant, except perhaps for storms crossing the wakes of previous storms. (In hindsight, the model used by Chang and Anthes had too coarse a resolution and was integrated for too short a period to see appreciable effects from ocean feedback) Interest in ocean feedback was renewed after publication of a series of papers by Isaac Ginis, Alexander Khain, and George Sutyrin, beginning in 1984. These researchers used advanced coupled models to demonstrate that ocean feedback has a first-order effect on hurricane intensity. By the end of the century, it had been demonstrated that the intensity of many hurricanes could be accurately predicted using even a very simple atmospheric model coupled to an essentially one-dimensional ocean model, as long as storms remain unmolested by adverse atmospheric influences such as environmental wind shear, which has been shown by Mark DeMaria and John Kaplan to be a statistically significant predictor of intensity change. An example comparing coupled and uncoupled hindcasts of a hurricane is shown in Figure 2.

As of this writing, the skill of intensity forecasts using coupled models is closing in on that of statistical models, which are still superior at least for short forecast periods. Experience suggests that numerical guidance will eventually dominate, when the speed of computers makes it possible to simulate hurricanes using

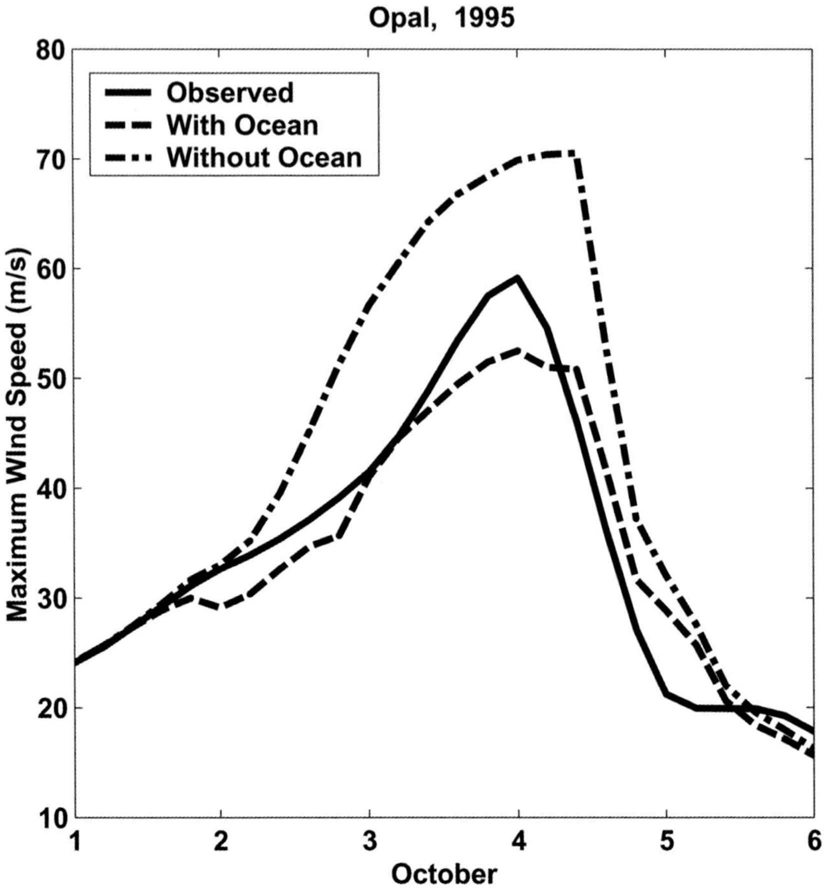

Figure 2. Evolution of the maximum wind speed in Hurricane Opal of 1995, compared to simulations with and without coupling with the ocean. From Emanuel [1999].

three-dimensional coupled models of the very high resolution necessary for simulating the fine-scale structure of the eyewall and rainbands.

HURRICANE MOTION

During the last three decades of the 20th Century, the physics of hurricane motion were refined and extended from the 1948 work of Rossby. Work on the barotropic theory of hurricane motion was advanced through the 1980s by Greg Holland, Johnny Chan, Mark DeMaria, Russell Elsberry, and Michael Fiorino, among others. In the early 1990s, Roger Smith and his colleagues developed analytic expressions for the motion of barotropic vortices. Hurricanes are baroclinic vortices, however, and often occur within large-scale environments that have vertical and lateral wind shear. This shear can have multiple effects on storm motion. Lloyd Shapiro emphasized the connection between shear and the background potential vorticity distribution, which can have a profound effect on the distribution and intensity of the so-called "beta gyres" that arise in pure form in barotropic simulations with uniform background vorticity gradients. On the other hand, Chun-Chieh Wu and I emphasized the effect of plumes of low potential vorticity swept downshear from the vortex at upper levels and later tried to find evidence of these and "beta gyres" in real data. The combined effect of a background potential vorticity gradient and the interaction between the upper level anticyclone and the lower level cyclone was also explored by Maria Flatau. Shear can also tilt the cyclonic potential vorticity anomaly associated with hurricanes, giving rise to complex motion, as shown by Sarah Jones and Roger Smith. All these effects are arguably contained in advanced numerical forecast models, such as the GFDL model, which in their performance have outrun the advance of physical understanding.

THE GENESIS PROBLEM

Perhaps no facet of hurricanes has proven more vexing than that of understanding and predicting their genesis. There is not even widespread consensus on the definition of the term, though it is usually applied to systems undergoing a transition from some non hurricane-like disturbance, such as an easterly wave or frontal trough, to a more symmetric, warm-core cyclone with a low pressure center at the surface. From the earliest post-war studies, it was recognized that all tropical cyclones originate in some independent disturbance. As described earlier, Tor Bergeron [1954] pointed out that groups of convective showers usually produce cool, anticyclonic outflow at the surface, thus preventing any spontaneous cyclogenesis near the surface. This is consistent with the analyses by William Gray in 1979 and John McBride in 1981, indicating that the secondary circulation in tropical cloud clusters exports moist static energy to the environment. When numerical models are initialized using environments in statistical equilibrium (as opposed to those with large convective available potential energy stored by virtue of a capping inversion), a finite amplitude disturbance is required to initiate intensification by wind-surface flux feedback, consistent with the observation that genesis must be triggered by an external disturbance.

While Erik Palmén [1948] had established that tropical cyclones form only over sufficiently warm ocean water, little further was done on delineating the character of environments conducive to genesis until the 1968 work of William Gray. Gray showed that genesis only occurs in environments characterized by small vertical shear of the horizontal wind and favors regions of large low-level vorticity. By the end of the 1970s, Gray had established a set of conditions that are apparently necessary (though by no means sufficient) for genesis. In addition to the two aforementioned factors, he argued that larger values of the Coriolis parameter—the heat content of the upper ocean (reflecting the depth of the ocean mixed layer) and the relative humidity of the middle troposphere—all favor genesis.

The early investigators, such as Riehl and Bergeron, debated about whether tropical cyclones arise on pre-existing troughs, such as fronts, or from disturbances in the upper troposphere. Later observations showed that there are many routes to genesis including nearly classic baroclinic development, interaction of easterly waves or other low-level disturbances with tropical upper tropospheric troughs, and possibly accumulation of wave energy in diffluent large-scale flow.

Another body of theory developed around the idea that patches of high vorticity associated with individual convective systems can, under certain circumstances, merge to form a more powerful incipient cyclone. This idea may have started with Fujiwhara's 1923[7] proposal that genesis involves the fusion of several small vortices, which has been revived recently by Elizabeth Ritchie and Greg Holland (see Chapter 12 in this volume for a more thorough discussion). Michael Montgomery and colleagues recently showed that small-scale patches of potential vorticity introduced into the flow field of a larger-scale vortex are quickly converted into the axisymmetric flow, in the process feeding their energy into the vortex. This suggests that mesoscale convective systems that develop outside the eyewall may help intensify the storm as a whole.

The central problem in genesis is the transformation of an existing disturbance into a system operating on the feedback between surface enthalpy fluxes and surface wind. Any complete theory must account for the fact that such transformations are relatively unusual and, in any event, only occur under the conditions reviewed by Gray in 1979. Bergeron [1954] had concluded that, under normal circumstances, convective downdrafts quench any nascent tendency for the boundary layer moist entropy to increase. He suggested that if, by some means, the surface cyclone could be made strong enough, the inward Ekman drift would overcome the anticyclonic outflow, leading to a positive feedback between surface enthalpy flux and wind and transformation into a warm core system. But it has become clear, from a series of numerical experiments and a field experiment that I conducted with several colleagues, that a necessary condition for genesis is the establishment of a mesoscale column of nearly saturated air in the core of the system so that cumulus convection rising into this air cannot produce low entropy downdrafts. Any environmental influence that disrupts the formation of such a saturated column will prevent genesis and weaken an existing system. Clearly, the ventilation of low entropy air in the middle troposphere through a nascent system will have this effect, as first pointed out by Robert Simpson and Herbert Riehl in 1958; this nicely explains why vertical shear is inimical to genesis. In axisymmetric models, the establishment of a mesoscale-saturated column

[7] Cited in Bergeron [1954], page 144, but no detailed reference is given.

appears also to be a sufficient condition for genesis. An important remaining question is how such a column can be established, and the various mechanisms, described in the previous paragraph, could all work in this direction. The same axisymmetric models show that a cyclonic circulation of sufficient strength at the surface can ultimately give rise to genesis, after a gestation period in which the friction and downdrafts weaken the circulation, while at the same time Ekman pumping and associated convection moistens a deep column in the core of the system. But detailed observations of Hurricane Guillermo in the eastern North Pacific in 1991 show that, in this development, the saturation was achieved by evaporation of rain falling from a stratiform anvil system associated with a mesoscale convective complex. In addition, downward advection of angular momentum in the mesoscale downdraft was an important ingredient in the subsequent genesis. It is by no means clear, however, that such a mechanism is at work in all or even most cases of genesis. This conclusion reflects the paucity of field experiments dedicated to the problem of tropical cyclogenesis.

EXTRATROPICAL TRANSITION

When tropical cyclones move into higher latitudes, they often regenerate as baroclinic systems, sometimes achieving great intensity and causing injury and damage. A celebrated case is that of Hurricane Hazel of 1954, which re-developed as it moved over the eastern U.S. and caused damage as far from the coast as Toronto. Erik Palmén studied this event in 1958 and concluded that both the high moisture content and circulation associated with the tropical storm were essential to the strong rejuvenation of the system. In 1990 Richard Anthes corroborated Palmén's findings through detailed numerical simulations of the event. Extratropical transition poses a formidable problem for forecasters, as even today's numerical models often miss or underpredict redevelopment due to the difficulty of correctly handling water vapor, cumulus convection, and the often very small scale of the circulation at low levels.

In recent years interest in extratropical transition has been rejuvenated. A number of studies by Lance Bosart, Patrick Harr, Russell Elsberry, Christopher Thorncroft, Sarah Jones, and others have cast the transition problem within a contemporary dynamical framework. This shows that at least part of the redevelopment results from the superposition of the intense, low-level potential vorticity anomaly associated with the former tropical cyclone with an upper tropospheric potential vorticity anomaly of extratropical origin. But there is large variation from case to case. In other circumstances, there may not be a strong upper level anomaly, but interaction of the tropical cyclone circulation with pre-existing gradients of potential vorticity at the tropopause (or with strong surface temperature gradients) sometimes plays an important role. Equally intriguing, but less understood, is the role of the water vapor anomaly associated with the tropical disturbance, as first alluded to in 1958 by Palmén. In the 1990s, several researchers note the importance of a deep column of nearly saturated, conditionally neutral air in the redevelopment. Others have shown, using sensitivity experiments with a numerical model, that the pre-existing, low-level potential vorticity anomaly associated with the tropical cyclone may not be a major factor in redevelopment, suggesting the importance of some other factor or factors associated with the tropical system. The role in extratropical transition of the deep, mesoscale column of nearly saturated air invariably associated with tropical cyclones needs to be better understood.

HURRICANES AND CLIMATE

> *Another problem, of much more far-reaching consequences, presents itself. What kind of secular changes may have existed in the frequency and intensity of the hurricane vortices of the Earth? And what changes may be expected in the future? We know nothing about these things, but I hope (to) have shown that even quite a small change in the different factors controlling the life history of a hurricane may produce, or may have produced, great changes in the paths of hurricanes and in their frequency and intensity. A minor alteration of the surface temperature of the sun, in the general composition of the earth's atmosphere, or in the rotation of the earth, might be able to change considerably the energy balance and the balance of forces within such a delicate mechanism as the tropical hurricane. During certain geological epochs, hurricanes may have been just as frequent as the cyclones of our latitudes, or they may have occurred all over the oceans and within all coastal regions, and they may have been even more violent than nowadays. During other periods they may have been lacking altogether. In studying paleo-climate and paleo-biological phenomena, especially along the coasts of previous geological epochs, it may be wise to consider such possibilities.*
>
> *-Tor Bergeron, 1954*

After Bergeron's striking insight a half century ago, interest in climatic influences on hurricane activity increased through the end of the century. As early as the mid 1950s, Jerome Namias showed that there were strong correlations between interannual fluctuations of Atlantic hurricane activity and changes in the general circulation, particularly in the pattern of long waves. Interest remained focused on interannual fluctuations of Atlantic tropical cyclone activity for another three decades. In 1982, Lloyd Shapiro published a pair of papers demonstrating a clear connection between interannual fluctuations in Atlantic hurricane activity and other climate signals including a phase measure of the quasi-biennial oscillation and sea surface temperatures west of Africa. Two years later, William Gray showed a strong connection between Atlantic hurricane activity and El Niño, and by the early 1990s Christopher Landsea and William Gray found a strong correlation between hurricane activity and rainfall in sub-Saharan Africa. In 1997, Saunders and Harris showed that the particularly active 1995 Atlantic hurricane season was related to exceptionally warm sea surface temperatures. However, Lloyd Shapiro and Stanley Goldenberg noted that most of this relationship might result from a connection between sea surface temperature gradients and vertical wind shear, with only a relatively small part due to the direct influence of sea surface temperature on cyclone formation. In 1996, Stanley Goldenberg also argued that the physical link between Atlantic hurricane activity and phenomena such as El Niño and sub-Saharan rainfall is through the effect of these phenomena on vertical wind shear over the genesis regions of the North Atlantic. By the 1980s, the Colorado State University group had started using these signals in an attempt to provide seasonal forecasts of Atlantic tropical cyclone activity. In the last decade of the century, a strong relationship between Atlantic hurricane activity and the North Atlantic Oscillation was discovered, giving some hope that decadal variations in hurricane activity might be partially anticipated.

While hurricanes are usually considered to respond passively to climate changes on many times scales, it may be that they are active players in phenomena like the North Atlantic Oscillation. I have argued that much of the thermohaline circulation is actually driven by global tropical cyclone activity. If this proves to be the case, then the variation of hurricane activity with climate may be integral to the physics of such phenomena as the North Atlantic Oscillation.

By the end of the 1980s, it was becoming apparent that average global surface temperature was increasing, perhaps in response to anthropogenically induced increases in greenhouse gases, and the effect of global warming on hurricane activity became a concern. In 1987, I argued that increasing greenhouse gases altered the energy balance at the surface of tropical oceans in such as way as to require a greater turbulent enthalpy flux out of the ocean, thereby requiring a greater degree of thermodynamic imbalance between the tropical oceans and atmosphere. Using a single-column radiative-convective model, I argued that the potential intensity of tropical cyclones would increase by about 3.5 m/s for each 1°C increase in tropical sea surface temperatures, and supported that with a calculation of potential intensity from a global climate model subjected to a doubling of atmospheric CO_2. Subsequent work by Thomas Knutson and colleagues, using a high-resolution tropical cyclone model given boundary conditions from a global climate model, gave a similar estimate for the increase in hurricane intensity that might arise as a consequence of global warming. But the problem of how global climate change might affect the frequency of hurricanes remains largely unsolved. Global climate models appear to give disparate results. For example, a 1992 study, using the GCM run by the British Meteorological Office, shows an increase in both the intensity and frequency of tropical cyclones, but an analysis using the GFDL model shows ambiguous results, with an increase in tropical cyclone activity if cloud-radiation feedback is not included and a decrease in activity otherwise. It is clear from all these studies that, even were the worst scenarios of global warming realized, the expected changes in tropical cyclone activity would be small compared to those arising from the kind of natural variability described in the previous paragraph, at least for the next 50 years. On longer time scales, the effects of global warming might be more serious.

Perhaps the most promising route to understanding the relationship between long-term climate change and changes in hurricane activity is through examination of the geological record, as Bergeron had suggested. The budding new field of paleotempestology is just beginning to show how hurricane activity varies on times scales of centuries to millennia. For example, Kam-Biu Liu has taken a series of sediment cores from near-shore lakes and swamps along the U.S. Gulf coast and deduced the timings of strong landfalling storms by observing sand layers in the cores and carbon-dating them from the surrounding organic matter. He has, in the last decade, developed a record of strong landfalling storms going back more than 3000 years at some sites. Work like this, if extended in space and further back into the past, may reveal how hurricane activity changes on very long time scales, offering an opportunity to understand how such activity is related to global and regional climate change on a variety of time scales.

SUMMARY AND A LOOK AHEAD

The 20th Century was a time of rapid progress in understanding and predicting hurricanes. Most of this progress came after World War II, when an influx of scientific talent, coupled with rapid advances in technology, catalyzed observational and theoretical development. Much of our current understanding of the genesis and intensification of hurricanes was developed or foreshadowed during the 1950s; the advent of CISK in the following decade was arguably a setback from this earlier work. Numerical modeling revealed the sensitivity of hurricanes to processes such as cumulus convection and surface fluxes and to factors in the environment but did not, in and of itself, lead immediately to improved physical understanding. In some cases, such as in the simulation of storms in highly unstable environments but without surface fluxes, numerical simulations simply served to codify bad theory. Notwithstanding this, models eventually proved their utility in track prediction. But the characterization of hurricane structure and behavior improved steadily through the post-war years, thanks to skillful analyses and the development of new technologies for measuring the atmosphere. Radar, reconnaissance aircraft, and satellites were all critical to the new synthesis that emerged in the second half of the century.

In spite of enormous progress, many facets of hurricane behavior remain poorly understood and predicted. We have not advanced to the point that forecasters can skillfully predict which tropical cloud clusters will become tropical cyclones and which will fade away; nor can we make reliable predictions of extratropical rejuvenation. Forecasts of hurricane intensity change remain notoriously unreliable, reflecting no doubt a poor understanding of the environmental and internal influences on intensity. Predictions of hurricane rainfall remain highly uncertain, and because rain-induced flooding is a major cause of death and injury (as in the Hurricane Mitch tragedy), a better grasp of this problem is essential. We are just beginning to understand the great sensitivity of hurricane activity to interannual and interdecadal climate fluctuations. Understanding the physical connection is central to the scientific problem of hurricanes as well as to forecasts of statistical storm activity on seasonal to decadal time scales. And while the scientific community debates when, if, and how much global warming might result from anthropogenic increases in greenhouse gases, it is important for hurricane researchers to understand how climate change in general affects hurricanes. Paleotempestology offers the hope of reconstructing changes in storm activity over thousands and perhaps millions of years; this would be a big step forward toward understanding the relationship between hurricanes and climate.

We have made great strides forward in scientific research; however, as a community, we have been less successful recently in conveying to our political representatives both the steps that are necessary for continued progress and the benefits to mankind they would bring. As a result, we bring somewhat less computational firepower to bear on numerical weather prediction than do our European colleagues, who suffer far less in the way of severe weather, while many of our most important observation networks fall into decline. Part of our great success in reducing injury and loss of life from hurricanes is due to the excellent cooperation that has developed among forecasters and emergency managers at the federal, state, and local levels. But we have largely failed to export this expertise to the developing world, which continues to suffer devastating losses from tropical cyclones. And closer to home, we have not been very successful in helping to bring about changes in the construction and insurance industries or in the formulation of zoning restrictions, which would mitigate the greatly increasing economic costs of hurricanes. As scientists, we are not positioned well to bring about these changes. But the responsibility for communicating the nature of the problem to the public lies with us.

REFERENCES

Bergeron, T., The Problem of Tropical Hurricanes. *Quart. J. Roy. Meteor. Soc.*, 80, 131-164, 1954.

Charney, J. G. and A. Eliassen, On the Growth of the Hurricane Depression. *J. Atmos. Sci.*, 21, 68-75, 1964.

Emanuel, K. A., Thermodynamic Control of Hurricane Intensity. *Nature*, 401, 665-669, 1999.

Hawkins, H. F. and S. M. Imbembo, The Structure of a Small, Intense Hurricane-Inez 1966. *Mon. Wea. Rev.*, 104, 418-442, 1976.

Malkus, J. S. and H. Riehl, On the Dynamics and Energy Transformations in Steady-State Hurricanes. *Tellus*, 12, 1-20, 1960.

Ooyama, K., Numerical Simulation of the Life-Cycle of Tropical Cyclones. *J. Atmos. Sci.*, 26, 3-40, 1969.

Palmén, E., On the Formation and Structure of Tropical Hurricanes. *Geophysica*, 3, 26-39, 1948.

Riehl, H., *Tropical Meteorology*. New York: McGraw-Hill, 392, 1954.

Yanai, M., Formation of Tropical Cyclones. *Rev.Geophys.*, 2, 367-414, 1964.

Chapter 10

A Century of Progress in Tracking and Warning

Improvements in Observations, Models, and Forecasts

H. E. Willoughby

CONTEXT

As described in *Isaac's Storm* by Eric Larson, Isaac Cline faced the most serious meteorological threat of his life based upon two pieces of information. The first was the climatological expectation that September hurricanes in the western Gulf of Mexico move northward. The second was a telegram from Washington predicting that the hurricane, which had pounded Cuba and Key West earlier in the week, would recurve across the southeastern U.S. into the Atlantic. Notwithstanding this prediction, Cline's only reliable information about the storm came from within the scope of his horizon: a low barometer, freshening wind from the north, and a long-period swell – well known symptoms of a hurricane's approach to a coastal area.

Within a decade after the Galveston disaster, telegraphed land-station reports were augmented by observations radioed from ships underway. During the 1930s, the first few balloon-borne upper air observations began to come in. Wartime developments in the 1940s brought aircraft reconnaissance and radar reflectivity observations, augmented in the 1950s by statistical track forecasting guidance and output from the coarse-resolution barotropic numerical models. During the 1960s, forecasters began to employ imagery from polar orbiting satellites. While the satellite images could be used to locate the position of the storm and even estimate its

Hurricane! Coping with Disaster
This paper not subject to U.S. copyright; published in 2003 by the American Geophysical Union
10.1029/055SP13

intensity, only scattered *in situ* observations could then provide quantitative measurements needed to initialize models. During the 1970s, geosynchronous satellite coverage became more extensive. At the same time, numerical models became multi-level and so could include baroclinic processes. Thermodynamic soundings from space-borne radiometers and winds estimated by cloud tracking were the first remotely sensed quantitative hurricane data. Assimilation of these observations—with their unsure accuracy, asynoptic timing, coarse resolution, and uncertain vertical position—posed significant challenges, but their existence provided many more observations over previously data-void oceans. Regrettably, these improvements tended to be accompanied by reduction of *in-situ* observing capability. Another key advance of the 1970s occurred when Chester Jelesnianski developed a storm surge model to predict the extent and depth of the wind-driven rise of the sea based upon a specified landfall position and hurricane intensity.

As discussed in Chapter 4, the track forecast error (the great circle distance between the forecast and actual storm positions) is the most commonly cited metric of hurricane forecast performance. It has been tabulated since 1954 by C. J. Neumann, Colin McAdie, and coworkers through careful verification of forecasts against the post-season "best track." The average error for the crucial 24-h lead time decreased from an initial value of 124 nmi to 80 nmi in 2000. Between 1970 and 1998, errors for forecasts to 48 h (initiated in 1961) and 72 h (initiated in 1964) decreased even more rapidly. The rate of improvement varies from 1% a year at 24 h to nearly 2% at 72 h, with some acceleration during the last decade.

By contrast, intensity forecasts have only about half as much skill (relative to climatology or extrapolation of past trends) as track forecasts (Chapter 4). This is a serious shortcoming because the most intense hurricanes account for a disproportionate fraction of the damage. For example, 244 hurricanes and tropical storms made U.S. landfall between 1925 and 1995 (Figure 1). Only ten were Category 4 hurricanes (4%), but they caused 45% of the hurricane-related property losses. The 40 Category 3 hurricanes (16%) caused another 35% [Pielke and Landsea, 1998]. It is for this reason that hurricanes in Categories 3 or greater, with winds stronger than 50 m/s, are termed "major hurricanes." Preparations for landfall could be more appropriate and in many cases less disruptive, if it were possible to identify major hurricane landfalls reliably in advance. A review of past hurricanes shows that essentially all major hurricanes become major through a process of "rapid deepening" that can transform a Category 2 hurricane into Category 4 overnight. Thus, rapid deepening is an important forecast challenge—both because of its suddenness and because it is the way that the worst hurricanes reach their destructive potential.

The last time more than 100 U.S. residents died in a hurricane was Agnes in 1972. The largest mortalities since Agnes were caused by Hurricane Floyd of 1999 and Tropical Storm Allison of 2001, each with fewer than half as many deaths. In both cases, as in Hurricane Mitch's tragic impact in Central America, the predominant cause of death was drowning in fresh water floods produced by torrential hur-

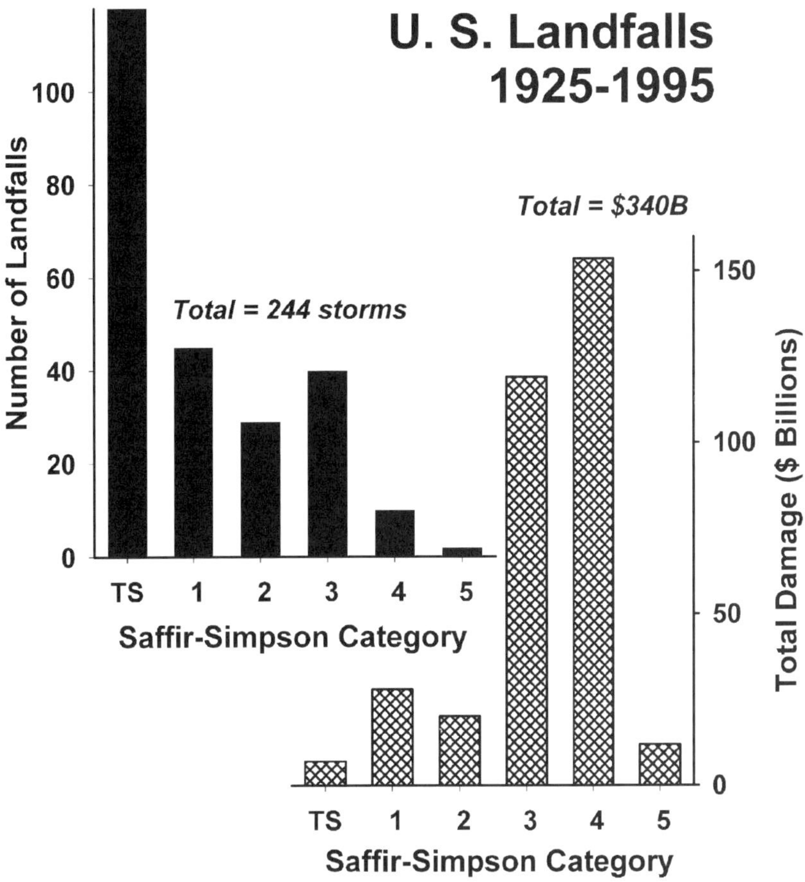

Figure 1. Histograms of U. S. tropical cyclone landfalls 1925–1995, showing the number of tropical storms and hurricanes by category, and the total damage for each category normalized for increased population, greater individual wealth, and inflation. The 50 Category 3 and 4 landfalls account for 80% of the total damage. Reproduced from Willoughby [2000], based upon Pielke and Landsea [1998].

ricane rainfall. Ed Rappaport found that 60% of U.S. deaths since 1970 were due to drowning in fresh water. This situation stands in stark contrast with earlier U.S. experience and present conditions in the developing world where storm surge is the main threat. As a result, human casualties from hurricanes in the U.S. decreased to an average of 25 deaths a year during the last three decades of the 20th Century.

Property losses, on the other hand, have increased to an annual expected toll of nearly $5 billion in step with the compounded effects of growing population, greater individual wealth, and inflation [Pielke and Landsea, 1998]. The annual average of 70 U.S. deaths between 1930 and 1969, multiplied by the 314%

increase in coastal population between Texas and Virginia since 1950, yields an expectation of 219 deaths annually if our responses to hurricanes had not improved since mid-century (Figure 2). Thus, a tangible benefit of modern forecasting and emergency response is the reduction of hurricane-related mortality by approximately 200 lives annually. The main reason for this result was the virtual elimination of drowning through prediction by storm surge models of inundation around the forecast landfall point and evacuation of the surge zone.

The errors in track forecasts at 24 h are somewhat larger than with the width of the damaging hurricane core. To avoid surprises, warnings must typically cover an area three times larger than the core. Jerry Jarrell and Mark DeMaria showed that the size of warning areas has actually increased by about a third since 1967—despite the decrease in track forecast errors—because increasing coastal populations require more lead time for evacuation. Evacuation and other preparations cost several hundred thousand to a million dollars per mile. Given an average of 400 nmi of coastline warned in each of two or three landfalls a year, preparation costs can approach a billion dollars annually. The cost of forecasting, emergency management, and supporting research is about an order of magnitude smaller than what is spent on evacuation.

In strictly scientific terms, the challenges facing the forecast enterprise are: (1) developing means to forecast rapid deepening; (2) continuing the trend toward more accurate track forecasts; and (3) providing neighborhood-level forecasts of local impacts: wind, rainfall, and storm surge. In more general terms, the value of accurate predictions lies in the ability to respond appropriately to the threat. Coastal residents can currently expect to prepare for a hurricane several times for every hurricane that they experience. It is the once-a-decade or once-a-century event, when the worst imaginable storm strikes the most vulnerable coastline, which will cause the greatest economic and human losses in the 21st Century, just as during the 20th. Most of the time institutions and individuals overreact; but when the "big one" comes, they are virtually certain to under-react.

MEASUREMENTS

Hurricanes' high winds and torrential rainfall have mesoscale spatial dimensions. They are typically concentrated within 100 nmi from the storm center. Through radar studies of Hurricane Daisy, Joanne Malkus and colleagues demonstrated that the hurricane's dominant time scale is synoptic, in contrast with other mesoscale systems. Significant changes require about a half pendulum day[1] (28 h at 25° latitude). This time scale also matches that of broadcast news so that tropical cyclones are natural media events. Hurricanes are easier to fore-

[1] A pendulum day is the time required for a freely-suspended pendulum to complete an apparent rotation about the vertical. It is a function of latitude, varying from 24 h at the poles, 48 h at 30°, and approaching infinity at the Equator.

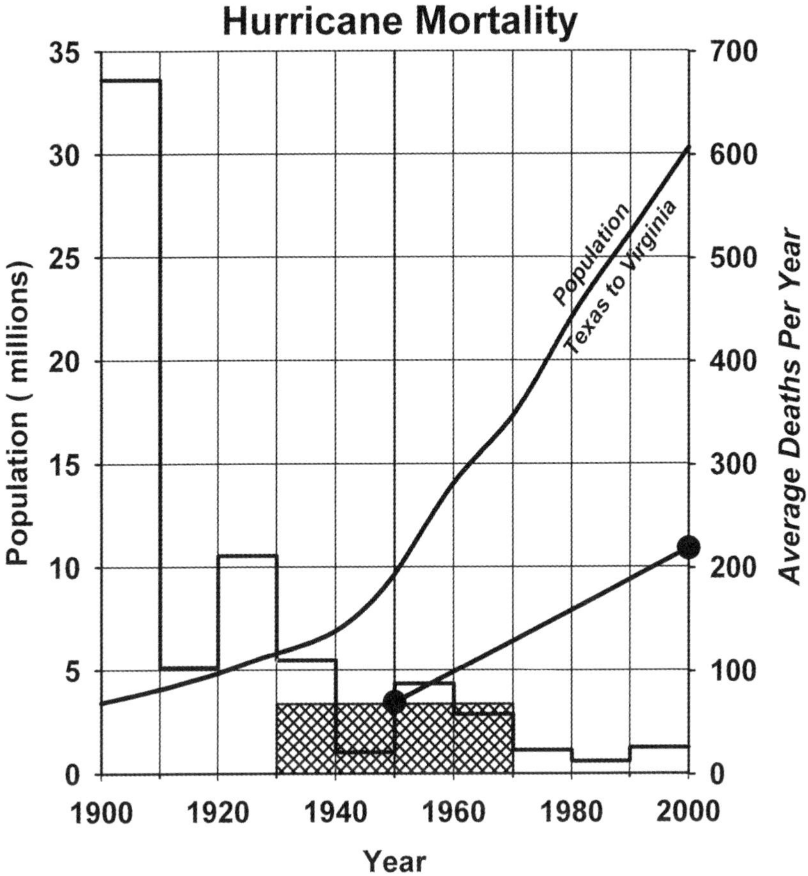

Figure 2. Histogram of U.S. hurricane mortality in the 20th Century plotted with a graph of Texas–Virginia coastal population. The shaded rectangle corresponds to the average mortality at mid century 1920–1969, which is scaled by the 314% increase in coastal population during the last half of the century to yield an estimate of 210 deaths annually if society's response to hurricanes had not improved since 1950.

cast than most mesoscale systems but just as hard to observe. To a great extent, hurricane forecasting has concentrated on storm position and intensity. Consequently, hurricane-observing platforms are generally tailored to spatially compact but slowly evolving phenomena.

For more than a half century, instrumented aircraft have provided position and intensity data. From the end of World War II through the 1960s, reconnaissance airplanes would find the center of the cyclone by radar or simply by flying inward perpendicular to the wind. During this period, naval aircraft flew below cloud base in order to estimate wind speed and direction from the appearance of the sea surface. Upon reaching the calm at the center of the eye, the flight crew would determine the minimum surface pressure by hydrostatic extrapolation or dropsonde and take a navigational fix. Because wind estimation was so subjective, forecasters routinely approximated the strongest winds statistically from the lowest sea-level pressure. The result of a successful aircraft sortie flown before the mid-1970s was one or more "center fixes" including position; central pressure; an estimate of maximum wind; and guesses at the radii of gale-force, hurricane, and 100 kt winds.

The first meteorological application of satellites was the production of visible or infrared images transmitted from spacecraft in inclined low-Earth orbits. If properly referenced to geography, these images allowed positioning of tropical cyclones and subjective estimation of their intensity from the distribution of clouds through techniques pioneered by Vernon Dvorak. Satellite imagery provided essentially the same information as aircraft reconnaissance but was somewhat less accurate. Use of remotely sensed data as an equivalent to *in-situ* observations, rather than something new and different with unique properties, was a recurring theme. For a time, advocates of satellites argued that they were a cost-effective replacement for aircraft. By the late 1980s, the U. S. military had stopped aircraft reconnaissance in the Western Pacific, although they continued to fly routinely in the Atlantic and intermittently in the Eastern Pacific.

As examples of unique satellite capabilities, imaging microwave radiometers operating in the 85 GHz band can capture the convective structure of tropical cyclones despite darkness or obscuration by high clouds. It is straightforward to recognize changes in the eyewall or environmental shear patterns through observations of rainfall by the Special Sensor Microwave Imager (SSM/I) or TRMM Microwave Imager (TMI). But these instruments provide a great deal more than pictures. The data also support quantitative retrievals of rainfall rate and thus of latent heating, which can be correlated with evolution of tropical cyclones' structures and intensities [Rodgers et al., 1998]. The Advanced Microwave Sounding Unit (AMSU) uses radiances sensed at 20 different wavelengths to observe temperature and water vapor soundings, precipitable water, and cloud water. Despite the relatively coarse (48 km) resolution, AMSU data can support estimates of wind in thermal balance with the remotely sensed temperature fields.

Pictures continue to play a vital role, nonetheless. Since the 1970s, satellites in geosynchronous orbits produced frequent images with superior time continuity. Animated images communicate the threat of an impending landfall dramatically and graphically. Isaac Cline's observations of freshening wind from an unusual direction and low barometer seem abstract compared with moving images of a giant vortex spinning across the ocean. It is easy to imagine that the low hurricane-related mortality in the U. S. since 1970 stemmed, at least in part, from the visual impact of broadcast satellite animations.

The observations discussed so far characterize the vortex as an isolated entity. From the earliest days, hurricane forecasters have tried to relate tropical cyclone motion to wind or pressure in the synoptic-scale environment. Greg Holland established clearly that tropical cyclone motion is a superposition of advection by the prevailing winds around the storm and the "beta effect," a poleward and westward propagation caused by the poleward increase of the component of the Earth's rotation parallel to the local vertical. As numerical modeling assumed a greater role in track forecasting, emphasis shifted to improvement of the initial conditions for simulation of the storm environment. A steady reduction of track forecast errors during the last 30 years is largely the result of better specification of the environmental flow. Idealized experiments with a baroclinic numerical model by Leslie et al. [1998] indicate that the lower bound on the errors in deterministic forecasts may be as small as 52, 83, and 121 nmi at 24, 48, and 72 h lead times. Thus, at the current rate of improvement, track forecasts could approach a deterministic limit some time before the middle of the 21st Century. Improvement beyond that point will require statistical treatment of a number or ensemble of forecasts that begin with small variations in the initial conditions. Another promising avenue is more careful analysis of propagation, which appears to be more complicated than the predominant beta effect.

In addition to their role in center reconnaissance, airplanes can be used for surveillance of the hurricane's environment to improve initialization of numerical track forecasting models. Experiments conducted from 1982 through 1993 demonstrated a 16–30% reduction in forecast error given aircraft measurements of the storm's environment [Burpee et al., 1996]. The success of these experiments led NOAA to commission a new Gulfstream IV airplane in 1997 specifically to do surveillance operationally. Sim Aberson and James Franklin found encouraging preliminary results with the new airplane, but owing to difficulties with model initialization, the second year's results were less conclusive. Nonetheless, correction of the initialization problems and rational targeting of observations to areas where ensemble forecasts diverge hold the prospect of even greater forecast improvements.

Satellites also sense winds by tracking motions of clouds or water vapor features. Obscuration by the central dense overcast prevents application of this strategy to observe vortex structure, but it is effective for initialization of the flow around the cyclone that determines motion. However, cloud winds are possible only where the clouds are—in the upper and lower troposphere—not at "steering level" in the mid troposphere where the winds correlate most strongly with storm motion. Nonetheless, proper assimilation of these data can reduce track forecast errors by 14% at 24 h and 12% at 48 and 72 h [Goerss et al., 1998].

The foregoing account summarizes an impressive range of *in situ* and remote sensors with potential to improve hurricane forecasts. Despite many solid successes, new sensors were too often developed in response to engineering opportunities instead of the expected effect on forecasts. Disappointing initial results were a frequent experience for first trials of innovative sensors. Analysis schemes tended to reject new observations, or the observations had little effect on simulations. Sometimes they even degraded forecasts. Numerical models are complicated entities in which many unforeseen interactions occur. Generally, each model has been adjusted to work well with the data sources available when it was introduced. Both the model and analysis scheme need to be adjusted to accommodate new kinds of data. Finally, the biases, random variations, and correlations of the data errors are often not clear or they are misunderstood. This situation is particularly vexing for remotely sensed data where the quantity measured is not directly related to the models' representations of the atmosphere. For example, the AMSU instrument observes radiances, which are complex functions of the model-predicted variables of temperature and water vapor. Thus, a key challenge is to use great quantities of diverse data in the best possible way.

DATA AND MODELS

Numerical models represent the atmosphere as arrays of meteorological quantities distributed on a geographic grid—a checkerboard of latitude-longitude boxes with winds, pressures, temperatures, humidities, and other variables assigned to the squares. These numbers depict the state of the atmosphere at an instant in time. The model computes the future state of the atmosphere by plugging the current state into the equations that describe atmospheric motions and extrapolating the resulting rates of change into the future. To start the calculation, one needs the best possible depiction of the state of the atmosphere (the "initial condition") expressed on the grid, but based upon observations with (often) unknown errors and scattered in space and time. This process is termed the analysis.

The simplest type of analysis is interpolation of the observed data onto the nodes, or grid points geographically near the position where each observation was taken. The analysis process often incorporates a background field, typically a short-range (e.g., 6 h) forecast from a previous analysis. The analysis scheme may also operate on differences between the background field and observations rather than on the observed quantities themselves. The analysis may then be adjusted by incorporating dynamical relations such as hydrostatic or geostrophic balance.

Modern analysis schemes include an *assimilation* of observations into the numerical model during the early stages of the forecast. When the observations are scattered in time as well as spatially, nearby nodes can be adjusted, or "nudged," toward observations during the course of an "assimilation run" that ends at the start of the new forecast cycle. In the early days of numerical weather prediction, analysis and initialization, though recognized as a necessary precondition to successful forecast, often received less sophisticated attention than did the models themselves.

A more systematic approach to initialization uses the formal process of optimum estimation, where the background and observations are combined in such a way that each affects the initial state in inverse proportion to its inherent error. Roger Daley's 1991 book provides an informative account of the meteorological application of these ideas.

The goal of the observations, analysis, assimilation, and forecasts is to arrive at a state of the model that corresponds closely with reality. This "state vector," is a list of winds, pressures, temperatures, etc. that may add up to thousands of values for a large-domain, fine-resolution model. As we have seen, the temporal changes may be computed from the state vector using the mechanical and thermodynamic equations that govern atmospheric motions. Extrapolation of these changes generates the state of the model after a short interval, Δt. A prediction for a time $k\Delta t$ in the future requires k extrapolation cycles. The resulting forecast is one source of information about the state of the atmosphere.

Observations are the other source of information, again expressed as a list of quantities such as wind, temperature, humidity, or electromagnetic radiance. The observations are almost never located on the geographic grid nodes. They may even be different physical quantities from those used in the model. However, they are useful because they may be expressed as functions of the state variables. This relationship is called a "forward" model. It includes biases and random errors and embodies physical relationships between the observed and state variables. Examples of forward models include the radiative transfer equations or spatial interpolation to the model's computational grid. The number of observations is almost always different from the number of state variables. If there were exactly the same number of observations as state variables, it would be possible to solve the equations of the forward model for the state variables exactly. This strategy would not be a good idea because it would throw away the information in the forecast for the initial time and retain all of the errors in the observations.

Thus, both models and observations provide useful information about the state of the atmosphere. The Kalman filter (Figure 3) is an elegant, though computationally expensive, mathematical scheme to combine the model results and observations such that each contributes to the ultimate estimate in proportion to its information content. The Kalman filter works by adjusting the background state variables to make estimates of the observed quantities calculated with the forward model agree more closely with the actual observations. Its statistical properties are designed to make greater adjustments to achieve agreement with accurate, representative observations and smaller ones to accommodate less accurate or less representative data. A key requirement is detailed knowledge of the error characteristics of each kind of observation. The genius of this approach is that numerous observations, even if they are individually inaccurate, will generally improve the analysis; whereas if the forward model were simply solved to get estimates of the state variables, they might degrade the analysis.

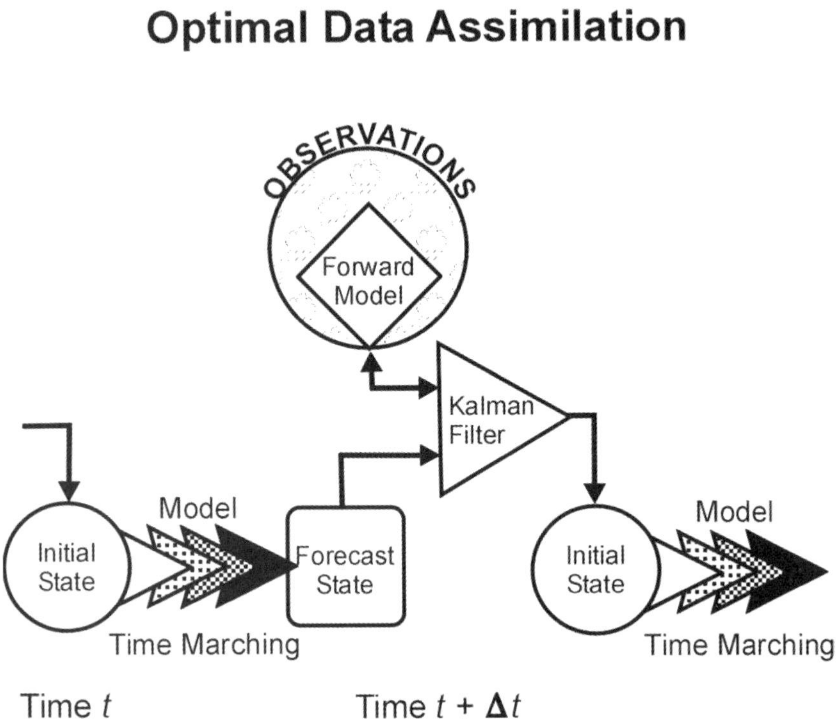

Figure 3. In the process of optimal data assimilation, the model predicts the state variables at time $t + \Delta t$ based upon the state variables at time t. The Kalman filter adjusts the state vector to minimize the differences between the observations and the observed quantities computed from the adjusted state variables, which then serve as the initial condition for the next cycle of time marching.

Forward models depend upon the statistical properties of the model and sensors. As new sources of observations become available over the lifetime of a model, each requires its own forward model that leads to changes in the properties of the analysis. The effort and physical insight entailed are significant obstacles to effective incorporation of new measurement technologies into operational forecasting.

A dramatic acceleration in the rate of improvement in track forecast errors followed operational implementation of the GFDL Hurricane Prediction System in 1995 [Kurihara et al., 1998]. This model began as a state-of-the-art research model in the early 1970s. It has a moving computational mesh that keeps the finest resolution (1/6° latitude or 18.5 km) centered on the vortex core. It uses a global model from the National Centers for Environmental Prediction to provide initial and boundary conditions. Although more sophisticated models existed even when it was first introduced to operations, the great achievement of the

GFDL group was to make a research model work effectively within operational time constraints. At about the same time, the Navy Operational Global Atmospheric Prediction System (NOGAPS) and the global model run at the United Kingdom Meteorological Office also became competitive hurricane forecasting models, although the latter was not available to forecasters within operational time lines. These improvements were immediately apparent to forecasters because of the rigorous protocol pioneered by C. J. Neumann that compares models with each other and assesses forecast difficulty relative to a climatology and persistence baseline.

Although the GFDL hurricane model provided spectacularly good track forecasting guidance, it consistently overpredicted intensity, ostensibly because it did not account for the reduction in energy available to the hurricane as a result of storm-induced cooling of the sea. Coupling of the atmospheric model with the Princeton Ocean Model substantially removed this bias, leading to improved performance [Bender and Ginis, 2000]. Clearly, future operational intensity forecasting models will need to represent a two-way interaction between the tropical cyclone and the sea. Studies of sensitivity to drag and thermodynamic transfer coefficients, interpretation of instrumented buoy observations of near-surface thermodynamics, and the uncertain role of spray evaporation show that a deeper understanding is essential for proper representation of sea-air interaction in a new generation of forecasting models.

Future tropical cyclone models will have resolutions of a few hundred meters, fine enough to resolve individual convective cells. Based upon the experience with the coupled GFDL model, accurate intensity forecasting will also require coupling with a prognostic ocean model. How elaborate the ocean model needs to be is less clear, but a potential exists for much greater precision in forecasting. A likely framework for development of future models is the Weather Research and Forecasting (WRF) Model. WRF is coded in modules and designed to run across a range of computer architectures in both research and forecasting settings. A key objective is quick, easy transition of innovations originating in universities and government laboratories into forecasting.

SUMMARY

The United States Weather Research Program's (USWRP) Fifth Prospectus Development Team's report [Marks et al., 1998], *Hurricanes at Landfall*, outlines ambitious but attainable goals for hurricane forecasting in the early 21st Century along with a well-thought-out plan for reaching them. Guiding principles for hurricane research in the early 21st Century include:

1. In the late 20th Century, the hurricane forecasting enterprise appears to be saving ~200 lives a year through evacuation of the storm-surge zone. The economic impact of reduced mortality is \$1–2 billion. The reduc-

tion of property losses is less easily documented because the number of properties at risk is increasing exponentially.

2. Hurricane impacts in the 21st Century will be determined by a few extreme events, predominately major hurricanes striking vulnerable, densely developed coastlines.
3. Foremost scientific priorities are: development of skillful intensity forecasts; continuation of the momentum toward more accurate track forecasts; and initiation of neighborhood-level forecasts of impacts such as wind, rainfall, and storm surge, which will require a new generation of fine-resolution, nonhydrostatic numerical models.
4. Scientific advances and engineering skill have produced an impressive array of sensors and platforms for operational observation of tropical cyclones. The key to realizing the benefit from these sensors lies in the development of forward data assimilation models.
5. The ocean has a pivotal role in intensity change. The new generation of models will also need two-way interactions with a dynamic representation of the ocean. Improved understanding of physical processes at the sea-air interface will prove essential to the design of these models.

REFERENCES

Bender, M. A. and I. Ginis. Real-Case Simulations of Hurricane-Ocean Interaction Using a High-Resolution Coupled Model: Effects on Hurricane Intensity. *Mon. Wea. Rev., 128*, 917–946, 2000.

Burpee, R. W., J. L. Franklin, S. J. Lord, R. E. Tuleya, and S. D. Aberson. The Impact of Omega Dropwindsondes on Operational Hurricane Track Forecast Models. *Bull. Amer. Meteor. Soc., 77*, 925–933, 1996.

Daley, R. *Atmospheric Data Analysis*. Cambridge, UK: Cambridge University Press, 457 pp, 1991.

Goerss, J. S., C. S. Velden, and J. D. Hawkins. The Impact of Multispectral *GOES-8* Wind Information on Atlantic Tropical Cyclone Forecasts in 1995. Part II: NOGAPS Forecasts. *Mon. Wea. Rev., 126*, 1219–1227, 1998.

Kurihara, Y., R. E. Tuleya, and M. A. Bender. The GFDL Hurricane Prediction System and its Performance in the 1995 Hurricane Season. *Mon. Wea. Rev., 126*, 1306–1322, 1998.

Leslie, L. M., R. A. Abby, and G. J. Holland. Tropical Cyclone Track Predictability. *Meteor. Atmos. Phys., 65*, 223–231, 1998.

Marks, F. D., L. K. Shay, and the Fifth Prospectus Development Team. Landfalling Tropical Cyclones: Forecast Problems and Associated Research Opportunities. *Bull Amer. Meteor. Soc., 79*, 305–323, 1998.

Pielke, R. A. and C. W. Landsea. Normalized Hurricane Damages in the United States. *Wea. Forecasting, 13*, 621–631, 1998.

Rodgers, E. B., W. S. Olson, V. M. Karyampudi, and H. F. Pierce. Satellite-Derived Latent Heating Distribution and Environmental Influences in Hurricane Opal (1995). *Mon. Wea. Rev., 126*, 1229–1247, 1998.

Willoughby, H. E. People, Property and Hurricanes. *Climate Report, 1,* 4, 2–6, Winter, 2000.

Chapter 11

The Burgeoning Role of Weather Satellites

C. Velden, J. Simpson, W. Timothy Liu, J. Hawkins,
K. Brueske, and R. Anthes

INTRODUCTION

The invention of weather satellites in the 1960s has virtually ended the problems associated with hurricane detection. With current technology, which allows for the detection of hurricanes in any ocean basin from space, a "surprise" hurricane, such as the Galveston event in 1900, is highly unlikely. Before the 1960s, it was very possible to "miss" a hurricane that was situated or forming away from shipping routes. By the 1970s, high-resolution geostationary satellite images from the visible (VIS) and infrared (IR) spectra were available at three- to six-h intervals over the hurricane-prone regions of the world's oceans. Numerous experts collaborated to develop subjective cloud pattern recognition techniques and apply them to satellite-detected hurricanes in order to estimate maximum winds. Today, the major global tropical cyclone forecast centers rely heavily on meteorological satellite surveillance. In many cases, the satellite is the only available method for estimating motion and intensity.

This chapter is not an exhaustive treatise on weather satellites or the radiative transfer theory on which satellite remote sensing is based. It is primarily focused on the more recent satellite sensors relevant to the study and forecasting of hurricanes. Section Two of this chapter describes the pioneering work with satellite imagery up through the mid-1980s and the rapid advancements in the use of geostationary satellite VIS and IR products since then. Section Three describes the emergence of new information contributed by satellite-borne instruments that

Hurricane! Coping with Disaster

10.1029/055SP14

observe in the microwave end of the spectrum. Section Four provides a summary and an outlook on the potential value of anticipated new satellite instrumentation toward improving future hurricane analysis and forecasting.

APPLICATIONS OF VISIBLE-INFRARED (VIS-IR) SATELLITE IMAGERY AND DERIVED PRODUCTS FOR HURRICANE DIAGNOSTICS AND FORECASTS

Satellite instruments receiving energy in the wavelengths of the visible spectrum see much like our own eyes. What these receivers sense derives mainly from reflection and scattering of solar radiation from cloud tops. Hence, the VIS imagery is not available during the nighttime hours. On the other hand, radiation sensed in the IR wavelengths comes from emission by the clouds, so that the energy emitted by the cloud tops can be received by a radiometer in space around the clock. The cloud-top temperatures (and altitudes) can be estimated if the IR sensors are properly calibrated and an atmospheric temperature profile is available nearby. It should be pointed out that because the IR sensors are "blind" to radiation below opaque clouds, the imagery is often confined to a depiction of the storm canopy and the "exhaust" clouds associated with the upper-level outflow of the hurricane. In conditions where the storm is characterized by a central dense overcast (no apparent eye in the upper-level cloud structure), it is often not an easy task to infer the exact center location and particularly the maximum surface wind speed.

Pioneering Use of Visible and Infrared (VIS/IR) Imagery

The first weather satellites made it possible to identify hurricanes without ship reports, which would have saved many, if not most, lives in Galveston had they been available in 1900. In 1960, the era of space-based reconnaissance began with the first successful launch of a weather satellite (TIROS 1). It had five scanning channels with variable VIS and IR ground resolution of 30-55 km, which was crude but showed promise of great results to come. In 1964, Nimbus 1 flew more advanced sensors with VIS and IR resolutions of about 5 km at perigee and 11.6 km at apogee. These polar orbiters were not only used to locate mature storms, but also to estimate maximum wind intensity from the portrayal of cloud organization in the images. Aircraft reconnaissance direct measurements showed that these initial satellite-based intensity methods were accurate to only plus/minus 20 kt, which translates to a one category difference in the Saffir-Simpson damage scale.

A major advance in hurricane detection from space came with the development of geostationary weather satellites. The first, ATS-1, was launched in 1966. Positioned over the Equator, it provided a view every 20 min of the Earth's sunlit disk between 52° North and South latitudes. This remarkable achievement was made possible by Professor Verner Suomi's (University of Wisconsin) famous spin-scan cloud camera. Its ground resolution was better than 4 km at the subsatellite point. Scientists at the NOAA Environmental Satellite Service continued efforts to estimate hurricane intensity from these improved images.

By the 1971 hurricane season, two satellite meteorologists were permanently assigned to the National Hurricane Center (NHC). These satellite specialists produced two hurricane classifications per day in support of forecasters during the 1971-1973 seasons. Each classification included the storm center location and estimate of maximum winds. During this period, NOAA scientist Vern Dvorak was developing his subjective image pattern recognition technique, which was first published in 1975 [Dvorak, 1975]. This method, although since refined by researchers such as NOAA scientist Ray Zehr and operational satellite meteorologists in practice, is still in use today as the standard for estimating tropical cyclone intensity around the globe.

The Dvorak technique is based on principles of cloud pattern recognition. The amount of organization is directly related to hurricane intensity (Figure 1). Trained satellite analysts determine the cloud pattern type and relate it to storm development stages. As satellite IR sensors became more sophisticated in the 1970s, Dvorak expanded his techniques to include special enhancements. The Dvorak enhancement curve (Plate 1) was developed to allow satellite specialists to focus on the convective vigor in the eyewall. He and other pioneering colleagues developed a statistical relationship between the IR-depicted eye (warm) temperature and the surrounding eyewall cloud (cold) temperature. From this relationship, intensity could be inferred. This enhanced IR (EIR) technique was added to the Dvorak scheme (Figure 1). Further details on this landmark study in hurricane meteorology can be found in Dvorak [1984].

Recently, university scientists collaborated to automate the Dvorak technique (Plate 2). This computer-based algorithm was designed to minimize human judgment in cloud pattern typing. While experienced satellite analysts are very much required, my colleagues and I [Velden et al., 1998b] showed that the Objective Dvorak Technique (ODT) is competitive. This algorithm has been transitioned into NHC operations, where it is part of the suite of satellite-based guidance tools.

The 1980s: Geosynchronous Satellite Information Comes of Age

During the 1980s, computer technology was advancing rapidly and just in time. The newer geostationary satellites were now delivering images at 30-min intervals. Advanced image processing and visualization systems were being developed that could ingest, enhance, and animate satellite imagery into colorful movie loops in real time. The prototype was designed at the University of Wisconsin-Space Science and Engineering Center. Called McIDAS (Man-computer Interactive Data Access System), this computer-driven system was installed at the NHC in

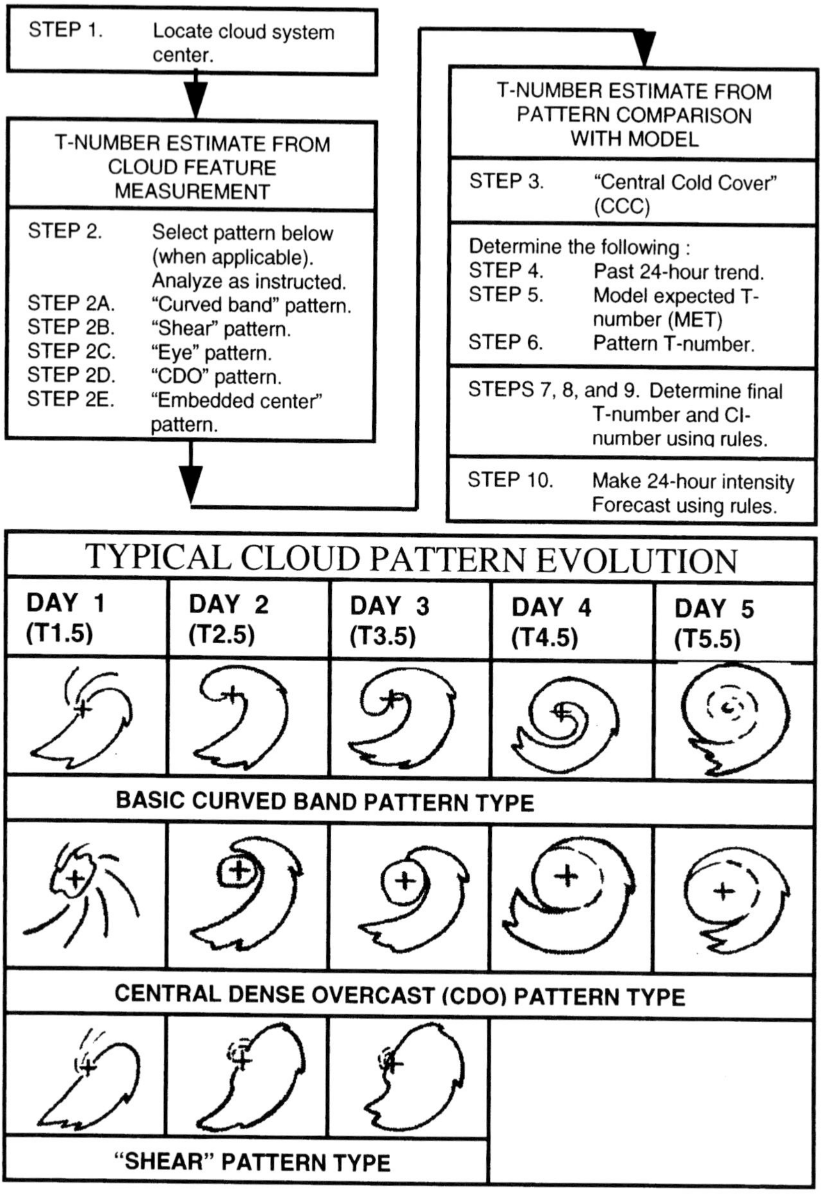

Figure 1. The Dvorak satellite-based hurricane intensity scheme. Top: Flow chart used by satellite analysts to refine the hurricane intensity estimation. Bottom: Cloud patterns are related to broad categories of hurricane intensity [taken from Dvorak, 1984].

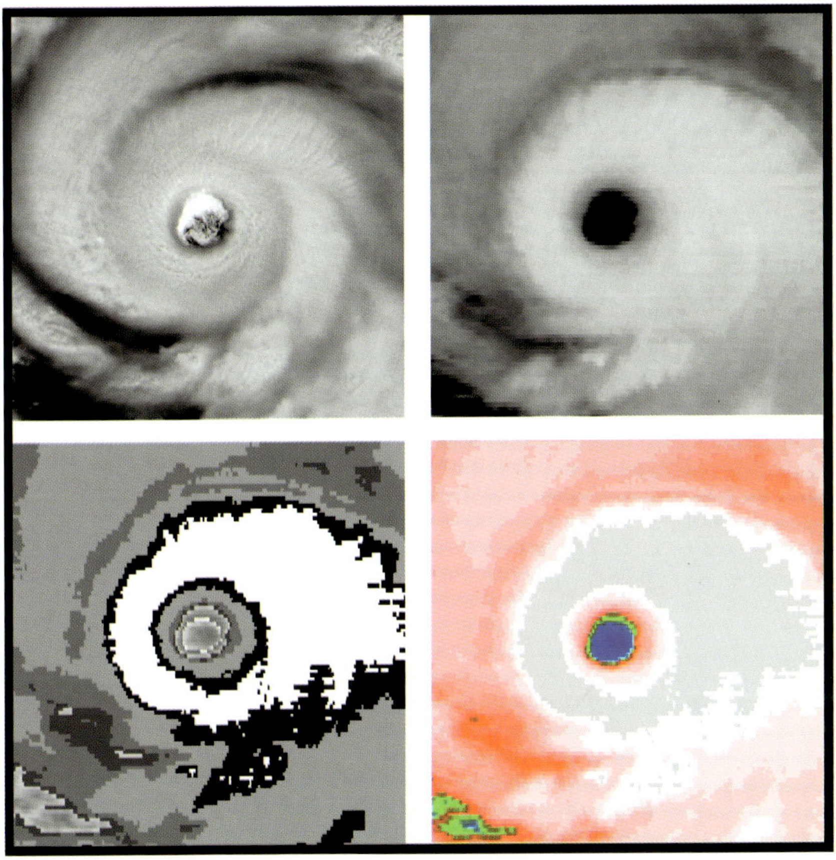

Plate 1. Examples of satellite imagery for Hurricane Floyd in 1999. Top left: Visible. Top right: Infrared. Bottom left: Dvorak enhancement curve. Bottom right: Color IR enhancement.

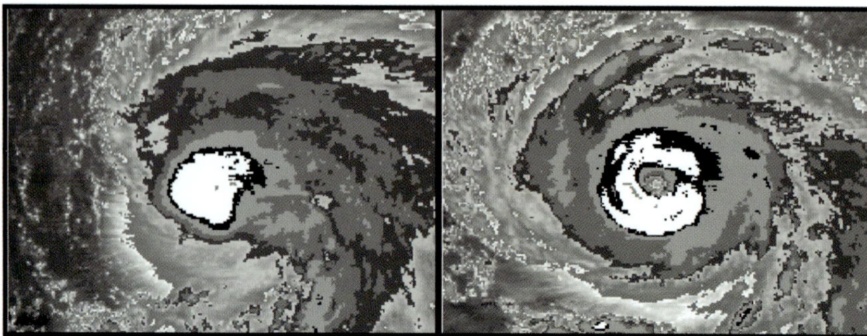

Date	Time	Intensity CI	MSLP	T-No Raw	Temperatures (C) Eye	Cloud	Scene Type	Step9 Flag
1999SEP11	164500	4.6	977.2	4.5	-59.2	-65.2	CDO	OFF
1999SEP11	181500	4.6	977.2	4.5	-51.8	-58.4	CDO	OFF
1999SEP11	201500	4.6	977.2	4.5	-29.4	-46.2	EMB	OFF
1999SEP11	221500	4.6	977.2	4.9	-15.8	-55.0	EYE	OFF
1999SEP12	14500	4.6	977.2	4.8	-29.6	-53.7	EYE	OFF
1999SEP12	34500	4.6	977.2	4.5	-46.5	-58.4	CDO	OFF
1999SEP12	64500	4.8	973.6	5.4	-16.8	-64.0	EYE	OFF
1999SEP12	84500	4.8	973.0	4.5	-33.4	-62.5	CDO	OFF
1999SEP12	104500	5.5	960.0	6.3	12.4	-65.6	EYE	OFF
1999SEP12	124500	5.7	955.2	6.0	-0.1	-66.1	EYE	OFF
1999SEP12	144500	5.9	950.4	6.1	12.9	-62.1	EYE	OFF
1999SEP12	164500	6.0	948.0	6.1	12.1	-62.5	EYE	ON
1999SEP12	204500	6.2	942.8	6.7	16.7	-70.4	EYE	ON

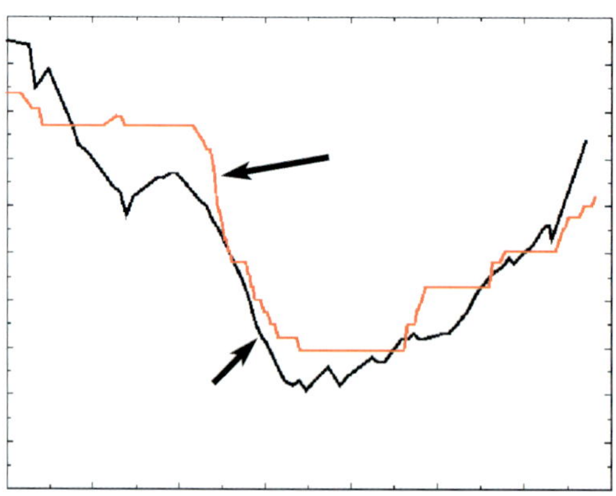

Plate 2. The Objective Dvorak Technique (ODT). This example shows a case of rapid intensification during Hurricane Floyd in 1999. Top: two geostationary satellite images show the Dvorak enhancement curve on September 11 and 12, respectively, and the development of an eye. Middle: ODT intensity estimates (every two hours between the two images) from objective, computer-based image analysis methodology. Bottom: ODT performance (red line) during Floyd relative to reconnaissance aircraft measurements (black line).

the early 1980s. It quickly became the primary satellite data display tool for NHC analysts and forecasters, who could now take advantage of the frequent sampling to detect circulations and storm motion. NHC director, Neil Frank, began to use the color-enhanced and animated displays in live TV interviews to educate and help warn the public about approaching storms.

Another satellite tool was introduced in the late 1970s and 1980s. The "water vapor channel" was added to IR sensors on the geosynchronous weather satellites deployed by European nations and the United States. This channel focused on long wave radiation absorbed and re-emitted by water vapor in the 6-7 micron frequency range. In contrast to the IR imagery, it was no longer necessary for clouds to be present, as this channel could depict moisture structures in the atmosphere as never seen before (Figure 2). Animation of this imagery revealed circulations and synoptic features in the hurricane environment that would affect the motion and track forecasts. Today, water vapor imagery is a prominent tool for use in analyzing the hurricane environment.

McIDAS was also used to quantify the information provided by the frequent satellite imagery. Once the rapid data acquisition and animation tools became available, it was clear that clouds and water vapor features could be tracked in successive images to infer wind fields (Plate 3, top). Given the scarcity of con-

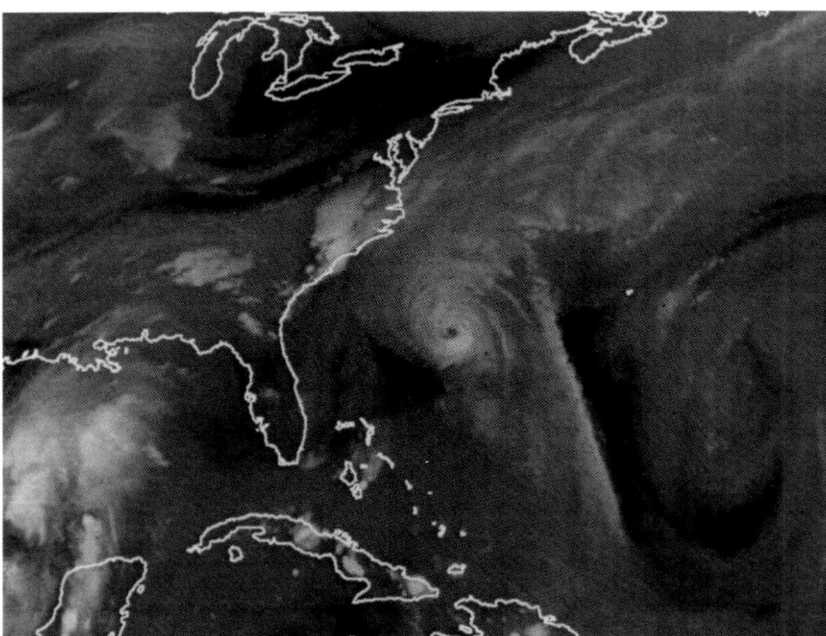

Figure 2. Geostationary weather satellite water vapor imagery during Hurricane Danielle in 1998. Moisture structure in cloud-free regions can be seen in the darker gray shades.

ventional observations over the oceanic regions, NHC quickly embraced the satellite-derived winds. Despite the limitation of imprecise height assignments, the data were used to help determine the environmental flow field that accounts for much of the hurricane steering and forecast motion. Today, automated techniques have been developed that track features in multispectral full disk imagery from multiple geostationary platforms to provide global wind sets around the clock. These data sets have proven useful in real-time analysis [Velden et al., 1998a] and in reducing hurricane track forecast error in numerical models [Goerss et al., 1998; Soden et al., 2001]. Special rapid-scan schedules from the GOES weather satellite are now standard and can sample at a 1-5 minute frequency over targeted hurricanes and, thus, provide very detailed wind fields (Plate 3, bottom).

Other Important Applications of IR Satellite Data

Another use of IR information is the determination of sea surface temperatures (SST). Hurricanes depend on ocean heat content and moisture fluxes to drive the convection necessary to sustain their low pressures, and it is well recognized that ocean surface temperatures greater than 26°C are usually necessary to generate and/or sustain hurricane vortices. Satellite IR radiometers are able to sense ocean skin temperatures in the absence of clouds. This information is particularly important as strengthening or weakening could depend on the SST gradients in the path of a hurricane. NHC forecasters can now monitor features such as the warm Gulf of Mexico loop current, the warm Gulf Stream and associated eddies, and cool-water "wakes" left over from the heat loss and mixing associated with previous storms (Plate 4). They are working toward incorporating this information in intensity forecasts. Recent studies show that height variations in ocean surface sensed by satellite altimetry can provide additional information on the ocean heat content (energy available in the upper layer of the ocean).

A major concern associated with landfalling hurricanes is the rainfall and flooding potential. IR imagery is used as a guidance tool to infer rainfall rates in hurricanes. Studies have been directed towards calibrating satellite IR-based rainfall estimation techniques to radar estimates at landfall. While these IR methods can provide a generally good (within a factor of two) bulk estimate of rainfall in the absence of other data, precise observations of heavy rain bands and localized accumulations are difficult from the IR alone. Improved multi-spectral satellite techniques are being developed that utilize the advantages of rapid temporal sampling from geostationary IR imagery and the cloud-penetrating properties in the microwave spectrum from polar-orbiting instruments.

IR and VIS satellite observations have become a mainstay of hurricane detection and analysis. The limiting factor of passive remote sensing in these spectra is the inability to sense through clouds. The next section will address this limitation with the introduction of new satellite sensors that can penetrate clouds and provide a unique view of hurricanes.

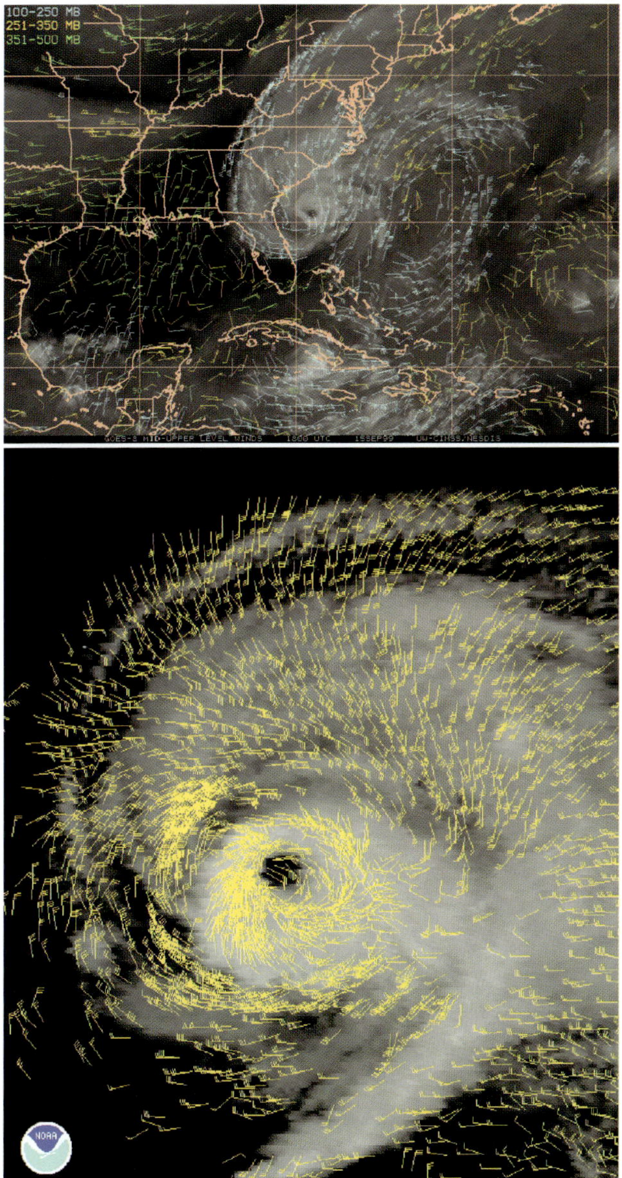

Plate 3. Top: Large-scale view of wind barbs determined by tracking upper-level cloud motions and water vapor in successive 30-min satellite imagery during Hurricane Floyd in 1999. Bottom: Zoomed-in view of high-density wind barbs determined by tracking upper-level cloud features in successive 5-min rapid-scan imagery during Hurricane Luis in 1996.

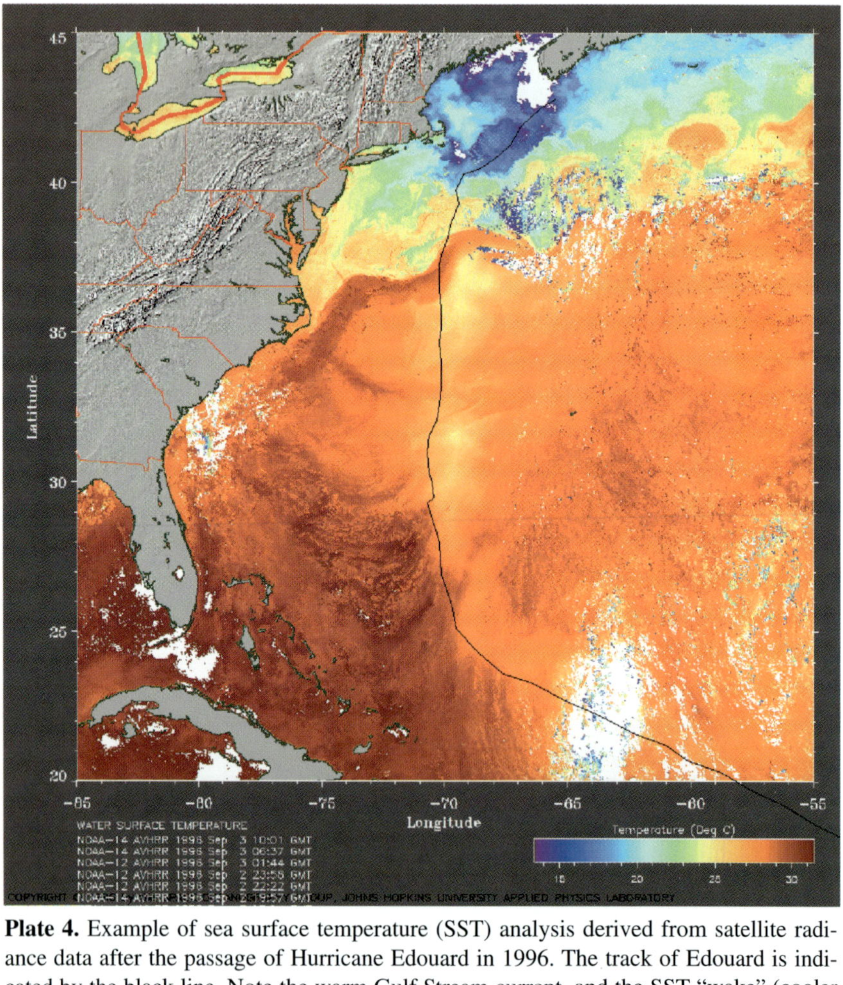

Plate 4. Example of sea surface temperature (SST) analysis derived from satellite radiance data after the passage of Hurricane Edouard in 1996. The track of Edouard is indicated by the black line. Note the warm Gulf Stream current, and the SST "wake" (cooler temperatures from mixing/upwelling) leftover along the path of Edouard. [Image courtesy of Space Oceanography Group, Johns Hopkins University Applied Physics Laboratory.]

THE EMERGENCE OF OBSERVATIONS IN THE MICROWAVE

Many scientists have shown that satellite-based passive microwave imagery greatly enhances our ability to monitor key hurricane characteristics [see Hawkins et al., 2001 and Chapter 12]. Passive microwave data ameliorates the prime limitation inherent with VIS/IR data by being able to sense through non-raining clouds and mapping organization not otherwise detectable. The longer wavelengths associated with microwave radiation are not sensitive to ice clouds and can, therefore, penetrate the storm's cirrus canopy to reveal important structure underneath. Unfortunately, the longer wavelengths require larger receivers that currently limit the instrumentation to low-altitude, Earth orbiting satellites. Therefore, very frequent sampling (an advantage of geostationary satellites) and image animation is not yet possible.

Applications of Satellite Microwave Observations

The concept of estimating tropical cyclone structure, intensity, and intensity change using polar-orbiting weather satellite passive microwave observations spans over two decades of active research. The first demonstration of the potential for microwave channels to delineate unique hurricane properties came in the late 1970s. Scientists used the first meteorological space-borne microwave sounding unit flown on the NOAA series of Earth orbiters to show how the characteristic warm core of hurricanes could be observed and, via hydrostatic approximations, related to storm intensity. Radiation sensed by oxygen emission in the 55 GHz region was used to derive upper-tropospheric temperatures in the hurricane core. Several factors limited these initial intensity estimation techniques including coarse spatial resolution and the ever-changing position of the hurricane within the instrument observation swath between successive observations. These factors combined to reduce the accuracy of the tropical cyclone inner-core thermal anomaly estimates.

Within the last decade, progress continues in understanding, treating, and reducing the undesirable effects of instrument-related warm core sampling limitations. This progress has been partly due to advanced microwave sounding instrumentation on the most recent NOAA Earth orbiters. Hurricane Floyd 1999 provides an excellent example of this new improved capability (Chapter 12). On September 14, 1999, Stan Kidder and colleagues [2002] used the NOAA-15 Advanced Microwave Sounding Unit (AMSU) data to obtain temperature profiles in Floyd's eye. From these they constructed vertical cross-sections of the hurricane's core region warm anomaly (Plate 5).

The ability of the AMSU to detect changes in a tropical cyclone warm core structure is vitally important as it yields information on storm intensity and intensity change. The magnitude and distribution of hurricane force winds are directly linked to the tropical cyclone sea-level pressure, which is governed by the

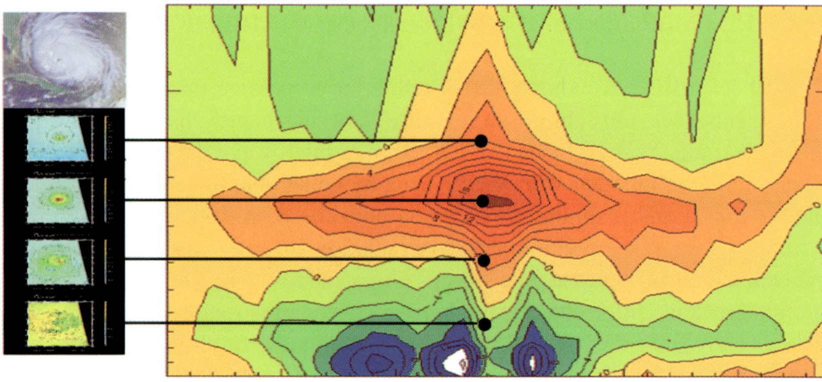

Plate 5. NOAA-15 AMSU vertical temperature anomaly (°C) for Hurricane Floyd at 1238 UTC September 14, 1999 (right). Floyd's warm core is clearly evident with temperature anomalies in excess of 18°C near 250 hPa. Strong cooling (>-16 °C) below the convective eyewall region is fictitious and due to the lack of explicit correction for scattering in the sounding retrieval package. AMSU 55.5 GHz, 54.94 GHz, 54.4 GHz, and 53.6 GHz limb corrected brightness temperatures (T_b, top to bottom) with corresponding peak radiance sampling levels (horizontal lines) are shown with coincident color-enhanced VIS image inset (left).

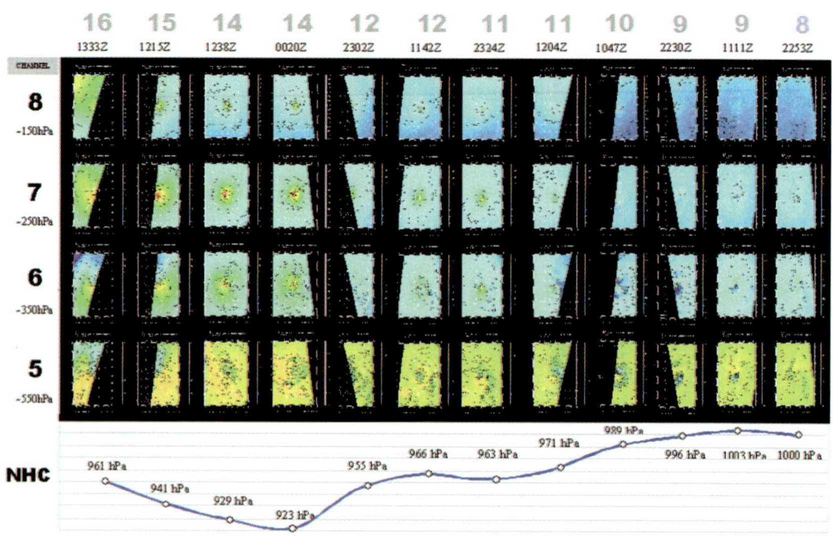

Plate 6. A time series comparison of NOAA-15 AMSU limb corrected brightness temperatures (°C) and aircraft reconnaissance observations of MSLP for Hurricane Floyd, September 8-15, 1999. AMSU channel and peak emission levels are indicated on the left: Ch. 8 (55.5 GHz ~150 hPa), Ch. 7 (54.94 GHz~250 hPa), Ch. 6 (54.4 GHz ~ 350 hPa), and Ch. 5 (53.6 GHz ~ 550 hPa). Aircraft reconnaissance MSLP estimates (hPa) are indicated below. The development of Floyd's warm core is clearly evident in the temperature anomaly trends depicted by the upper-tropospheric sensing channels (green to red indicates warming).

three-dimensional thermal structure of the overlying atmosphere. The qualitative inverse relationship between the size and vertical distribution of the warm core and tropical cyclone intensity is well illustrated by a time series of NOAA-15 AMSU observations of Hurricane Floyd from September 8-16, 1999 and near-coincident aircraft reconnaissance sea-level pressure estimates (Plate 6). Floyd's sea-level pressures (SLP) are at their maximum values (>1000 hPa) during periods of weakly organized warming within the inner core region, whereas minimum SLP values (<930 hPa) occur during periods when the warm core is well defined — particularly when upper-tropospheric temperature anomalies are at their maximum values. Chapter 12 discusses some of the physical processes causing Floyd's warm core to develop and strengthen.

In most cases, even the higher resolution AMSU measurements cannot fully resolve the true warm anomaly present in the eye. Brueske and Velden [2002] address this and find promise in further quantifying the relationship between AMSU-observed upper-tropospheric warming and hurricane intensity. They model the explicit correction of scan-geometry-related warm core sub-sampling. More precise intensity estimates are then possible, as shown by the promising preliminary results presented in Plate 7. Starting in 2001, experimental AMSU-based tropical cyclone intensity estimates were made available in near real-time for evaluation by forecasters at the NHC and the Joint Typhoon Warning Center (JTWC). We expect that these techniques will be transitioned into operations in 2002.

Another unique capability of the AMSU is its ability to provide important information on the structure and distribution of tropical cyclone gale-force winds using thermodynamic and dynamic constraints. Stan Kidder and colleagues describe this method in detail [2000]. We present only a brief summary here. A statistical algorithm is used to determine the temperature as a function of pressure from the AMSU radiances. The hydrostatic equation is then integrated downward from 50 hPa (mb) to the surface using an upper-level boundary condition provided from a Numerical Weather Prediction (NWP) analysis. Prior to the hydrostatic integration, an empirical correction is applied to the temperature profiles to adjust for attenuation of the AMSU radiances by liquid water and scattering by ice. Once the height field is determined as a function of pressure, the wind field is calculated assuming gradient balance. The height field is azimuthally averaged relative to the storm center for this purpose.

Plate 8 shows an example of the retrieved gradient wind as a function of radius and height for Hurricane Gert. This figure shows that without the water/ice correction (top panel), the gradient winds below about 3 km and outside a radius of about 200 km are not very realistic. With the correction, the retrieved wind field shows a cyclonic circulation through a deep layer within a radius of about 200 km, and an anticyclonic circulation at upper levels outside of about 200 km.

This figure also shows the limitation of AMSU spatial resolution. The maximum retrieved wind in the low levels is only about 50 kt at a radius of about 175 km. At the time of this analysis, the maximum winds were estimated by the

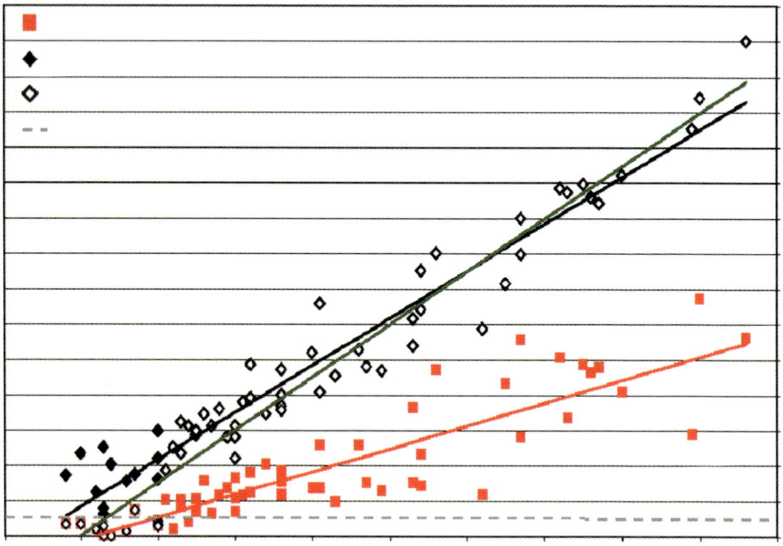

Plate 7. A comparison of AMSU 54.94 GHz-derived tropical cyclone upper-tropospheric warm anomalies (UTWAs) and aircraft reconnaissance MSLP estimates. The red squares (red linear least squares line) represent raw UTWA observations, the black diamonds (black linear least squares line) are the same sample corrected for scan-geometry dependent sub-sampling [Brueske and Velden, 2002]. The hollow diamonds (green linear least squares line) are also from the same sample and are also corrected for scan-geometry dependent sub-sampling with the exception of raw UTWA values <= 0.5°C. The results demonstrate that tropical cyclone UTWA are indeed sub-sampled by the AMSU and exhibit a strong linear relationship to MSLP after explicit correction of scan-geometry related warm core sub-sampling.

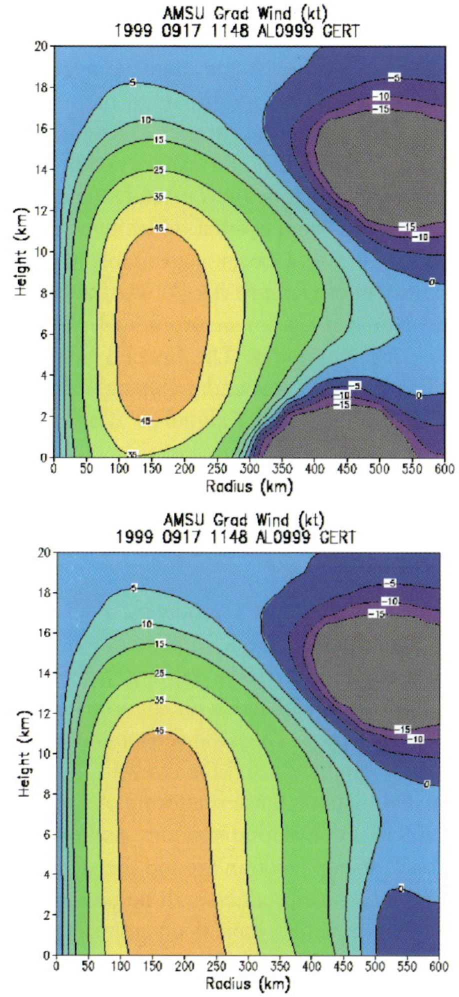

Plate 8. Radial/height gradient wind (kt) cross-sections for Hurricane Gert on September 17, 1999 near 1200 UTC. The figure above (below) is without (with) the empirical correction for cloud liquid water and ice [taken from Demuth et al., 2000].

National Hurricane Center to be 125 kt at a radius of about 45 km. Thus, the AMSU data does not have adequate resolution to fully sample the inner core structure of intense tropical cyclones. To help correct for this problem, a statistical algorithm is being developed to estimate the maximum winds and the radii of 34, 50, and 64 kt winds from properties of the AMSU analyses. The inputs are analysis properties such as surface pressure drop from 600 km to the storm center, the maximum retrieved gradient wind, and the average cloud liquid water near the storm center. The output is an estimate of the observed maximum wind and wind radii. This promising algorithm is currently under development. Demuth and colleagues describe a preliminary version [Demuth et al., 2000]. Studies are also underway to apply the corrective measures of Brueske and Velden [2002] to this gradient wind estimation technique.

Another important satellite tool became available to forecasters in the late 1980s. The U.S. Department of Defense (DoD) was the first to fly a passive multifrequency microwave imager on a meteorological satellite in 1987, and this program of satellites continues today. The Special Sensor Microwave Imager (SSMI) channels, unlike the NOAA sounder channel data described above, penetrate to the Earth's surface unless the signal is attenuated by precipitation or large aerosols. The first SSMI showed some of the potential of this new imaging tool, but the potential was not fully realized until multiple satellites were operating and providing more than one view per day. By the 1990s it was typical to have two to four active satellites that were capable of providing five to seven images of a storm per day.

One of the SSMI channels senses at a frequency of 85 GHz, with a spatial resolution of 13-15 km. This channel is able to penetrate non-raining clouds due to the small size of ice crystals aloft. However, larger, frozen hydrometeors (e.g., hail, graupel) and raindrops associated with vigorous convection dramatically scatter radiation at this frequency. Thus, the sensor can easily detect intense rain associated with hurricane rainbands and the eyewall due to the lowered brightness temperatures created by the intense scattering. A time series of 85 GHz data can reveal a storm's internal convective structure and evolution by mapping the organization and vigor of the convection around the storm center.

Examples of SSMI microwave products will be presented later in the chapter. It is important to note that these data have been enthusiastically used not only by the DoD sponsored JTWC in Pearl Harbor, Hawaii, but by virtually all of the World Meteorological Organization (WMO) tropical cyclone warning centers over the globe.

The TRMM Satellite: A Success Story

Building on the success of the DoD program, NASA launched a special satellite aimed at measuring meteorological quantities over the tropics using passive and active microwave sensors. The Tropical Rainfall Measuring Mission

(TRMM) satellite completed five years of successful observations in November 2002. With the change in altitude from 350 km to 400 km made in August 2001, it has enough fuel to continue measurements another year and possibly longer[1]. TRMM is a low Earth-orbiting satellite with a unique orbit, which oscillates about the Equator between roughly 35°N and 35°S. A joint mission between the U.S. and Japan, TRMM's primary stated goals were to: (1) improve the understanding of crucial links in climate variability due to the hydrological cycle, (2) improve the large-scale numerical models of weather and climate through assimilation of TRMM data, and (3) advance our understanding of cloud ensembles and their impacts on larger scale circulations. Shortly after launch, scientists recognized that TRMM would also provide essential new information on hurricanes in all stages of development.

The primary TRMM sensors and their scanning patterns are shown in Figure 3 and Table 1. Table 1 shows that the Precipitation Radar (PR) scans only over a relatively narrow swath of 215 km, limiting the sampling ability of the PR. TRMM's PR is an active sensor and the first successfully deployed civilian rain-rate monitoring radar to operate from space. The PR can provide 3-D profiles of precipitation through the storm cloud patterns. Plate 9 shows two cross sections at right angles through the core of Hurricane Floyd (1999). Floyd's early stages are discussed in Chapter 12. Chris Kummerow and colleagues [2000] provide more detail on the TRMM instruments, algorithms, and a wide range of early results.

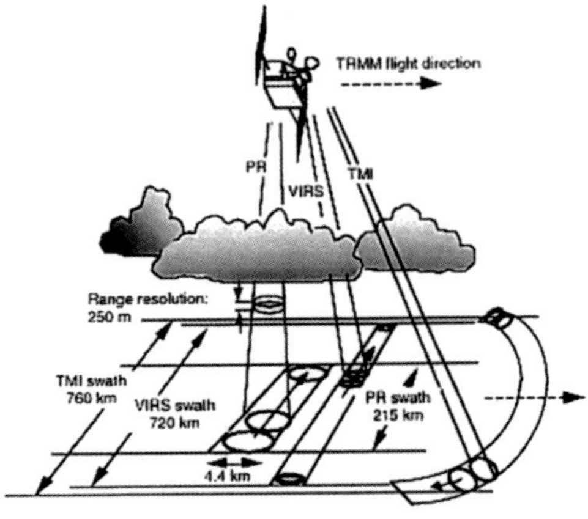

Figure 3. Schematic of the TRMM satellite over-flying clouds. The scanning patterns of the Precipitation Radar (PR), TRMM Microwave Imager (TMI), and Visible-Infrared Radiometer (VIRS) are shown on the surface.

[1] The drag on the spacecraft, owing to the denser atmosphere associated with the solar maximum in 2002/2003, has reduced earlier hopes of continuing until 2007. If the satellite is to have a controlled re-entry to an empty oceanic target, the extra fuel required for that will end the mission in late 2003.

TABLE 1. TRMM Sensor Summary

Microwave Radiometer (TMI)	Radar (PR)	Visible/Infrared Radiometer (VIRS)
10.7, 19.35*, 21.3, 37, 85.5 GHz (dual polarized) 21 km resolution at 19 GHz	13.8 GHz 4.3 km footprint 250 m range resolution	0.63 μm VIS and 10.8 μm IR, also 1.61, 3.75 and 12 μm at 2.2 km resolution
760 km swath	215 km swath	720 km swath

The passive microwave radiometer (TMI) is a multi-channel, dual-polarized, conically scanning passive microwave instrument similar to SSMI. The purpose of the VIS/IR instrument (VIRS) is to enable TRMM to be a "flying rain gauge." The TRMM satellite radar and radiometer combination is intended to obtain high quality vertical profiles of precipitation as well as surface rainfall estimates. TRMM's physically based rain-rate observations from the combined radar and passive microwave instruments allow the calibration of empirical rain estimates from the IR sensors. As a result, uncertainty in tropical rainfall has been greatly reduced from earlier space-based estimates. Currently, these techniques are being applied to better estimate hurricane rainfall.

TRMM observations are also being used to provide better initial conditions for numerical models. Florida State University professor T. N. Krishnamurti and colleagues [Krishnamurti et al., 2000] developed a complex modeling approach using TRMM data that is showing promise to improve three-day hurricane forecasts of both track and intensity. Since, as shown in Chapter 4, the skill of intensity forecasts is considerably less than that of track forecasts, these new approaches are of great importance. The approach uses multiple analyses and multi-models to create a "super-ensemble" forecast. Slightly different analyses for model initialization come from applying products from the different TRMM and SSMI algorithms. The selected numerical models are all those in use at major forecast centers and a global model developed at Florida State University. Using supercomputers, hundreds of test runs are made using past global weather. By comparing resulting forecasts with validation (observations of variables such as rain, wind, temperature), poor models are discarded, and the final forecast is a weighted average of all the resulting model runs. The weights are determined by statistical relations between forecasts and observations of each component model. Figure 4 shows an example for Hurricane Georges in 1998. In this particular case, the "super-ensemble" 72 h intensity forecast is superior to other model forecasts, none of which caught the full intensification. While this result

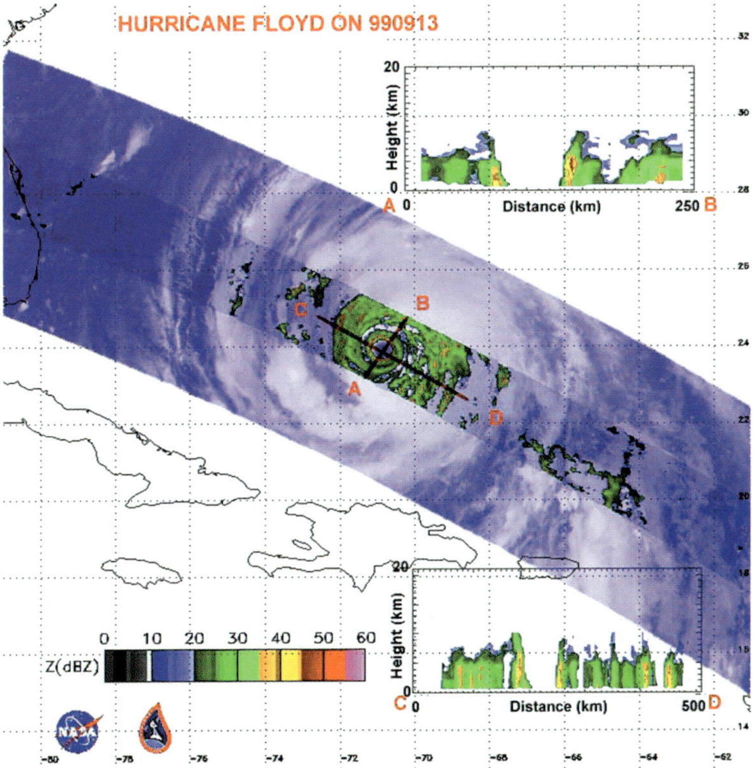

Plate 9. A TRMM overpass of Hurricane Floyd (1999). The swath of the PR is shown, along with two cross sections at right angles through the core indicating the three-dimensional precipitation profiling capabilities of the space-borne radar.

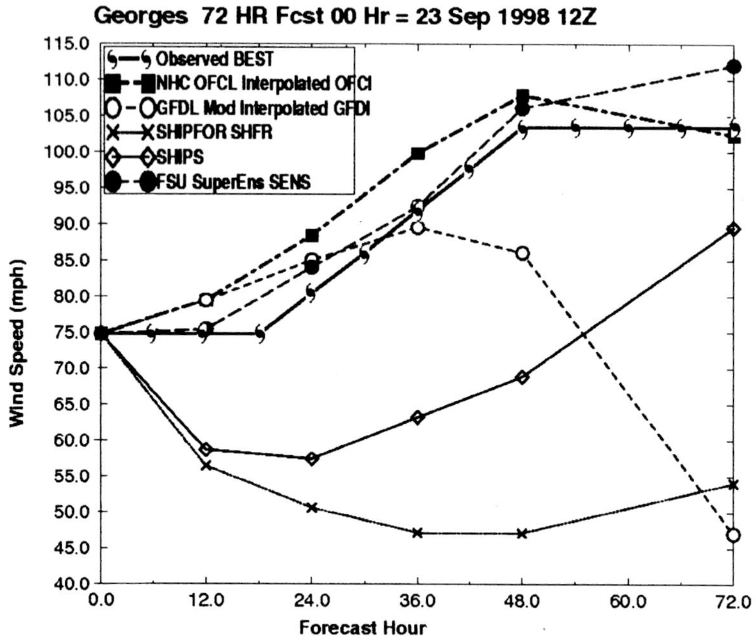

Figure 4. Seventy-two hour forecasts of intensity of Hurricane Georges in 1998 made at 0000 UTC September 23. The line with the hurricane symbol is the best observed maximum wind, while the dashed line with the squares is the official forecast made by NHC [Courtesy of Professor T.N. Krishnamurti].

has not yet been proven typical in a rigorous statistical evaluation over many cases, new methods that attempt to assimilate satellite information are showing great promise. This "super-ensemble" model approach is also proving valuable in improving large-scale weather forecasts.

TRMM data are being provided in near real time to hurricane forecasters. The high-resolution imagery is being used to help locate hurricane centers and to assess convective organization trends. The value of TRMM to the forecast and research communities is evidenced by its wide usage in operational tracking and forecasting of tropical systems, along with the increased understanding of the global water cycle.

More Examples of Passive Microwave Imagery Monitoring of Tropical Cyclones

The TRMM TMI complements the SSMI data by filling in temporal gaps and providing superior spatial resolution. The TMI 85 GHz ground footprints are 5-7 km, which is two to three times better than the SSMI, and the TMI 37 GHz channel can penetrate deeper into the hurricane to reveal additional details about the storm organization.

The ability of passive microwave data to sense through high-level cirrus is especially relevant to detect lower-level features that define the circulation center, which can lead to more accurate storm center fixes. Plate 10 illustrates how an exposed low-level circulation center (LLCC) is readily apparent in the SSMI product, while the cirrus shield in the enhanced IR image obscures this feature. Use of the SSMI and TMI can help reduce large center relocation errors that negatively affect numerical model initialization and resultant track forecasts.

Passive microwave data are also quite valuable when trying to understand the organization of systems undergoing rapid intensity changes. Intensifying hurricanes are often characterized by double (or concentric) eyewalls. Double eyewalls are rarely seen in VIS/IR imagery since upper-level clouds almost always obscure these structures. However, the passive microwave sensitivity to ice and rain scattering in the inner and outer eyewall typically allows this important structural signature to be observed. A good example of this phenomenon is illustrated in Plate 11, which shows coincident TMI and geostationary satellite IR data during Hurricane Gert as it underwent a double eyewall formation. The IR data caught the formation of the eye due to the dissipation of high clouds from subsiding air in the eye. The outer eyewall is clearly shown in the TMI image, but is obscured under the cirrus canopy in the IR image. Trained forecasters can observe trends in these data (a strong and constricting inner eyewall generally means intensification; a collapsing inner eyewall in favor of the outer eyewall generally implies decreasing intensity) and use these trends as a proxy for the evolution of eyewall dynamics. This information can serve as important guidance in their intensity forecasts.

SSMI scalar wind speed values can be derived from ocean surface backscatter and are used operationally to help diagnose the surface wind field surrounding hurricanes. The SSMI passive microwave frequencies used to create the wind speed product are susceptible to contamination when the field of view contains very high water vapor values and/or rain. Wind speed retrievals are, thus, not usually reliable in a hurricane's inner core region. These data have their biggest impact in specifying the radius of gale force winds. Although the correlation between rain and gale winds is high, many storms have quadrants that are relatively rain free due to shear and other storm cloud/rain asymmetries. SSMI wind speeds can then be used to update gale radii in conjunction with all other *in-situ*, satellite and climatology information.

The TRMM TMI sensor can also be used to provide surface wind speeds and has the advantage of an additional lower frequency at 10 GHz. This channel permits wind speed retrievals closer to rain and high winds and, thus, typically detects some higher winds. The 750-km swath is a limitation, but the wind speed information can be superior to SSMI values due to this inherent capability.

Large precipitation particles affect each microwave channel differently, with the scattering effect particularly large at the highest frequency of 85 GHz. This

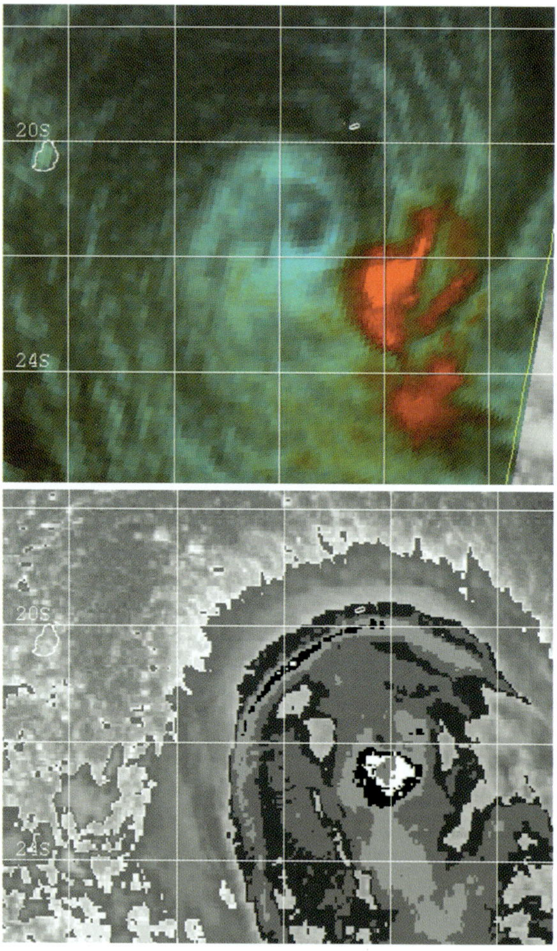

Plate 10. Comparison of color-enhanced SSM/I product (above) with the standard Dvorak IR enhancement (below) for tropical cyclone 05S on January 16, 2001 at 0149 UTC. Both images have 2° latitude/longitude markings and reveal the ability of the SSM/I to detect sheared systems with exposed low-level circulation centers (LLCC) not viewable in IR.

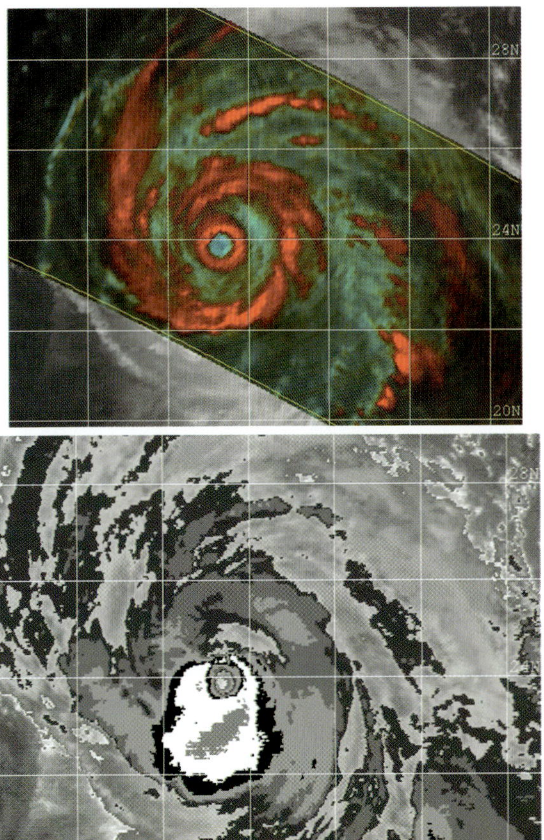

Plate 11. TRMM 85 GHz color product detecting inner and outer eyewall for Hurricane Gert (above), while Dvorak IR enhancement shows little evidence of outer eye (below) due to obscuration by upper-level clouds.

instrument response offers an immediate candidate for hurricane rain rate estimation techniques and does offer some advantages over IR-based methods. Combined microwave and IR techniques are being developed and calibrated with radar estimates in cases of landfalling hurricanes. Experimental versions are showing promise and should lead to improved operational hurricane rainfall estimates in the near future. The importance of accurately estimating hurricane rain rates cannot be understated, as inland flooding from landfalling hurricanes is a leading cause of human fatalities (see Chapter 8).

Progress in Scatterometer Applications

A scatterometer sends microwave pulses to the Earth's surface and measures the power scattered back from the surface roughness. Over the ocean, the roughness elements producing the backscatter are largely due to small waves on the surface, which are believed to be in equilibrium with the local wind stress. The backscatter power depends not only on the magnitude of the wind stress, but also on the wind direction relative to the direction of the radar beam. Wind vector retrievals from backscatter measurements are based on empirical relations. The relation between backscatter and ocean surface winds under the strong wind and rainy conditions of a hurricane is not well established because of the lack of validation data. The ability to estimate surface wind speed and direction under both clear and cloudy conditions day and night is what makes the scatterometer unique.

Since 1978, NASA and the European Space Agency (ESA) have launched five scatterometers into space on Earth orbiters. In the past decade alone, the daily global coverage from scatterometers has increased from slightly over 40% to well over 90%, and the spatial resolution has increased from 50 km to 12.5 km. Historically, ESA scatterometers use C-band (5 GHz), while the NASA instruments use Ku-band (14 GHz). Ku-band is more sensitive to wind variation at low wind speeds but is also more susceptible to rain effects.

The first scatterometer on a European Remote Sensing (ERS-1) satellite was launched in 1991 and was followed by an identical sensor launched on ERS-2 in 1996. Early studies using these data demonstrated the ability to determine the winds and subsequently the central pressure in typhoons, which contributed to the improved estimation of the intensity and the location of the storms. ERS data also allowed researchers to estimate the relative vorticity of tropical cyclones and compare the sizes of Pacific and Atlantic systems. The positive impact in predicting and describing hurricanes by assimilating ERS scatterometer winds into numerical weather forecast models has been demonstrated in numerous studies. The two ERS scatterometers provided nine years of continuous wind data.

The success of the earlier scatterometers prompted the rapid deployment of a new instrument by NASA in 1999. Called QuikSCAT, the conical scan of this sensor's pencil-beam antennas provides a continuous 1800-km wide swath and unprecedented global ocean coverage. Wind fields from QuikSCAT are available in near-real time to most hurricane forecast centers. The standard wind product has 25 km spatial resolution, but special products with 12.5-km resolution have been produced to monitor hurricanes. Tim Liu and his colleagues [2000] showed that the surface wind and fresh water flux of a high-resolution operational numerical weather prediction model could not resolve the rain bands of Hurricane Floyd (1999). By simply replacing the surface level model wind divergence with QuikSCAT data, the fresh water flux becomes more realistic compared with observations by the TRMM PR. With QuikSCAT winds, the vertical profiles of heating and rainfall rate revealed the eye and rainbands, in agreement with the PR observations. Kristina Katsaros and her colleagues [2001] examined QuikSCAT data for all of the tropical systems in the Atlantic during the 1999 hurricane season. They found that, for the majority of them, closed circulations with intensities meeting the criteria of a tropical depression were observed by QuikSCAT before their classification by the National Hurricane Center. For example, NHC declared Floyd a tropical depression on September 7, 1999, east of the West Indies. QuikSCAT data were used to track the surface vortex all the way back to the African coast on September 2nd (as illustrated in Chapter 12). Plate 12 illustrates an example of QuikSCAT wind data during Floyd.

In examining thousands of co-located QuikSCAT and ocean buoy measurements, researchers have found that there is almost no mean difference (bias) in wind speed and only 6° variance in wind direction. However, there is an ongoing debate in the tropical and satellite research communities over the accuracy of scatterometer wind retrievals under extreme wind and rain conditions associated with hurricanes. Because of the lack of *in-situ* wind measurements, the accuracy of the wind retrievals under high wind conditions (over 15 m/s) and in heavy precipitation is more uncertain. There are clear indications that heavy rain affects the wind retrieval, but there is currently no proven threshold of rain intensity at which the wind retrieval using existing algorithms becomes invalid. This is an area of intense research. Currently, wind vectors under suspected areas of rain contamination are flagged, and the forecaster must use these data with caution. Nonetheless, the satellite scatterometer surface wind data have become an integral part of the hurricane analysis tools now utilized by operational forecast centers. Tim Liu [2002] provides a review of scatterometry and science applications.

Use of GPS Radio Occultation Methods for Hurricane Research and Forecasting

A promising new technique for measuring the temperature, pressure and water vapor in the environment of hurricanes is the radio occultation (RO) technique, which was developed by scientists at Stanford University and the Jet Propulsion Laboratory (JPL) for the remote sensing of planetary atmospheres. The Global Positioning System (GPS) is a satellite-based system developed for navigation and positioning. The GPS today consists of 24 satellites that transmit L-band

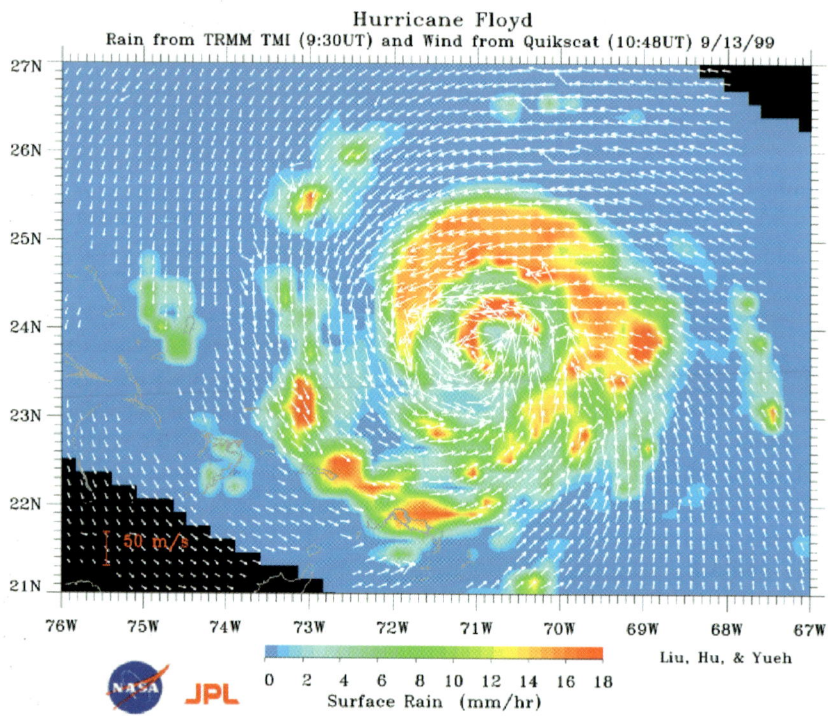

Plate 12. Example of QuikSCAT surface wind data overlain on TRMM rainfall signatures during Hurricane Floyd in 1999 [taken from Liu et al., 2000].

radio signals (19 and 22 cm wavelengths) to a wide variety of users for navigation, time transfer and relative positioning as well as an ever-increasing number of scientists in geodesy, atmospheric sciences, oceanography, and hydrology.

Atmospheric soundings are obtained through the RO technique, in which satellites in low-Earth orbit, as they rise and set relative to the GPS satellites, measure the change in frequency of the GPS dual-frequency signals. The Doppler-shifted frequency measurements are used to compute the bending angles of the radio waves, which are a function of atmospheric refractivity. The refractivity is a function of electron density in the ionosphere and temperature, pressure and water vapor in the neutral atmosphere.

In the stratosphere and troposphere, the bending angle-derived refractivity profiles are a function of temperature (T in degrees absolute, K), pressure (P in mb), and water vapor pressure (e in mb),

$$N = 77.6\ P/T + 3.73 \times 10^5\ e/T^2 \tag{1}$$

Given measurements of refractivity N, the relationship among refractivity, temperature, pressure and water vapor (Eq. 1), together with the hydrostatic equation, comprise a system of two equations with three unknowns. Equation (1) may be solved for either temperature (T) or water vapor pressure (e) given independent estimates of the other variable.

Where water vapor is negligible, such as in the stratosphere and upper troposphere, pressure and temperature may be obtained directly from the refractivity (e=0 in (1)). Temperatures retrieved with the assumption of zero water vapor are termed "dry temperatures" (T_{dry}); neglecting the water vapor implies that T_{dry} is always less than or equal to T.

The GPS/MET (Global Positioning System/Meteorology) Program demonstrated the relevance of this technique for accurate sounding of the Earth's atmosphere. Rocken et al. [1997] present statistical validation of GPS/MET data by comparing over 1,200 GPS/MET soundings of the troposphere and stratosphere to a variety of correlative data sets. Figure 5 shows comparisons of GPS/MET T_{dry} profiles with radiosonde temperatures in the tropics. There is excellent agreement between T_{dry} and the radiosonde temperatures between 10 and 35 km. Below 10 km, the presence of water vapor causes T_{dry} to be less than the radiosonde temperature. Temperatures retrieved from observed N and values of e obtained from other independent sources are termed T_{wet}. As shown in Figure 5, T_{wet} computed from observed N and e from the NCEP analysis agrees closely with the radiosonde temperatures.

GPS/MET demonstrated atmospheric limb sounding from low-Earth-orbit with high vertical resolution, high accuracy, and global coverage in all weather situations. As a result, GPS/MET technology is now recognized as a potential candidate of the global observing system in support of weather analysis and prediction, including hurricanes. For a detailed summary of the radio occultation

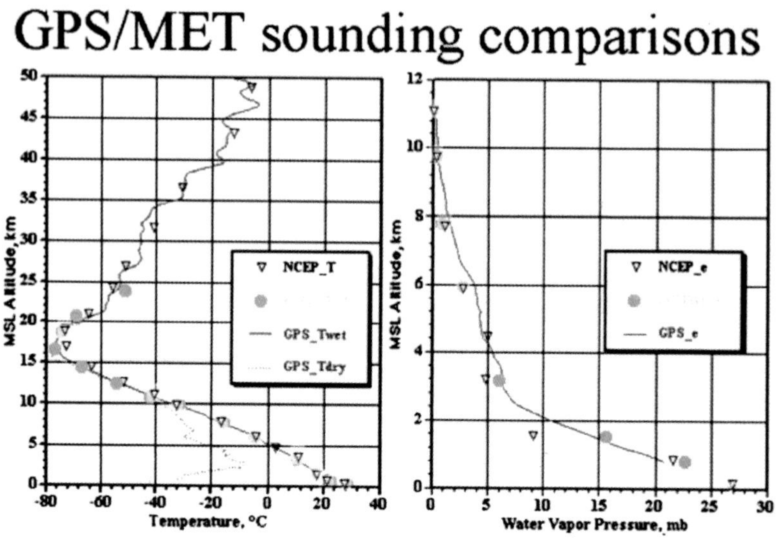

Figure 5. GPS/MET occultation profiles from 1140 UTC on June 22, 1995 at 9°S, 172°W compared with global analyses from NCEP and ECMWF [Rocken et al., 1997].

sounding technique and the results from GPS/MET, the reader is referred to a special edition of Terrestrial, Atmospheric and Oceanic Sciences [TAO, 2000].

SUMMARY AND OUTLOOK

The last half of the 20th century bore witness to "the burgeoning role of weather satellites" in monitoring hurricane activity. From simple images to more sophisticated techniques for extracting quantitative information, satellites have become the predominant platform for monitoring hurricanes. Analysts have developed methods to interpret the imagery to infer intensity and predict short-term motion. Newly available passive microwave imagery has provided unique insights into the organizational structure of hurricanes. The ability to observe beneath the hurricane cloud shield has enabled researchers and forecasters to increase their understanding of fickle hurricane tendencies.

Yet the fact remains that the satellites yield only remotely sensed data. They do not "feel" the hurricane elements, but "see" them from afar. Despite recent advances in instrumentation, satellite data must be augmented by conventional *in-situ* observations to achieve an optimal monitoring scenario. Reconnaissance aircraft still provide the highly reliable inner core data in Atlantic basin hurricanes needed for precision forecasts. An unfortunate consequence of improving satellite technology was the cost-cutting discontinuation of aircraft reconnaissance missions into western Pacific typhoons in 1986. Not only was this a loss of valuable real-time data for prediction purposes, it also cost researchers an important data set for validation studies. Over time, advances in remote sensing will hopefully reduce the need for risky and costly aircraft flights into hurricanes. But until weather satellite sensors can achieve the precision of *in-situ* measurements, it is vitally important that aircraft-based hurricane reconnaissance missions (manned airplanes and/or unmanned drones) fly the Atlantic basin.

The success of the Tropical Rainfall Measuring Mission (TRMM) has led to the formulation and high prioritization of a Global Precipitation Mission at both NASA and its Japanese counterpart NASDA. A tentative launch date of 2007/8 is being planned. The mission is envisioned to consist of a core satellite carrying a dual-wavelength radar and a TMI-like radiometer, and it is expected to achieve an incremental improvement in precipitation measuring accuracy from space observations. In cooperation with the international community, NASA plans a constellation of up to eight passive microwave instrument-bearing satellites flying at about 600 km in sun-synchronous orbits. With these programs, the sparse sampling of TRMM will be overcome, and more frequent observations of hurricanes will be possible. The obvious synergy between the core satellite, designed to improve our physical understanding of rainfall systems, and the constellations designed to improve sampling implies that the Global Precipitation Mission will deliver high quality measurements with the high temporal frequency needed by many applications requiring accurate rain rate data.

NOAA and the DoD have plans for merging the NOAA and DMSP series of satellites into one program. This new series will build upon the successes of the AMSU and SSMI measurements. These satellites will contain advanced instrumentation with improved spatial resolution. With an improved global hydrologic budget, the many and varied interactions between Earth's biosphere, atmosphere, oceans and cryosphere (ice portions) will be better understood and simulated in prediction models. Most directly important to the public, hurricane forecasts will be improved through earlier warnings.

Based on the success and scientific results of GPS/MET radio occultation studies applied to measuring atmospheric variables, a team of international partners is developing COSMIC (Constellation Observing System for Meteorology, Ionosphere and Climate), which is a follow-on project for weather and climate research, climate monitoring, space weather, and geodetic science. The COSMIC program plans to launch a small constellation of six low-Earth orbiting satellites in late 2005 or 2006. Each COSMIC satellite will retrieve up to 500 daily profiles of atmospheric thermodynamic properties from the tracked GPS radio signals as they are occulted behind the Earth limb. The horizontal spacing of COSMIC data on a daily basis varies from approximately 400 km near the poles to 500 km in the tropics. Thus, COSMIC data by themselves will not resolve the detailed structure of hurricanes with scales of typically several hundred kilometers. However, they will contribute significantly to the definition of the pressure, temperature and water vapor, and through data assimilation, the wind field in the large-scale environment surrounding the storm. These improved analyses and resulting numerical model forecasts will provide significant benefits to hurricane research and prediction.

Fortunately, the future appears bright for our space-based observing system. Advanced multispectral imagers, sounders (infrared and microwave), and scatterometers are planned for launch in the near future. Hyperspectral measurements from newly developed interferometers are expected to be flown experimentally by 2006. The information content will vastly exceed that of the current measuring devices. Instead of about a dozen viewing channels, these instruments will have nearly two thousand channels over a wide range of the spectra. The satellite data downloads are expected to exceed several terabytes (one trillion bytes) per day. Fortunately, communications and computing capacity are increasing at a rate that can, with strong pursuit, accommodate this explosion of data. Emerging new technologies (including the use of rapidly-developing visualization tools) will be employed. It is important that the evolving space-based observational system keep one step ahead of the demands being placed by advances in numerical weather prediction. It will be an enormous task and challenge to assimilate this wealth of data into meaningful parameters. However, the outlook is bright for unlocking the still-unresolved mysteries, and thereby improving our understanding and prediction of hurricanes.

REFERENCES

Brueske, K. F. and C. S. Velden, Tropical Cyclone Intensity Estimation Using NOAA-KLM Advanced Microwave Sounding Unit (AMSU) Data, accepted in *Mon. Wea. Rev*, 2002.

Demuth, J. L., M. DeMaria, J. A. Knaff and T. H. Vonder Haar, An Objective Method for Estimating Tropical Cyclone Intensity and Structure From NOAA-15 Advanced Microwave Sounder Unit (AMSU) Data, *Preprints, AMS 24th Conf. Hurr. and Trop. Meteorol.*, May 29-June 2, Fort Lauderdale, FL, 484-485, 2000.

Dvorak, V., Tropical Cyclone Intensity Analysis and Forecasting From Satellite Imagery, *Mon. Wea. Rev.,* 103, 420-430, 1975.

Dvorak, V., Tropical Cyclone Intensity Analysis Using Satellite Data, *NOAA Tech. Rep. NESDIS* 11, 47 pp., 1984. [Available from NOAA/NESDIS, 5200 Auth Rd., Washington, DC 20233.]

Goerss, J. S., C. S. Velden and J. D. Hawkins, The Impact of Multispectral GOES-8 Wind Information on Atlantic Tropical Cyclone Track Forecasts in 1995. Part II: NOGAPS Forecasts, *Mon. Wea. Rev.*, 126, 1219-1227, 1998.

Hawkins, J. D., T. F. Lee, K. Richardson, C. Sampson, F. J. Turk and J. E. Kent, Satellite Multi-sensor Tropical Cyclone Structure Monitoring, *Bull. Amer. Meteor. Soc.*, 82, 567-578, 2001.

Katsaros, K. B., E. B. Forde, P. Chang, and W.T. Liu, QuikSCAT Facilitates Early Identification of Tropical Depressions in 1999 Hurricane Season, *Geophys. Res. Lett.*, 28, 1043-1046, 2001.

Kidder, S., M. Goldberg, R. Zehr, M. DeMaria, J. F. W. Purdom, C. S. Velden, N. C. Grody and S. J. Kusselson, Satellite Analysis of Tropical Cyclones Using the Advanced Microwave Sounding Unit (AMSU), *Bull. Amer. Meteor. Soc.*, 81, 1241-1259, 2000.

Krishnamurti, T. N., C. M. Kishtawal, Z. Zhang, T. LaRow, D. Bachiochi, E. Williford, S. Gadgil and S. Surendan, Multimodel Ensemble Forecasts for Weather and Seasonal Climate, *J. Clim*, 13, 4196-4216, 2000.

Kummerow, C., J. Simpson, O. Thiele, W. Barnes, A. T. C. Chang, E. Stocker, R. F. Adler, A. Hou, R. Kakar, F. Wentz, P. Ashcroft, T. Kozu, Y. Hong, K. Okamoto, T. Iguchi, H. Kuroiwa, E. Im, Z. Haddad, G. Huffman, B. Ferrier, W. S. Olson, E. Zipser, E. A. Smith, T. T. Wilheit, G. North, T. Krishnamurti, and K. Nakamura, The Status of the Tropical Rainfall Measuring Mission (TRMM) After Two Years in Orbit, *J. Appl. Meteor.*, 39, 1965-1982, 2000.

Liu, W. T., H. Hu and S. Yueh, Interplay Between Wind and Rain Observed in Hurricane Floyd, *Eos, Trans.*, 81,253-81,257, 2000.

Liu, W. T., Progress in Scatterometer Application, *J. Oceanogr.*, 58, 121-136, 2002.

Rocken, C., R. Anthes, M. Exner, D. Hunt, S. Sokolovskiy, R. Ware, M. Gorbunov, W. Schreiner, D. Feng, B. Herman, Y.-H. Kuo and X. Zou, Analysis and Validation of GPS/MET Data in the Neutral Atmosphere. *J. Geophys. Res.*, 102, 29849-29866, 1997.

Soden, B. J., C. S. Velden and R. E. Tuleya, The Impact of Satellite Winds on Experimental GFDL Hurricane Model Forecasts, *Mon. Wea. Rev.*, 129, 835-852, 2001.

TAO: Terrestrial, Atmospheric and Oceanic Sciences, Special Issue for Applications of the Constellation Observing System for Meteorology, Ionosphere and Climate (COSMIC), 11, 1, 2000.

Velden, C. S., T. L. Olander and S. Wanzong, The Impact of Multispectral GOES-8 Wind Information on Atlantic Tropical Cyclone Track Forecasts in 1995. Part I: Dataset Methodology, Description, and Case Studies, *Mon. Wea. Rev.*, 126, 1202-1218, 1998a.

Velden, C. S., T. L. Olander and R. M. Zehr, Development of an Objective Scheme to Estimate Tropical Cyclone Intensity From Digital Geostationary Satellite Infrared Imagery, *Wea. and Forec.*, 13, 172-186, 1998b.

Chapter 12

Present Day Satellite Technology for Hurricane Research

A Closer Look at Formation and Intensification

E. A. Ritchie, J. Simpson, W. T. Liu, J. Halverson, C. S. Velden, K. F. Brueske, and H. Pierce

INTRODUCTION

Tropical cyclones are the most damaging and destructive weather event on Earth. Happily, forecasting their motion and landfall has improved greatly in the years since the mid-1970s, in considerable part owing to satellite observations and computer models. However, intensity forecasts show much less skill, and the process of hurricane[1] formation involves many conflicting and unverified hypotheses. A major problem has been the lack of observations, particularly observations that resolve mesoscale systems[2]. For example, the Doppler radar

[1] Hurricanes (eastern Pacific, Caribbean, North Atlantic) are known as typhoons in the western North Pacific, and tropical cyclones in the south Indian Ocean, South Pacific, and Australian regions.

[2] Mesoscale weather systems are weather systems that have dimensions of 25 km to 500 km and last for periods of a few hours to a few days. Typically, several data points are required to resolve a system adequately, e.g., to resolve a 25-km feature, observations would need to be less than 5-km resolution.

Hurricane! Coping with Disaster

10.1029/055SP15

network on the U. S. coastline can detect mesoscale systems when they are close to shore. When in mid-ocean, however, fleets of research aircraft, with radar, dropwindsondes and many other remote and *in-situ* instruments have been necessary to detect mesoscale features in developing storms. Deployments of research aircraft are expensive and only rarely do they "catch" a hurricane in its formation or early deepening stage. Additionally, few nations under the threat of tropical cyclones can afford such technology. As a result, in this chapter we use satellite tools that are capable of measuring or deducing events on the mesoscale to help clarify issues associated with hurricane formation and intensification. Because such technology can be made available to many nations via the Internet, it is a potentially valuable source for improving tropical cyclone warning and prediction worldwide.

In general, numerical analyses enable scientists to examine the larger-scale environment involved with hurricane formation. Along with the National Hurricane Center (NHC), we define formation this way: When a west wind appears to the south of center in a tropical wave in the easterly trade-wind flow and a closed low-pressure area exists, the system is called a Tropical Depression (TD). Here we describe two cases of African origin, both of which reached TD designation. The first, after struggling for a week, finally intensified to become destructive Hurricane Floyd of 1999. The other was an initially stronger tropical depression but did not develop the necessary warm core and gradually dissipated.

We present the material as follows: a background review of the present study allows us to make physical connections that cannot yet be routinely validated by direct observation. We show some prior results that help clarify the present study. We then describe our approach and data used. Two case study results follow, along with discussion of the physical mechanisms as evidenced and hypothesized. We conclude with summary remarks and plans for future research. Because available satellite tools for research and forecasting are described in Chapter 11, we discuss them only as they directly affect future mesoscale research, as we foresee it.

BACKGROUND

In 1974, Frank and Hebert showed that waves or *seedlings* moving westward off Africa are the source of more than half the tropical cyclones in the Atlantic. A 30-year summary is even more revealing: From an average 61 Atlantic tropical waves per year, only 11 become TDs of African origin, of which just over half become named storms. At the same time, the distinctive physical conditions that separate developing from non-developing disturbances have not been resolved, despite numerous research efforts. It is known that the large-scale environment needs to be favorable for a hurricane to develop. These conditions include well-mixed ocean temperatures of greater than 26°C to a depth of about 60 m, weak vertical wind shear[3], conditional instability[4] of the air, and high midlevel humidity [Gray, 1998]. Many of these conditions are met much of the time over large regions of the tropical oceans, whereas hurricanes form only rarely.

In the 1950s, it was understood that a mechanism to develop the intense high-level warm core of 10-18°C, as measured by aircraft, was necessary to support the intense low-level pressure gradients and high wind speeds of the hurricane. It was recognized that ascent of the mean tropical environment would support only a modest warm core associated with a central pressure fall to about 990-995 mb. The use of nearly undiluted "hot towers" to provide the warm core of the *mature* hurricane heat engine[5] was well understood and quantitatively documented beginning in the 1960s, with the first aircraft flights into Hurricane Daisy [Riehl and Malkus, 1961]. What remained a puzzle, however, was how the hurricane heat engine actually got started. Joanne Malkus (Simpson) and Herbert Riehl postulated that there must be an extra heat source from the ocean to *start and accelerate* the hurricane engine. Because cloud clusters with tall convective towers flourish in quantity over the tropical Atlantic, but fewer than ten per year become hurricanes, "Why are there so few hurricanes?" became the vital question.

Using *in-situ* aircraft observations from hurricane reconnaissance missions in the late 1960s through early 1980s, Bill Gray [1998] identified the presence of a "wind surge," a region of higher low-level winds that developed within approximately 200 km of the center of the disturbance during genesis. He postulated that these wind surges are the result of small-scale imbalances in the pressure-wind balance. He further proposed that the wind surges result in externally forced convergence into a pre-existing mesoscale vortex, resulting in the formation of long-lasting deep convective towers — the "hot towers" identified earlier by Malkus and Riehl as necessary to develop the hurricane warm core. It is possible that these wind surges provide the missing "kick start" for the hurricane heat engine for which they were searching (see Chapter 1).

In the early 1980s, Wayne Schubert and colleagues [1980] and Vic Ooyama [1982] demonstrated that it was necessary to increase the "inertial stability" of

---

[3] Vertical wind shear is present when the horizontal wind changes with height. Strong shear can disrupt the vertical cloud organization in developing systems, or displace the warm core from the center of a mature tropical cyclone, which has detrimental effects on its structure and intensity.

[4] Conditional instability exists when the upward displacement of an unsaturated air parcel would undergo a downward restoring force, while a saturated parcel would be buoyant and, therefore, would continue rising.

[5] The hurricane acts like a very inefficient heat engine. It converts a very small percentage of the heat energy received from the ocean into the hurricane winds.

the atmosphere[6] to retain the heat released by the cumulus convection. Inertial stability can be demonstrated by dropping a small stone into a deep bowl of quiet water. Gravity waves are seen at the surface propagating outward. However, if the water is stirred strongly to create a vortex and a stone is dropped in again, the waves will be trapped and cannot propagate outward. In the atmosphere when the background vorticity is strong, the atmosphere becomes inertially stable in that region. Then, if there is strong convection, the released latent heat is trapped in a similar manner to the gravity waves in the bowl of water, and a warm core may accumulate. If the atmosphere is not inertially stable, the latent heat released by the clouds will propagate away as gravity waves. The Rossby radius of deformation, L_R, is a measure of the horizontal scale on which the atmosphere is inertially stable. In the ambient tropics L_R is large, about 2000 km, so the inertial stability is low. L_R must be decreased to about 300 km or so for a hurricane-sized vortex to be stable.

In the 1980s, Mike Fritsch and others [1994] studying long-lived convective cloud systems over the United States discovered that the stratiform anvil of such systems was an ideal environment for the growth of midlevel (~ 4-8 km up in the atmosphere) vortices of about 300 km in diameter. These cloud systems lasted long enough (12-24 h) for the vortex to stabilize so that, when the anvil dispersed, the vortex remained coherent. Chen and Frank [1993] hypothesized that the stratiform anvil provided a unique local environment where L_R was reduced, thus, significantly increasing the local inertial stability. In this atypical environment, the differential warming in the cloud and cooling below, due to precipitation processes, created a temperature gradient maximized near cloud base. Potential vorticity was, therefore, enhanced at cloud base, and the stretching due to mesoscale processes within the stratiform cloud resulted in stable vortices that were the size of the anvil region. I (Ritchie), with my advisor Greg Holland, documented in 1997 that the numerous similar cloud systems in the tropical Pacific also develop stable vortices. We thought it was probably due to similar processes. Many field experiments using aircraft to collect data in these tropical oceanic mesoscale convective systems provide evidence that *midlevel vortices in the stratiform anvil regions are the norm rather than the exception.* These mesoscale cloud systems provide a mechanism for developing vortices on a smaller scale than the ambient tropical L_R. Thus, they may provide important pre-conditions for the development of a hurricane. However, the circulation associated with these midlevel vortices is strongest at around 6 km up in the atmosphere and the strength diminishes with decreasing height. In addition, we deduced from the aircraft observations of the developing Typhoon Irving that, while numerous vortices may develop in the midlevel anvil regions, very few develop distinct circulations at the surface. For

[6] There are two basic types of stability used in meteorology. Static stability is obtained from a thermodynamic diagram to determine whether the atmosphere is favorable to cumulus cloud development. Inertial stability depends on the motion field alone.

example, in the case of Typhoon Irving, as many as four vortices were identified at 500 mb (~ 5 km) in one aircraft flight, whereas only one distinct vortex was identified at 700 mb (~ 3 km) from dropwindsondes [see Figure 10 from Ritchie and Holland, 1997]. Although the higher resolution of the aircraft flight-level data, compared with the dropwindsonde data, may account for some of this discrepancy, it is important to determine how midlevel vortices might develop a circulation at the surface, with enough strength to generate sufficient sea-air fluxes of heat and moisture to begin the intensification cycle.

Greg Holland, Joanne (Malkus) Simpson, and I recently developed the hypothesis that vortex mergers could be a way to develop the circulation down to the surface. Greg and I applied theoretical model studies, which showed that when two vortices of unequal strength interact, one could absorb vorticity from the other to achieve tropical cyclone formation. In applying our vortex models to real cloud systems, we found that when the background vorticity was sufficiently high, the observed midlevel vortices could live long enough to merge. Models suggested that the stronger vortex thickened in the vertical and was, thus, able to reach the surface while the weaker vortex stretched out laterally into a vorticity filament,[7] similar to a rainband (see Figures 1A-1F). Such a hypothesis could explain why numerous vortices might be observed at 500 mb but fewer at lower levels because they would be the products of vortex mergers. However, there are few datasets that can adequately resolve mesoscale vortices, or the convective processes that produce them, so that this hypothesis can be tested.

Joanne Simpson and some colleagues acquired a particularly exciting data set during the formation and intensification of Tropical Cyclone Oliver in a monsoon environment just off the east coast of Australia, during the 1993 TOGA COARE[8] field campaign. In fact, the analyses of that data set strengthened the evidence for the importance of mesoscale vortex mergers in the formation and intensification of tropical cyclones. Joanne Simpson and others [Simpson et al., 1997] documented the development and interaction of two large mesoscale convective systems seen first on satellite images in flight. These convective systems formed in association with midlevel vortices within a low-level monsoon circulation, as shown from using data from the two NASA aircraft in combination with satellite infrared (IR) imagery and a ground-based C-band radar located at Willis Island (Figure 1). While the aircraft documented the three-dimensional

[7] During merger, vortices move closer together and become distorted. Thin "filaments" of energy are drawn away from the main body of each vortex. These filaments dissipate so that the energy left in the merged system is reduced. As the inertial stability of the background environment increases, less energy is lost to filamentation during merger resulting in a stronger merged vortex.

[8] TOGA COARE stands for Tropical Ocean Global Atmosphere Coupled Ocean Atmosphere Experiment.

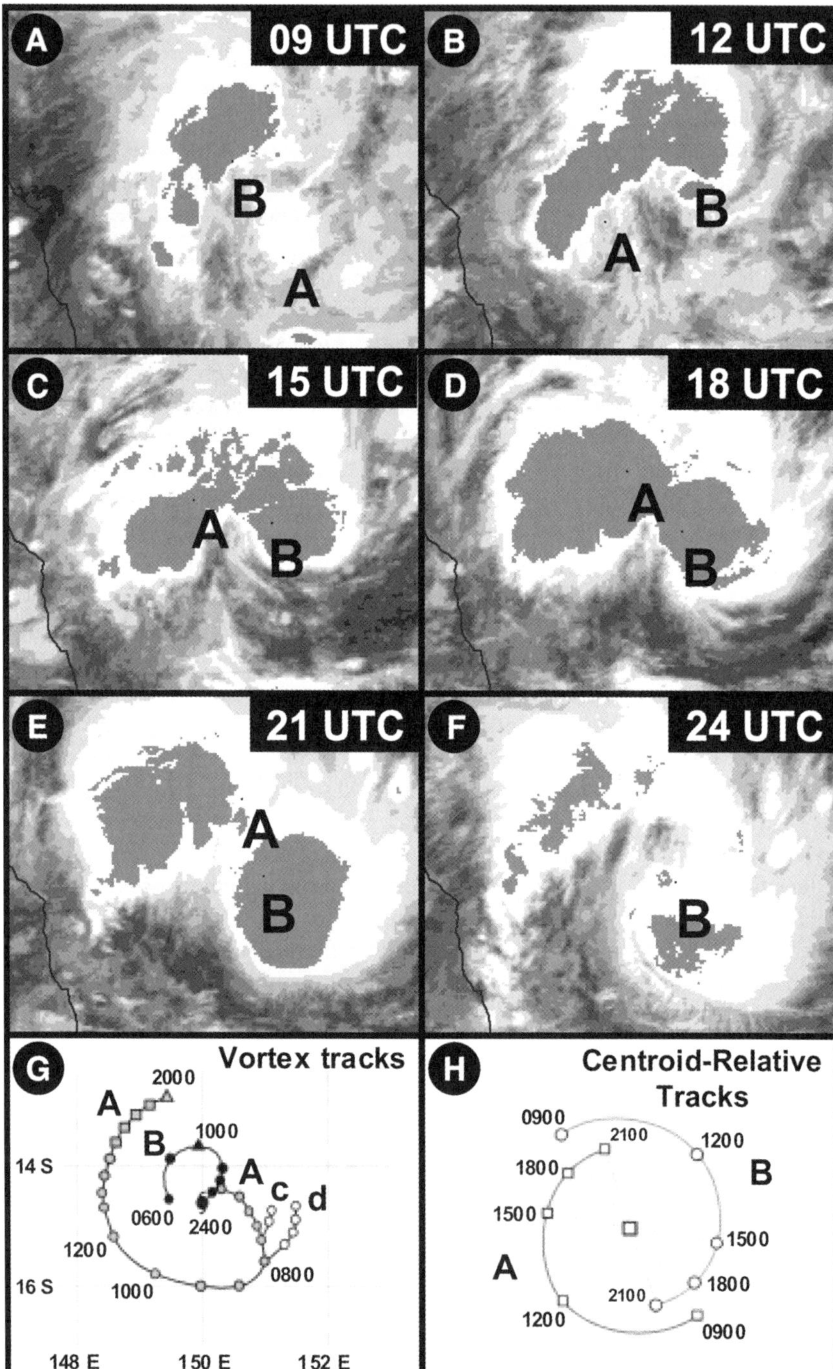
A
09 UTC
B
12 UTC
C
15 UTC
D
18 UTC
E
21 UTC
F
24 UTC
G
Vortex tracks
2000
1000
0600
2400
1200
1000
0800
14 S
16 S
148 E
150 E
152 E
c
d
H
Centroid-Relative
Tracks
0900
2100
1200
1800
1500
1500
1800
2100
1200
0900

structure of the vortices embedded within the strong low-level monsoon circulation, the evolution of the cloud and precipitation patterns during the formation period were documented using data from the Willis Island radar and IR imagery. When these separate data sets were combined, the results showed that two major vortex interactions were important in the development and intensification of TC Oliver. The first was an interaction between the two midlevel vortices associated with two spectacular mesoscale convective systems that preceded TC formation. The vortices were tracked by the precipitation returns from the radar (Figure 1G). When plotted in a centroid-relative frame of reference, a classic picture of two interacting vortices emerged (Figure 1H). As vortex A (the weaker vortex) rotated about the west and north side of vortex B (Figure 1G), the associated mesoscale convective system became sheared into a rainband-like feature similar to a filament (Figures 1E, 1F). During this time, vortex B executed a much tighter loop, strengthened, and moved to the center of the low-level monsoon circulation in which the entire system was embedded.

Model studies show that an interaction between unequal midlevel vortices will result in the stronger vortex executing a small inner loop while absorbing most of the circulation associated with the weaker vortex, stretching the remnants out into a long filament of vorticity. The victorious vortex will be strengthened by this interaction and thicken vertically. In addition, an interaction between a stronger smaller midlevel vortex and a weak surface monsoon-like circulation results in the monsoon circulation wrapping around the midlevel vortex [Ritchie and Holland, 1997]. This results in a reduction in size, an increase in strength of the surface circulation, and the development of a small, confluent region of stronger winds — a *wind surge* similar to that observed by Gray [1998]. In the case of Oliver, this second vortex interaction, between the victorious midlevel vortex and the low-level sheared monsoon circulation, was also important in the development of Oliver. As the pre-Oliver vortex tightened and strengthened, it probably accelerated the low-level inward flow in the monsoon trough, providing enhanced heat and moisture flux into the core of the developing storm. During the six hours the two NASA aircraft were in the storm, a partial eyewall developed on the south edge of the two interacting cloud systems. The deep convective towers indicative of forced low-level convergence, possibly due to a wind surge, were observed on aircraft instruments and satellite imagery while dropwindsondes from the aircraft recorded the concurrent warming in the nascent eye.

Figure 1. A) - F) The locations of the two mesovortices A and B during the development of Tropical Cyclone Oliver superposed on three-hourly satellite imagery for the period 0900 UTC February 4 to 0000 UTC February 5, 1993; G) Tracks of four of the vortices obtained from radar data. The positions are not evenly spaced and so times (in UTC) of some of the vortex positions are marked; H) three-hourly centroid-relative tracks of mesovortices A and B from 0900 UTC to 2100 UTC February 4th [Simpson et al., 1997].

The mesoscale vortex interactions associated with genesis in both Typhoon Irving and Tropical Cyclone Oliver occurred in strong monsoon trough regions. We wondered whether such a genesis mechanism is applicable only to monsoon environments. Would any similar interactions be found in the Atlantic region, where hurricane formation is associated with an easterly regime? In addition, the aircraft and radar data collected during the development of Typhoon Irving and Tropical Cyclone Oliver were vital in exploring vortex interaction associated with genesis. However, these data are not routinely available.

Thus, our next step was to inquire whether: (1) mesoscale vortex interactions may be important in tropical cyclone formation elsewhere; and (2) whether it is possible to identify key physical mechanisms associated with tropical cyclone formation using data from satellite technology without instrumented aircraft or surface radar. In order to proceed, two pieces of information reviewed in this section are used in the following sections to infer vortex behavior from cold high cloud behavior on satellite infrared imagery. The first is that missions flown into mesoscale convective systems in the Pacific during the TCM-92[9] and TCM-93 field campaigns as well as TOGA COARE revealed that midlevel vortices are commonly observed in association with mesoscale convective systems. The second is that midlevel vortex interaction and merger is a mechanism to maintain and strengthen vorticity in a cloud cluster, as documented in both Typhoon Irving and Tropical Cyclone Oliver. Because of these pieces of information, we will infer vortex behavior from mesoscale convective systems observed in satellite IR imagery, in combination with other types of remotely sensed data, such as scatterometer surface winds, AMSU brightness temperatures when available, and data from the Tropical Rainfall Measuring Mission (TRMM) satellite.

APPROACH AND DATA USED

Our approach is to build a time sequence of three-dimensional mental pictures of each hurricane seedling studied. The goal is to understand and document the processes by which the system undergoes the observed changes between one and the next time interval using all available observations. Global numerical model analyses are used to describe the storm environment and to tie the satellite observations together. For all satellites mentioned here, their sensors and products, the reader is referred to Chapter 11, which explains how they work and are used in tropical cyclone studies.

Geostationary Satellite Data

The new generation of geostationary satellites offers multispectral infrared (IR) and visible imagery at intervals of 30 minutes or less. These images can be

[9] TCM stands for Tropical Cyclone Motion. The numbers are the years.

sequenced to create movie loops, which can be animated to depict the temporal evolution of mesoscale structures of interest in this study. The IR images can be color-enhanced to depict the cold (high) convective tops. Movie loops of IR imagery from both GOES-EAST and METEOSAT-7 were created to follow the hurricane seedlings as they emerged off Africa. Together, these two satellites view the entire Atlantic Ocean and can be used to help identify mesoscale convective system activity (a.k.a. midlevel vortex activity) as well as the critical convective bursts (hot towers), which act to develop the warm core structure necessary for further development of the tropical disturbance.

Earth-Orbiting Satellite Data

Microwave Imagers. The advent of passive microwave satellite instruments provided the opportunity to see down into the rain and ice structure of the atmospheric cloud systems over oceans and determine how this structure evolves with time. This provides valuable clues to the differences between developing and non-developing cases. The two passive microwave imagers used in this study are the Special Sensor Microwave Imager (SSMI), flown on the Defense Meteorological Space Program (DMSP) satellite; and the Microwave Imager (TMI), flown on the Tropical Rainfall Measuring Mission (TRMM) satellite. The TMI is similar to SSMI except that it has an additional channel and has more than twice the resolution of that of SSMI because of a much lower satellite orbit. We use the microwave imager data mainly in connection with the data sets on the Naval Research Laboratory's (NRL) website (see below). Chapter 11 describes how the satellite instruments work.

Scatterometer Data. QuikSCAT has been a revolutionary and valuable tool in that it provides observations of winds at the sea surface of the ocean [Liu, 2002]. QuikSCAT is an active microwave sensor that observes the sea surface roughness to estimate the wind speed and direction. QuikSCAT has a continuous 1800 km swath that covers 93% of the global ocean in a single day with a standard wind product of 25 km spatial resolution. Wind vectors retrieved from QuikSCAT generally exceed accuracies of 2 m/s in speed and 20° in direction, except under heavy rain conditions. However, we do retain the potentially rain-contaminated data in this investigation of early development.

Microwave Sounders. The Advanced Microwave Sounding Unit (AMSU-A) is a next-generation passive microwave radiometer [Brueske and Velden, 2001]. By virtue of its increased horizontal resolution (48 km at nadir vs. 110 km for the original Microwave Sounding Unit) and improved radiometric accuracy, the AMSU-A provides improved measurement of tropical cyclone warm cores in the absence of traditional aircraft or radiosonde observations. The fifteen AMSU-A channels map the three-dimensional thermodynamic structure of the inner core region, with key information provided at approximately 550, 350, 250, and 150 mb by Channels 5, 6, 7, and 8, respectively. Because the AMSU-A scan swath is approximately 2100 km wide, the tropical cyclone warm core (compared with the environment) can usually be directly evaluated from the radiance data without recourse to model data. However, since most tropical cyclone eyes are less than 48 km across (i.e., less than the resolution of the instrument), the peak warm core values may be underestimated. Corrective measures are discussed in Chapter 11. Channels 5 and 6 are particularly useful for measuring hydrometeor-induced scattering (and associated cold anomalies) in regions of organized convection and stratiform clouds. The expected warm temperature anomalies (1-2 K) in the stratiform anvils are just above the noise limits of the AMSU-A instrument and so are not generally as well observed as the cold pools. These profiles provide evidence of thermodynamic structure in the mesoscale convective systems supportive of, and consistent with, the presence of midlevel vortices. They are presented as additional evidence to the IR imagery that midlevel vortices are probably present.

Some Useful Websites

Three websites were major sources of data for this study. The Cooperative Institute for Meteorological Satellite Studies (CIMSS) started a website in 1994. This website contains real time, and historic satellite images and movie loops of tropical cyclones around the world. In addition, CIMSS has pioneered the calculation of satellite-derived wind vectors that fill much of the data-void ocean regions. These images, as well as those from the AMSU instruments are available in near real-time.

The NRL tropical cyclone website makes available visible and infrared GOES data, together with TMI and SSMI, whenever microwave sensors pass over tropical cyclones or potential tropical cyclones. They produce collocated images on the same base maps, making it easy to assimilate several types of data at once. Combinations of imagery are made with the passive microwave sensors' high-resolution channel (85 GHz) to enhance the high towers and with the 37 GHz channel to bring out the low-level rainfall.

The National Hurricane Center of the National Oceanographic and Atmospheric Administration (NOAA) website provides archival information including the best track data and preliminary reports on individual tropical cyclones for the Atlantic and Eastern Pacific basins. During the hurricane season, this site can also be accessed for current information and forecasts of existing tropical cyclones.

The Navy Operational Global Atmospheric Prediction System (NOGAPS)

Gridded analyses from the NOGAPS model are used to represent the large-scale environment surrounding each African "seedling." These analyses of winds, pressure, temperature and humidity are created twice daily, using conventional data and satellite-derived wind data. The products are available on a 1-degree latitude, longitude grid at the surface and 19 pressure levels between 1000 mb and 10 mb, which is above the top of all hurricanes. Here we use the fields to calculate the temperature and moisture (combined into a variable called equivalent potential temperature or θ_e), the vorticity and divergence of the wind, vertical wind shear (the difference between the upper and lower environmental winds over the seedling), and the total wind field.

RESULTS AND DISCUSSION

Physical Conditions of the Atlantic Environment

In the Northern Hemisphere summer, the warm African continent draws in oceanic air from the southwest. At the northern edge of the southwest flow, a "monsoon trough" of low pressure is formed, often with easterlies to the north. A boundary with easterlies to the north and westerlies to the south has horizontal wind shear in the cyclonic (counterclockwise in the Northern Hemisphere) sense. Observations show that the southerly African monsoon flow often reaches 12°-14°N. In the two cases we present here (TD 2 of 2000 and Hurricane Floyd of 1999), the monsoon trough also extends westward out over the ocean as far as 30°W, more than 1500 km. Figure 2 shows the trough over water, perturbed by the early Floyd vortex with the northeasterly trade winds of the Atlantic at the left side of the panels. Both seedlings studied here originated as waves over Africa [Lau and Lau, 1990 and references therein] and were observed in QuikSCAT data as surface vortices just as they moved westward off the coast. Thus, our study breaks down into two parts. First, how does each system become a TD? Second, how does one of these TDs become a major hurricane and why does the other fail?

Figure 3 shows a profile of temperature and dewpoint plotted on a thermodynamic diagram from radiosondes released from the Cape Verde Islands. The sounding was made when the pre-Floyd vortex passed to the south of Cape Verde. There is a pronounced "inversion" of temperature, a region where temperature increases with height instead of the normal cooling with height, beginning at about 2 km (920 mb) above the surface. In an inversion, the dewpoint is much lower than the temperature, which means the air is dry and statically stable, and is unfavorable for cumulus-type clouds. Below the inversion, the temperature and dewpoint are equal, which means the air is saturated, and stratocumuli exist in such regions. The saturated layer from about 5-5.5 km (540-500 mb) in Figure 3 indicates another cloud layer higher up. The curved line on the right of Figure 3 shows the temperature profile that a saturated cloud parcel would have if it rose without being diluted by the surrounding drier air. If that curve, called a *saturated adiabat*, lies to the right of the air temperature sounding, it means that a saturated, undiluted air parcel will be warmer than the surroundings and, therefore, buoyant, so that it will accelerate upward. But if the curve is to the left of the air temperature sounding, the air parcel is not buoyant and will sink. This diagram, thus, shows two important features. First, the near surface air will be unable to rise through the inversion, unless it is given a strong upward "push." Second, the saturated parcel will not be able to rise very high because it will mix with the surrounding dry air in the inversion layer.

One factor, which nearly always has a pronounced effect upon hurricanes, is the sea surface temperature (SST). Tropical hurricanes rarely, if ever, form at SSTs lower than about 26.5°C. The reason is the nonlinear relation between temperature and the amount of water vapor the air can hold. SSTs below about 26°C do not

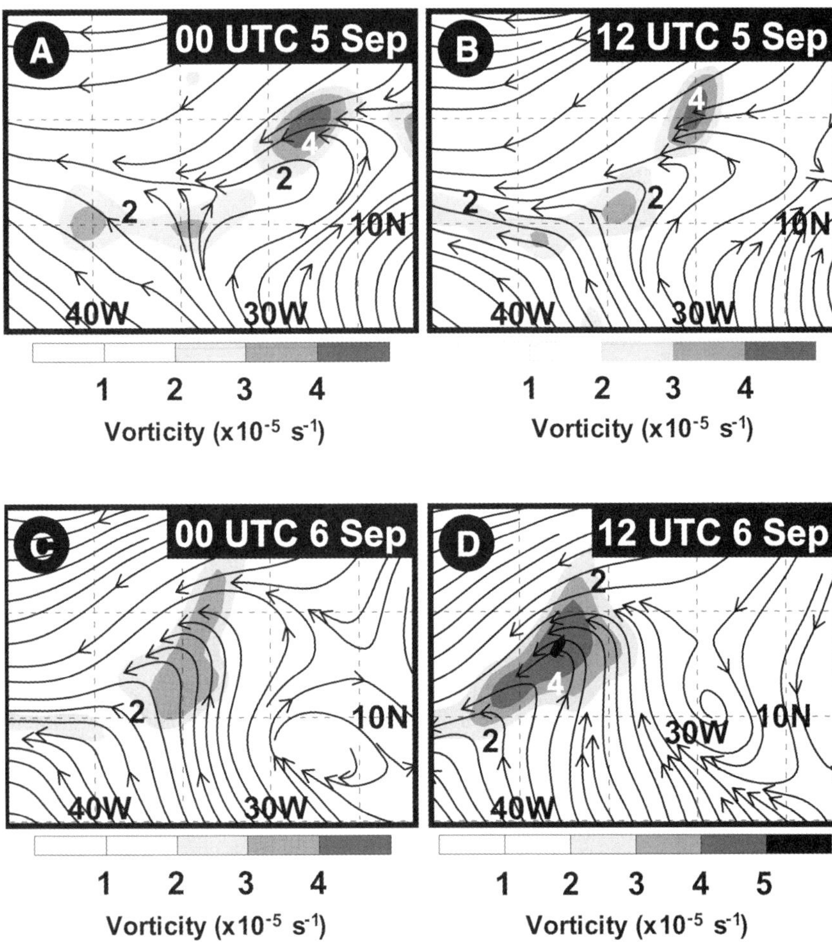

Figure 2. Relative vorticity plots at 700 mb (shading begins at 1×10^{-5} s^{-1}, increments of 1×10^{-5} s^{-1}) and 1000 mb streamlines from NOGAPS for A) 0000 UTC September 5; B) 1200 UTC September 5; C) 0000 UTC September 6; and D) 1200 UTC September 6.

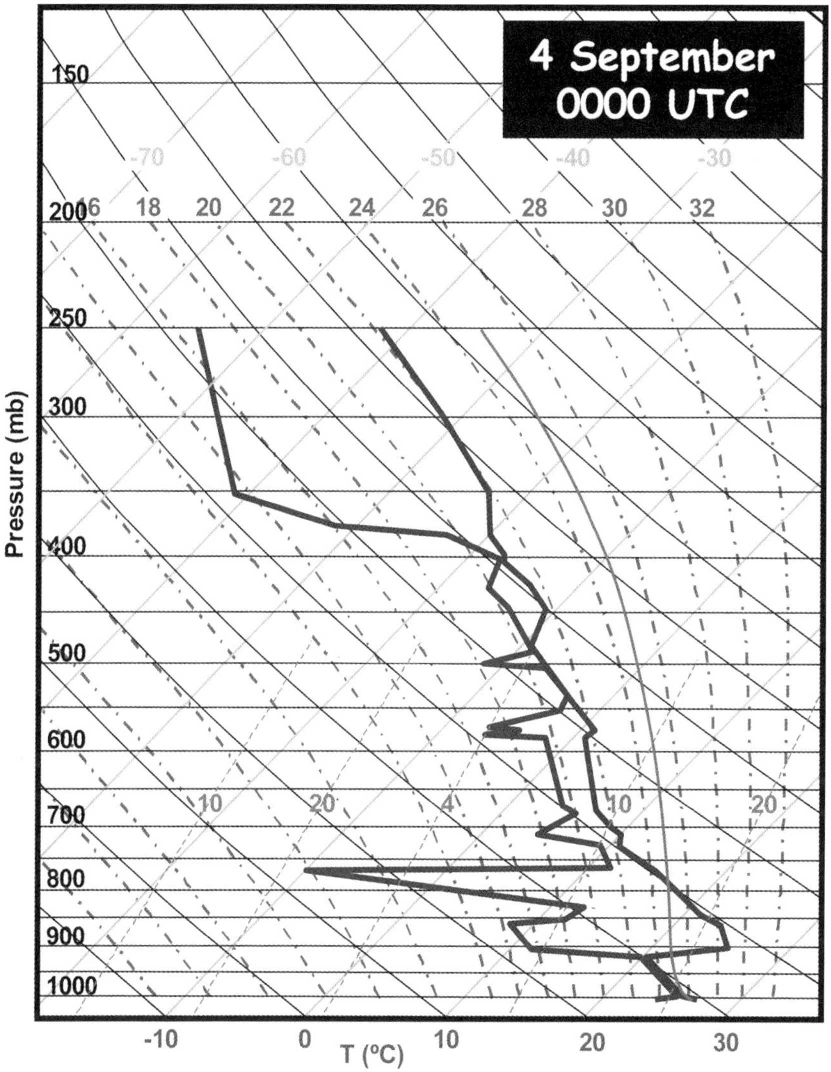

Figure 3. Cape Verde sounding for 0000 UTC September 4th, showing strong low-level inversion but a fairly moist layer from 760 mb to 400 mb. The thin solid lines that slope diagonally to the right are constant temperature lines. The black diagonal lines that slope to the left are dry adiabats — they show how the air temperature would change inside a rising or descending unsaturated air parcel. The curved dashed lines are called moist adiabats — they show how the air temperature would change inside a rising or descending saturated air parcel.

release enough latent heat of condensation for initiation of the hurricane heat engine. This is part of the reason why, all other factors equal, an increase in SST often leads to intensification, while colder water nearly always favors weakening. We will see that in early stages, even before a pre-hurricane disturbance develops a high-level warm core, higher or increasing SSTs favor mesoscale interactions that keep the pre-hurricane circulation alive for long periods over the ocean.

Related to the above, another factor that must be right for hurricane formation and intensification is the total energy content of the air at low and middle levels of the atmosphere. Meteorologists measure the total air energy content by a value called "equivalent potential temperature" or θ_e. This quantity is related to the temperature the air would have if all the water vapor it contains has been condensed out to warm the air. One of the main results from the National Hurricane Research Project in the 1960s was related to θ_e requirements for development. Not only must the surface layer θ_e be high enough, on the order of 352 K or more, but also the difference between θ_e at the surface and that at middle levels must be greatly reduced in comparison to the mean tropical air mass. In the Caribbean, the mean difference is about 12 K, and this must be reduced to about 8 K or less for hurricane formation. That is, the SST must be high enough and the middle levels must be moist enough for many undiluted convective clouds to penetrate upward through the entire troposphere. A sounding such as that shown in Figure 3 is marginal for both surface and middle level requirements.

Our first example, Floyd 1999, reached Saffir-Simpson Category 4 off the Bahamas, killed 57 people, and created a \$3-6 billion flood disaster on the U.S. coast, particularly in North Carolina. However, the initial development of Floyd, especially during the first week out of Africa, was characterized by a constant struggle for survival.

The Birth and Early Life of Floyd 1999, September 2nd-5th

The track of Floyd is shown plotted on underlying weekly average sea surface temperatures[10] in Figure 4A. The O's indicate where NHC declared Floyd a TD, which occurred halfway across the Atlantic at 1800 UTC on September 7th. The X's follow the center of the surface vortex observed in QuikSCAT combined with the IR movie. The pre-Floyd disturbance moved off Africa on September 2nd as a weak surface vortex observed in scatterometer data (Plate 1A). Moderate convection was organized into a mesoscale convective system to the northeast of the center.

During the first few days after pre-Floyd moved off the coast of Africa, the large-scale wind environment was unfavorable. The ambient vertical wind shear[11]

[10] Derived from satellite imagery discussed in Chapter 11.

[11] The ambient vertical wind shear was calculated from NOGAPS analyses by subtracting the 850 mb winds averaged over a 12° lat by 12° long grid centered on the disturbance from the 250 mb winds averaged similarly.

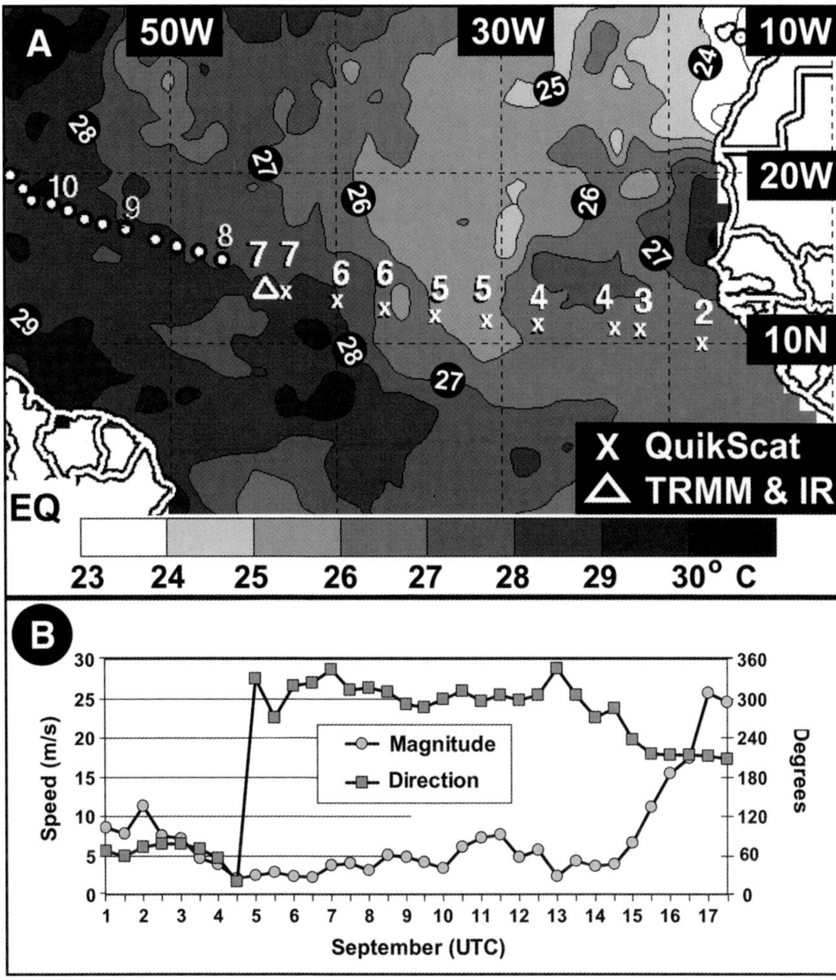

Figure 4. A) Sea surface temperatures with positions of Hurricane Floyd (September 7-17, 1999) superposed; and B) Time series of area-averaged shear over Floyd (September 1-17, 1999).

was greater than 7 m/s until 1200 UTC on September 3rd (Figure 4B). However, by early on September 4th, the vertical wind shear reduced to less than 5 m/s and ceased to be a problem for pre-Floyd. The sea-surface temperatures just off Africa were also marginal for development, less than ~27°C. Furthermore, soundings from Nouadhibou (21°N, 17°W), for the week prior to pre-Floyd's exit off Africa, indicate a deep and extremely dry layer of air exiting off the African coast (not shown). This dry air is indicative of the Saharan air layer or SAL, which has been observed to extend for many hundreds of kilometers across the Atlantic Ocean

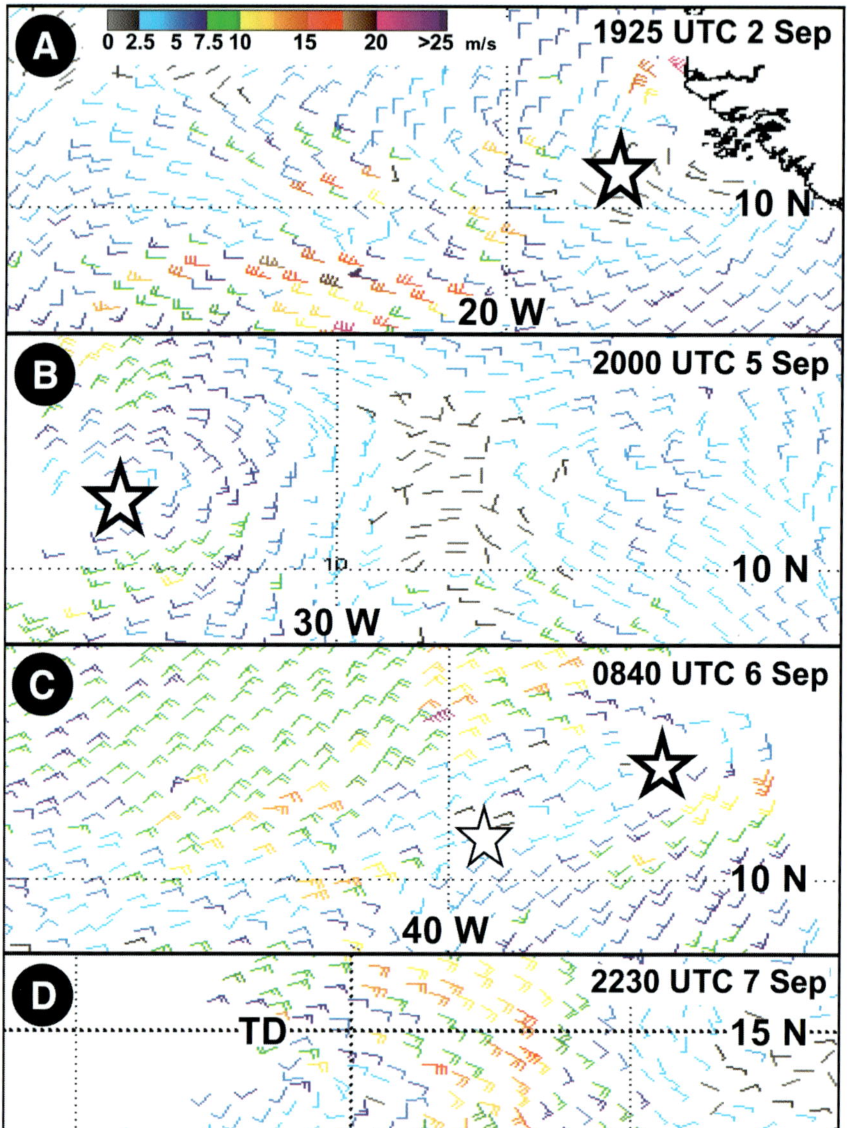

Plate 1. QuikSCAT scatterometer winds for A) 1925 UTC September 2nd, 1999; B) 1950 UTC September 5th; C) 0840 UTC September 6th; and D) 2230 UTC September 7th. A full barb = 5 m/s, and a triangle = 25 m/s. Black wind barbs are possibly rain contaminated.

and may be an inhibiting factor for hurricane development. New products are currently being developed at CIMSS to track this air layer via the dust that travels with it [J. Dunion, personal communication]. In the case of Floyd, this dry air and accompanying wind shear may also have been a factor in inhibiting deep convection during the early part of Floyd's development.

Despite the hostile environment, the vortex moving off Africa appeared quite deep, extending down from the middle levels of the atmosphere in NOGAPS analyses (not shown) with an associated surface circulation of 4-6 m/s indicated in the scatterometer data (Plate 1A). Some convergence and low-level inflow was associated with the surface vortex, so that the forcing was there for a few cumulus clouds to penetrate the inversion some distance (200 - 400 km in Figures 5A, 5B) from the surface circulation center. We speculate that this saturated air in cumulus clouds, mixed with the drier air of the inversion, gradually moistening the dry layer. An IR movie (not shown) suggests that some of these towers reached 12-14 km, but that they disappeared within 12 h. The rotational winds of the surface vortex seen in the scatterometer data weakened only slightly between September 3rd and 4th, although little convection developed immediately over the surface vortex center. This is important, as without convective forcing, some spin-down of the vortex due to frictional convergence would be expected to occur. However, Figures 5A-5D show a mesoscale convective system that developed or redeveloped each day in the same relative location north of the surface vortex center. Although there were no AMSU data to corroborate the thermodynamic structure within these cloud systems, the anvils were large enough (~300 km) to sustain midlevel vortices. These could be the cause of the sustained strength of the surface vortex against frictional spin down.

By September 5th, there were two midlevel vortices detectable in both cloud motion of the IR movie (Figure 5D) and in the NOGAPS analyses (Figure 2B). One was clearly associated with the mesoscale convective system to the north of the surface circulation center, and the other was a midlevel reflection of the surface circulation tracked in QuikSCAT. The two midlevel vortex centers appeared to merge during September 5th, and the surface vortex winds intensified by 2-4 m/s to 6-8 m/s (Plate 1B). In addition, pre-Floyd crossed an ocean area with SST less than 26°C (Figure 4A) under unfavorable conditions for penetrative convection. The IR movie showed no new convective towers during this period. However, despite the lack of centralized convection in the disturbance during the period September 2nd-5th, particularly on September 5th, the surface circulation did not lose rotation speed from friction, but in fact, its speed increased. The circulation probably was sustained and even strengthened by vortex mergers, including the one observed during September 5th in the IR movie. By late on September 5th, pre-Floyd again moved over increasing SSTs. A pool of warmer θ_e air appeared to the north of the system, partially built by the swirling flow around the east and north side of the vortex (not shown).

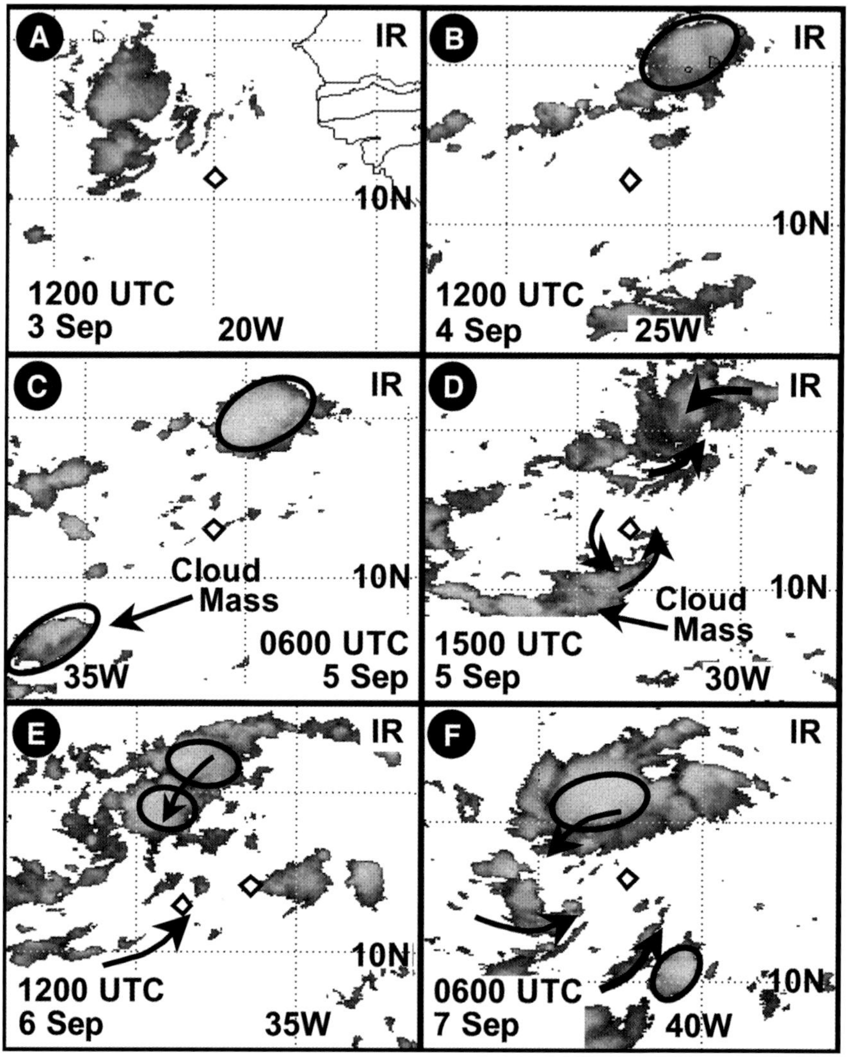

Figure 5A. A – F) Sequence of METEOSAT IR. The time of image is labeled on each panel. The estimated locations of surface centers based on QuikSCAT data are indicated by (◇). The thick black ovals indicate inferred vortices, and black arrows show where cloud rotation was observed in the IR movies. Grid lines are every 5 deg.

Intensified Convection and a Low-Level Inburst of High θ_e Air, September 6th-9th

Thus far, we have hypothesized that vorticity enhancement through vortex merger compensated for the lack of convection to sustain pre-Floyd's primary circulation from September 2nd-September 5th. However, no studies to date have

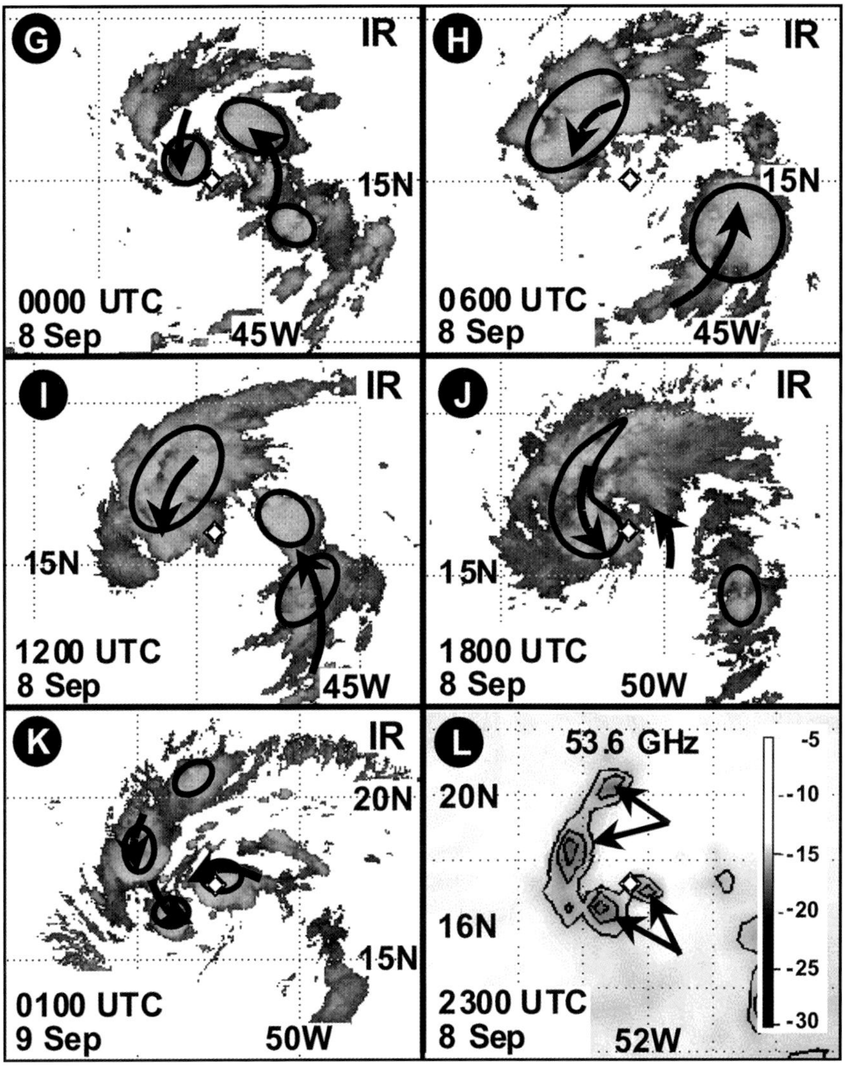

Figure 5B. G- K) As per Figures 5A, A–F, and L) AMSU-A 53.6 GHz (~550 mb) image corresponding to K). AMSU grid lines are every 2 deg.

shown that vortex merger alone can start the hurricane heat engine. The story of hurricane development is the story of the successful interaction of many different processes, each acting on a different spatial and temporal scale than the others. The technology that became available in the 1990s has enabled us to observe several of these interactions that previously had only fragmentary data, leaving many hypotheses difficult to test. To understand how Floyd got the hurricane "heat engine" started, we must look at how events of different scales relate to each other.

By September 6th, the pre-Floyd circulation had moved over sea surface temperatures of higher than 28°C, so that increasing fluxes from the sea raised the θ_e of the inflowing air. This warming, moistening, and destabilization of the air caused convection to be much more active, with multiple developments within about 4° latitude of the center of the surface circulation (Figures 5E, 5F). As existing mesoscale systems dissipated, new ones were already developing. In addition, the scatterometer image for September 6th, 0840 UTC (Plate 1C) indicates not one but two surface vortices embedded in a shear zone oriented E-NE to W-SW. Prior to this time, no double surface vortices were detected on scatterometer images. Since the eastern (right) vortex is consistent with the motion of the pre-Floyd surface vortex prior to this time, it is probably the same vortex QuikSCAT first detected on September 2nd.

Thus, we postulate that the western (left) vortex originated from multiple cloud developments to the southwest of the pre-Floyd (eastern) vortex. This cloud mass can be seen in Figures 5C and 5D, near 9°N, 35°W. Between Figures 5D and 5E, a 21-h period, the cloud mass develops several mesoscale convective systems and propagates northward through 13°N, 40°W. Evidence of midlevel vortices in this region is the multiple mesoscale convective system developments observed in the IR imagery. We speculate that through vortex interaction processes, the vorticity in this region became strong enough so that the circulation developed down to the surface, where it became the western circulation on Plate 1C. However, regardless of how it got there, from 1950 UTC on September 5th to 0800 UTC on September 7th, there are several pieces of evidence supporting a merger between these two surface vortices (e.g., Figure 2). First, there is a 1.3° northward "jump" in the track of pre-Floyd (e.g., Plates 1C, 1D, and Figures 5E, 5G) consistent with interaction with another vortex to its west. Second, by 0800 UTC on September 7th (Plate 1D), the QuikSCAT data show a single vortex with an area of increased and highly sheared winds on its eastern edge. This is a match with the merging vortex model [Ritchie and Holland, 1997], a classic picture of a sheared "losing" vortex being carried from the west of, around the south and east of, and then merged into the victor. It is likely that these vortex mergers produced the intensification of the surface swirling winds to greater than 8 m/s by late on September 7th, when Floyd was declared a TD by NHC.

Starting at about 0000 UTC September 8th, mesoscale systems began to develop, one to the east-southeast and one to the west-northwest of the center of the surface circulation. The cloud system to the east-southeast was forced by the low-level convergence (calculated from the QuikSCAT data) due to the increased winds from the sheared surface vortex (Plate 1D). In addition, Plate 2 shows the NOGAPS analysis field of low-level θ_e overlaid on QuikSCAT scatterometer fields with the storm motion subtracted. Whereas on September 7th, while the above-mentioned vortex mergers were occurring, the main surface vortex center

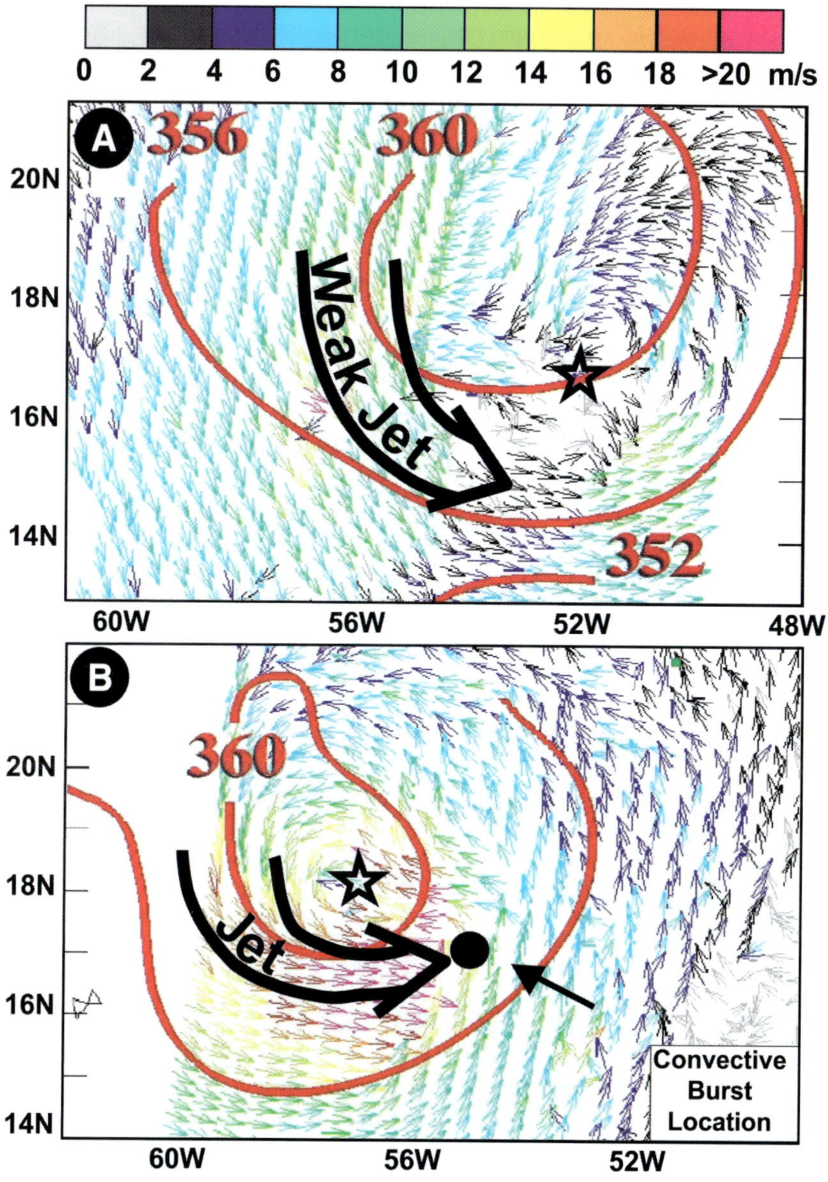

Plate 2. Storm-relative wind field calculated from the QuikSCAT scatterometer data with 1000 mb equivalent potential temperature (θ_e) calculated from NOGAPS analyses for A) 2155 UTC September 8, 1999; and B) 2130 UTC September 9, 1999.

was south of the high-energy air. By September 8th, the main surface vortex center was on the southwest edge of the high-energy air (Plate 2A). This region of high-energy air was also a region of strong easterly flow (greater than 10 m/s) just observable in Plate 1D north of the TD center location. The mesoscale convective system to the west-northwest of the center formed in this very favorable region. The two mesoscale convective systems rotated and merged over an 18-24 h period (Figures 5G-5K). They performed in a manner analogous to those for TC Oliver 1993 [Simpson et al., 1997]. During this time NHC upgraded the system to tropical storm strength, although they note a lack of a centralized core in their report.

There was a hint of a nascent eyewall at 1800 UTC on September 8th (Figures 5I, 5J), as the western clouds wrapped around the center of circulation in a hook pattern, also similar to that of Tropical Cyclone Oliver. Thermodynamic support for the presence of midlevel vortices is provided by the AMSU 53.6 GHz image in Figure 5L, which corresponds to Figure 5K. This image shows the precipitation cooling that would be expected below the stratiform base for four convective systems within the pre-Floyd disturbance. The midlevel vortices would develop above this layer, and so we indicate them in Figure 5K by the black ovals.

By late on September 8th, NOGAPS analyses show that the pool of 360 K air had moved in and to the north of the center (Plate 2A). The QuikSCAT storm-relative winds also showed that a weak low-level jet, about 100 km wide, had developed on the southwest side of the vortex (2A) and was sweeping the high-energy core θ_e air from the west-northwest side of the vortex core around to the southwest. The jet continued to intensify over the next 24 hours and was continuing to sweep the high θ_e air from the west of the vortex into a confluent region to the southeast of the center (not shown). We postulate that this injection of high-energy air, with its convergent forcing, set the stage for the first convective burst described in the next section. Twenty-four hours later this jet was at least 20 m/s in strength (Plate 2B). The cause of the observed low-level jet is still under investigation.

Floyd's Key Convective Burst

The rapid intensification period of Hurricane Floyd began early on September 9th with the sudden eruption of hot towers in the form of convective bursts (Figure 6). Two bursts were identified through inspection of combined GOES IR and passive microwave imagery, and also animations of half-hourly IR imagery. These convective bursts were essential for both the rapid deepening and eye formation in Floyd, and they provide a crucial link between the larger-scale forcing (i.e., convergence due to low-level wind surges) and vortex-scale intensification.

Figure 6 reveals that each burst was active for approximately 24 hours and followed in close succession during the period of most rapid vortex spin-up. Both of these bursts featured a cold, circular cirrostratus shield that expanded to 400-

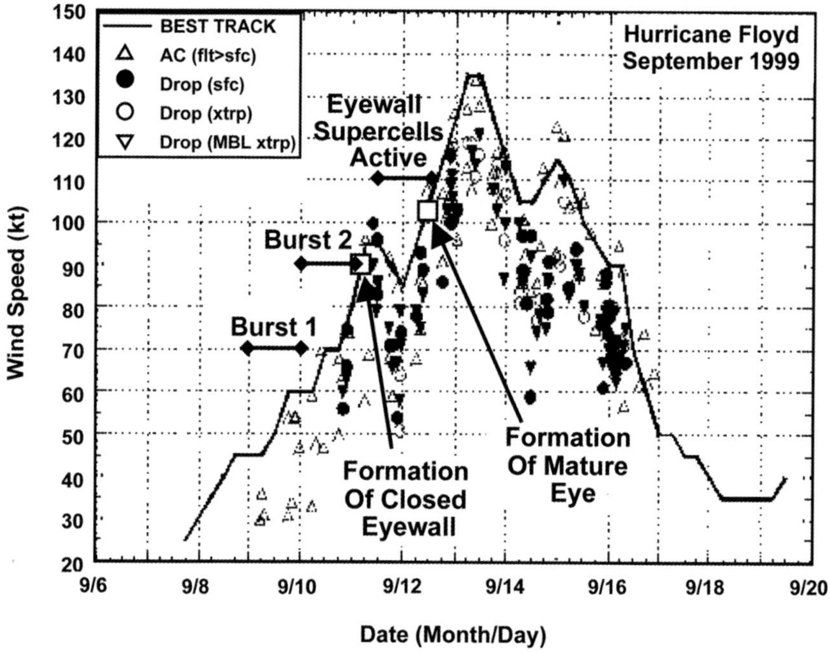

Figure 6. Evolution of convective burst events during the intensification and eye development stages of Hurricane Floyd.

500 km diameter during maximum extent, with repeated eruption of overshooting hot towers within the center of the cloudy mass. Minimum cloud top temperatures in the tallest overshoots approached -85°C. Formation of a closed low-level eyewall occurred during the latter stage of Burst 2, when the storm's maximum sustained winds approached 50 m/s on September 11th. Further intensification that followed on September 12th was associated with multiple eyewall supercells, culminating in the development of a mature eye in the infrared imagery by 1200 UTC, September 12th.

Plate 3 highlights important sub-mesoscale evolution within these two bursts. In these mosaics the top panel presents a color-enhanced infrared view of Floyd's cloud top features, while the bottom panel reveals the ice scattering signature of cloud tops using the 85 GHz channels of the passive microwave sensors. The first burst began before 0600 UTC on September 9th (Plate 3A). A small cluster of intense hot towers is identified by the bright ice scattering signature (PCT < 163 K) in the microwave channel. These high towers are within the zone of strong surface convergence set up by the "wind surge" (Plate 2B). They flanked the estimated center position of the vortex in the southeast quadrant. Two hours later (Plate 3B), the cold cirrostratus shield of the burst doubled in areal extent, while the area of active hot towers remained the same. This implies that the convective burst is a dynamic

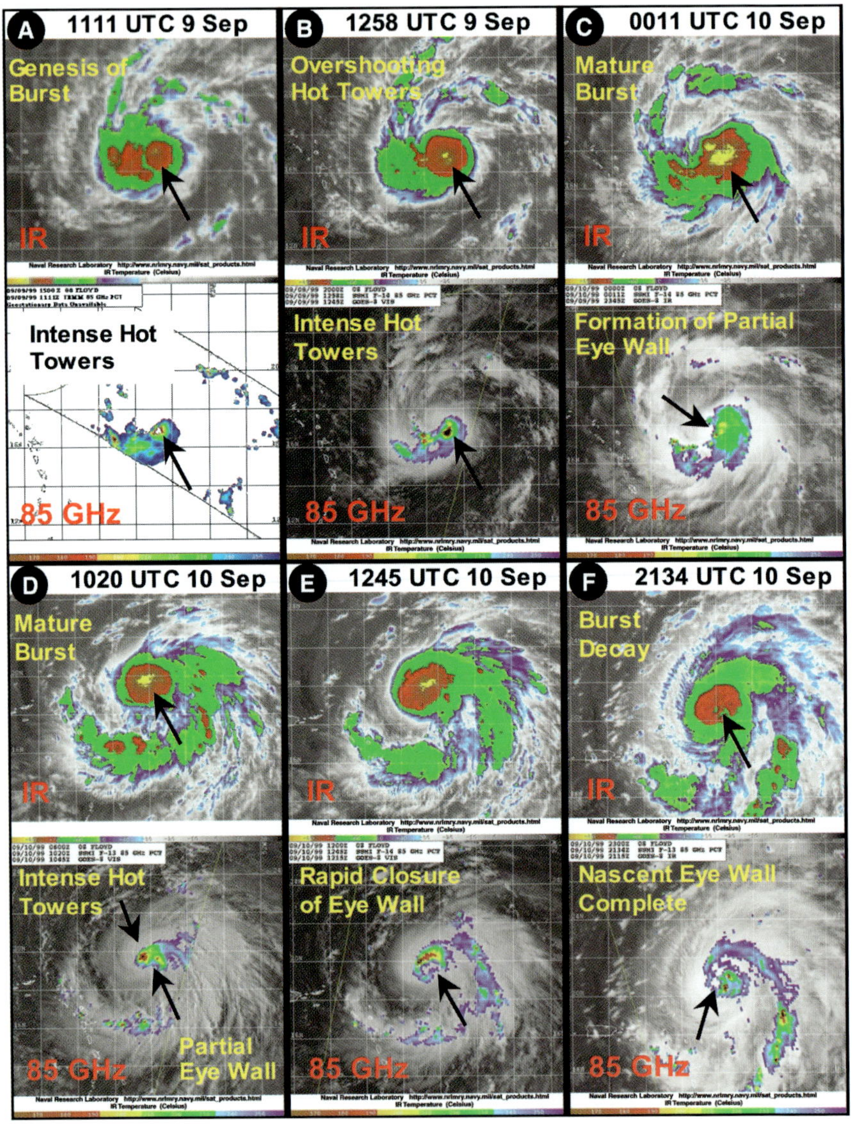

Plate 3. Floyd Convective Burst 1: A) 1111 UTC September 9th, 1999; B) 1258 UTC September 9th; C) 0011 UTC September 10th; and Convective Burst 2: D) 1020 UTC September 10th; E) 1245 UTC September 10th; and F) 2134 UTC September 10th. The black arrows point to the feature of interest, written in the upper right.

entity, with mass balance requiring substantial upward moisture flux and latent heat release in just a few undiluted conduits. By 0011 UTC on September 10th (Plate 3C), the burst was in its mature stage, with widespread cloud top cooling in excess of -80°C and the formation of a partial eyewall in the eastern quadrants. This initial burst then decayed by 0300 UTC on September 10th, with warming of cloud top temperatures and assumption of a ragged appearance.

A second major burst erupted around 0300 UTC September 10th in the northwest quadrant (i.e., downwind with respect to the swirling wind). It was within the pool of highest θ_e air (Plate 2B). This burst assumed similar areal extent and intensity levels as the first, reaching maturity by 1020 UTC (Plate 3D). Again, the area of active hot towers was relatively small compared to the diameter of the upper-level cloud canopy, implying that vigorous updrafts must be present. Just two hours later, the small cluster of hot towers in Plate 3D wrapped cyclonically around the vortex core, to form a nearly complete eyewall (Plate 3F). Such rapid spin-up of the vortex may, in part, be a consequence of the vigorous ascending motion stretching and amplifying the storm's vorticity. Decay of the second burst occurred late on September 11th with full closure of a shallow eyewall and first appearance of the eye's central warming in the infrared channel. At this time Floyd's maximum wind speed was approaching 55 m/s.

Warm Core Development

To us, one of the most surprising aspects of Floyd's development was the length of time it took to develop a warm core in the upper levels. An upper-level warm core is necessary to support both the low central pressure and the strong winds at the surface. Despite the continued development of convection, particularly through the period September 6th-8th, an upper-level warm core did not form. In fact, it wasn't until the early hours of September 9th, when the towers in the first convective burst carried the necessary high-energy air to high levels, that a warm core finally began developing[12]. Figure 7 shows the evolution of the warm core derived from AMSU-A sounding data. At 2300 UTC on September 8th, Floyd showed no significant warm core. By 1200 UTC on September 9th, after the first major hot tower eruption, a weak warm core appeared. A corresponding increase of surface winds then resulted in an increase in the ocean fluxes and, thus, the energy of the surface air that ascended in the second burst, so that by late on September 10th, the warm core had increased again (Figure 7). At that time, maximum winds had exceeded 36 m/s, and a complete nascent eyewall is seen in Plate 3F. Finally, by late on September 11th, at the end of the second burst, the AMSU data show a warm core anomaly of about 7°C at 250 mb. Our hydrostatic calculation showed that this warm anomaly would support a 31 mb reduction in the surface pressure relative to September 8th. This is

---

[12] Jason Dunion [personal communication] said Floyd had just moved out of the African dust by September 9th.

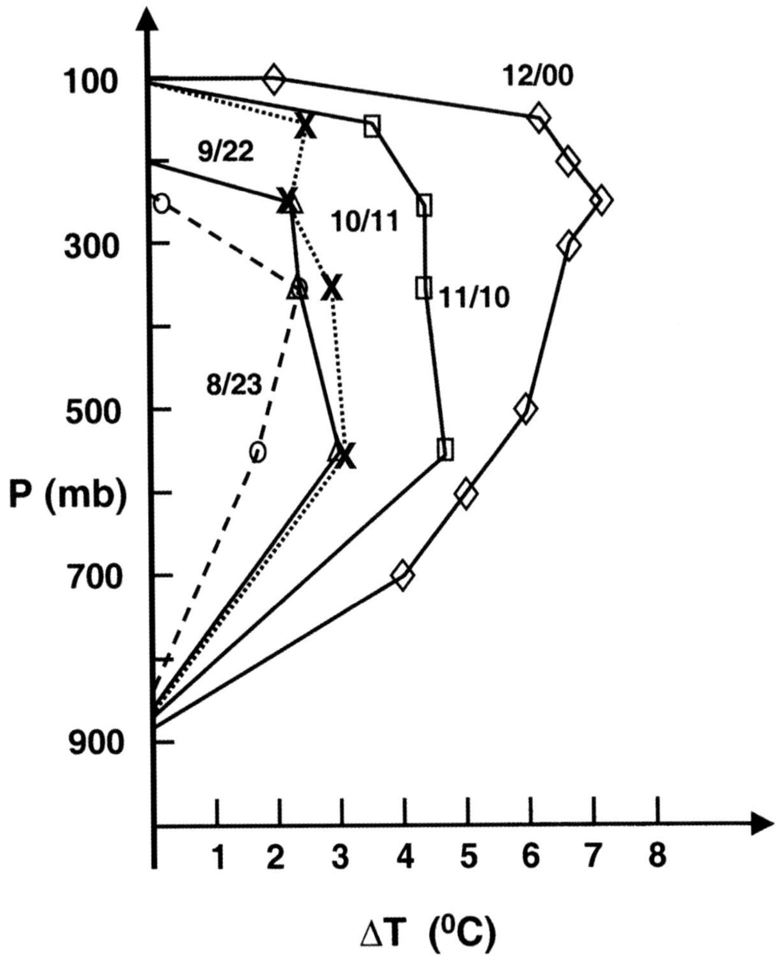

Figure 7. Vertical profiles of anomaly temperatures derived from the AMSU-A sounding instrument brightness temperatures for Hurricane Floyd, showing the evolution of the warm core prior to, during, and subsequent to the convective bursts on September 9 and 10, 1999. Numbers are the dates and times in UTC.

consistent with NHC's minimum central pressure data of 1000 mb on September 8th and 963 mb by the end of September 11th.

Discussion

The most interesting feature of Floyd's development is the importance of the mesoscale vortex mergers that maintained the rather weak African-origin vortex half way across the Atlantic Ocean. Whereas initially the water temperatures were not high enough to support the intensification of Floyd, by September 8th warm water was able to build up a deep enough layer of high-energy air to start the hurricane engine going through bursts of convection. A particularly intriguing event was that the source of air, or impulse for the first convective burst, seemed to originate with a 100-km wide wind jet, or injection of high-energy air into the vortex core. Bill Gray postulated the necessity for this type of wind surge some years ago [Gray, 1998]. Further research on other storms, with the help of mesoscale hurricane models, is in progress to determine the prevalence and physics of these wind surges.

The "Near Hurricane" That Failed: TD 2, June 22-25, 2000

Next we examine a vortex, which was stronger than pre-Floyd when it left Africa, but underwent an entirely different fate. Experience dictates that no African seedling develops into a hurricane in June because the water is too cold and the environment is too dry and stable. In fact, without QuikSCAT data showing rotational winds exceeding 25 m/s, it is questionable whether the disturbance which emerged from Africa on June 22, 2000 would have been designated a TD at all.

Early TD 2: June 22-23, 2000

During the period June 22-23, 2000, the large-scale dynamic environment off the west coast of Africa appeared favorable for development. The vorticity in the monsoon trough zone was weaker than that for Floyd at the surface. However, it was much stronger at 700 mb, adequate to permit 200-500 km mesoscale vortices to persist. The ambient vertical wind shear was weak — less than 6 m/s (Figure 8). Another important feature was a bulge of anomalously high sea surface temperatures above 28°C just off the coast of Africa, as shown in Figure 9. This warm ocean caused the low-level air to have unusually warm energy content, with a surface θ_e higher than 350°K (not shown). However, the low midlevel θ_e of about 335°K shown in Figure 10 and, particularly, the stable, dry atmospheric sounding at Dakar (Figure 11) indicate that in general the atmosphere was not conducive to the growth of tall convection such as was required in Floyd for intensification. Note that both storms were about 4° latitude south of Dakar and the Cape Verde Islands at the time the soundings were made. Thus, both storms would have been located in somewhat less hostile conditions than indicated by the soundings. However, it is the difference between the two that is significant. Figure 11 shows a stronger inversion, no saturated layers, and a much drier air mass than does Figure 3. Thus, a stronger upward forcing is required for penetrative clouds, and much more moistening of the environment would be necessary to pre-condition the atmosphere for a convective burst.

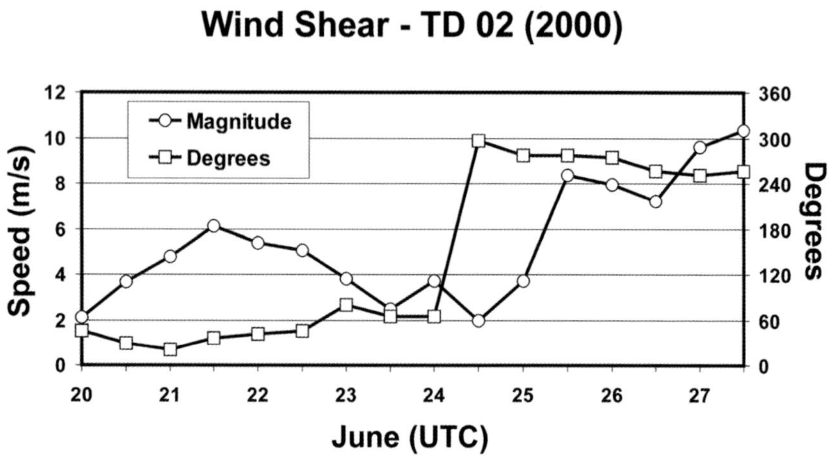

Figure 8. Time series of area-averaged shear over TD 2, June 20-27, 2000.

The disturbance moving off Africa appeared deep, extending downwards from 500 mb in NOGAPS analyses, with an associated surface circulation of 8-10 m/s (Plate 4A). There was a region of low-level convergence greater than -20×10^{-6} s^{-1} for June 23rd just ahead of the surface circulation. Another region with values greater than -10×10^{-6} s^{-1} was found just to the northeast of the surface circulation, as shown in the scatterometer data (Plate 4b). Despite the stronger inversion (Figure 11), the convergence ahead of TD 2 was strong enough to "push" the

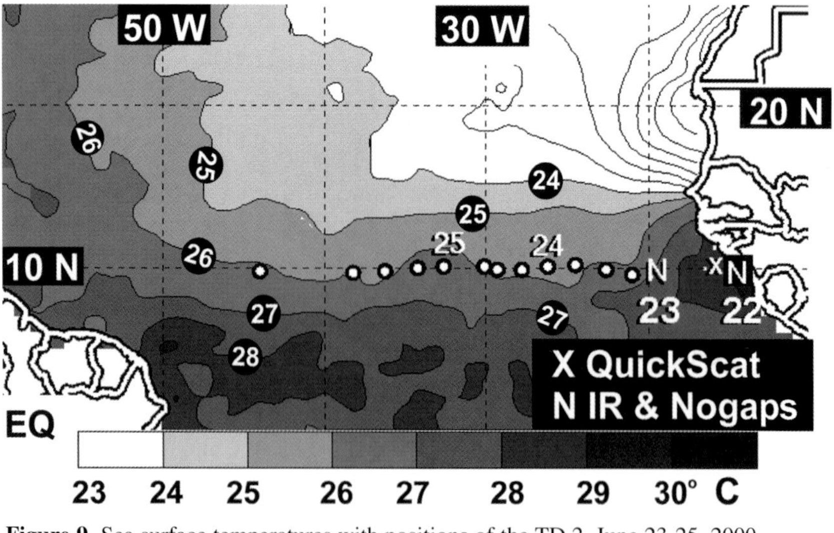

Figure 9. Sea surface temperatures with positions of the TD 2, June 23-25, 2000.

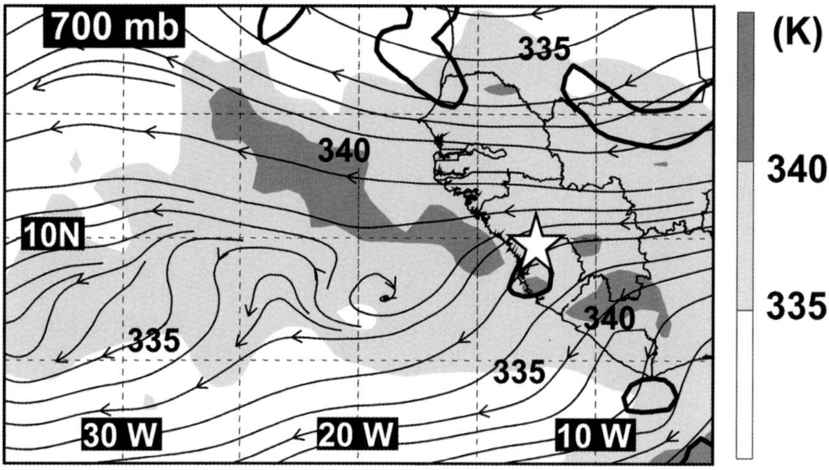

Figure 10. Equivalent potential temperature (shaded, increments of 5 K) and streamlines from NOGAPS at 700 mb for 0000 UTC June 22nd. Regions of convergence > 10 x 10^{-5} s^{-1} are surrounded by the thick black line, and shading starts at 335 K.

warm, moist low-level air upward to create several convective towers that reached higher than 14 km during June 22nd and the early hours of June 23rd. However, there was no evidence in the AMSU data for warming aloft as a result of this convection. Figure 12A shows a time series of the height of the coldest cloud tops observed in METEOSAT throughout the period June 22nd-23rd. No cloud tops higher than 15 km[13] were observed after 1200 UTC June 23rd, with most clouds topping below 13 km. Figure 12B shows the last high tower that was observed. After this time SSTs lower than 26°C underlay the storm. The main explanation for the diminishing strength and penetration of cumulonimbus towers is the decreasing energy (θ_e) of the low-level air, as the TD moved westward over the cooler ocean, which is typical for June in this region. By 0000 UTC on June 23rd, the θ_e of the underlying air had decreased by 8 K. In addition, the dryness of the midlevel air in June dilutes cloud towers, killing their buoyancy.

TD 2 Intensifies: Role of Mesoscale Vortex Interaction

The last series of convective towers that reached higher than 15 km occurred between 0500 and 0800 on June 23rd (Figure 12), and it was during this period, at 0700 UTC that the TD 2 vortex was observed to most closely approach hurri-

[13] The cloud top heights are estimated by locating the height of that temperature using a sounding (radiosonde), which provides a relationship between temperature and height. If the cloud top temperatures are not equal to that of the environment, a small error will result.

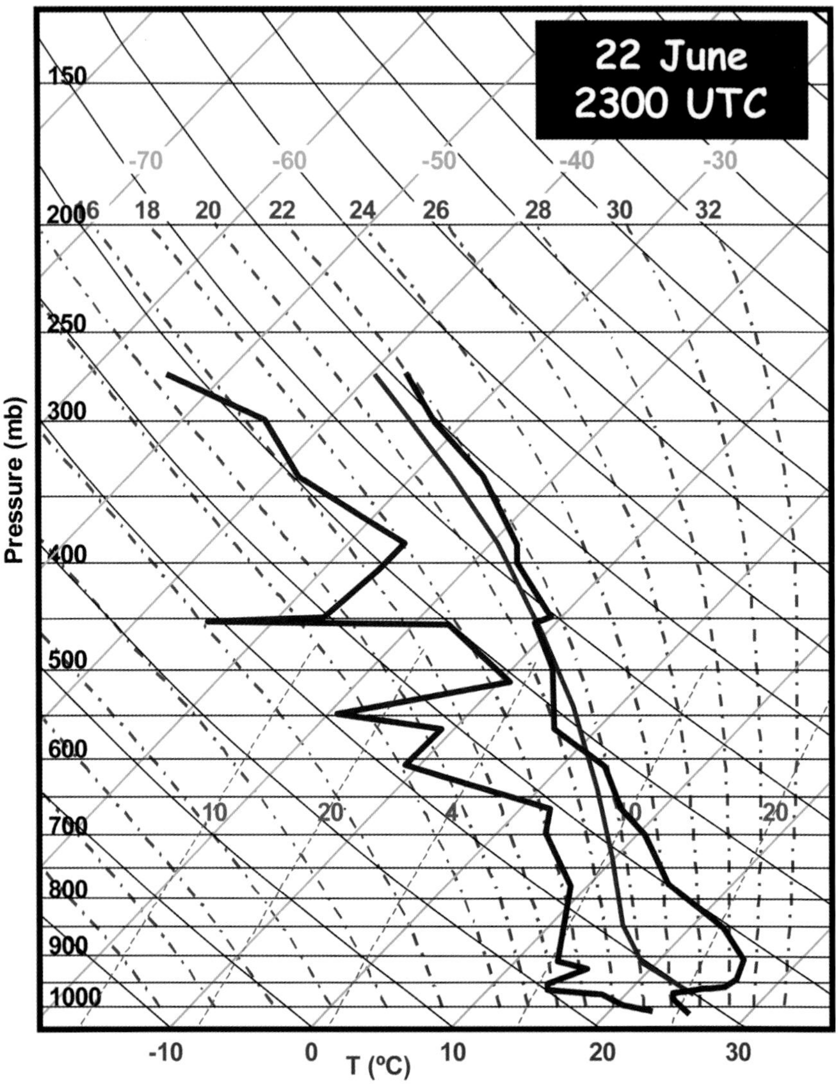

Figure 11. Dakar sounding valid for 2300 UTC June 22, 2000 showing a dry, stable layer throughout the troposphere. The thin solid lines that slope diagonally to the right are constant temperature lines. The black diagonal lines that slope to the left are dry adiabats.

cane wind speed and structure. Figure 13 shows how the merging of midlevel vortices may have assisted this strengthening and organization.

As TD 2 first moved off Africa on June 22nd (Figures 13A, 13B), convergence of low-level high-θ_e air resulted in the development of several mesoscale con-

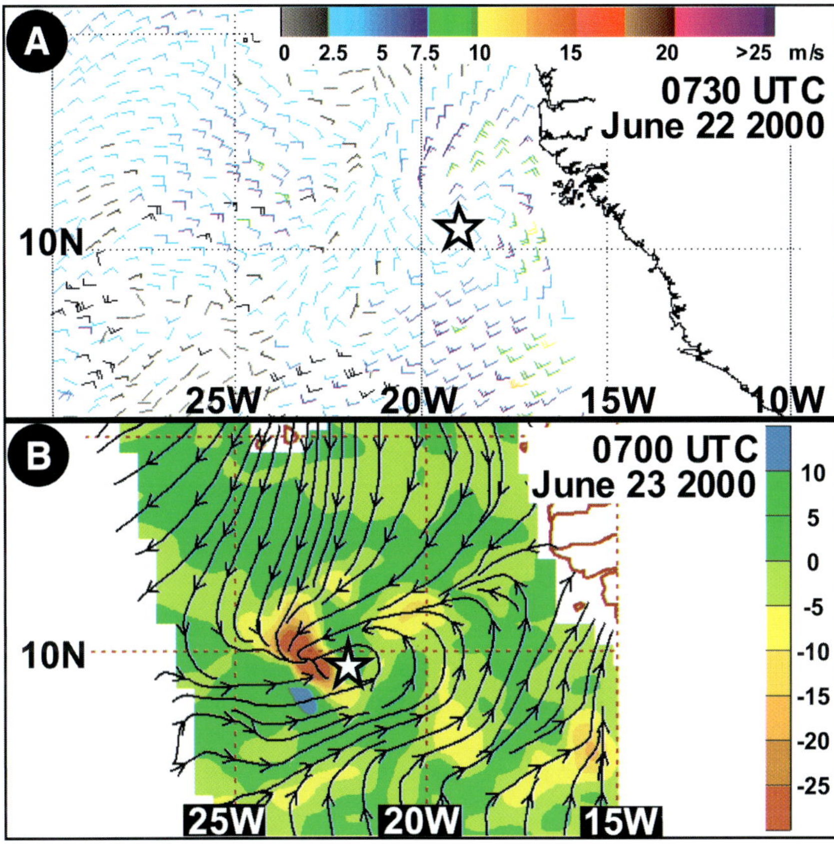

Plate 4. A) Surface winds from QuikSCAT scatterometer data for 0730 UTC June 22, 2000 showing the surface circulation of TD 2 just after it moved off Africa; and B) Divergence and streamlines calculated from QuikSCAT scatterometer data for 0700 UTC June 23, 2000. Note the strong convergence values (warm colors) of greater than -25×10^{-5} s^{-1} just ahead of the surface circulation center.

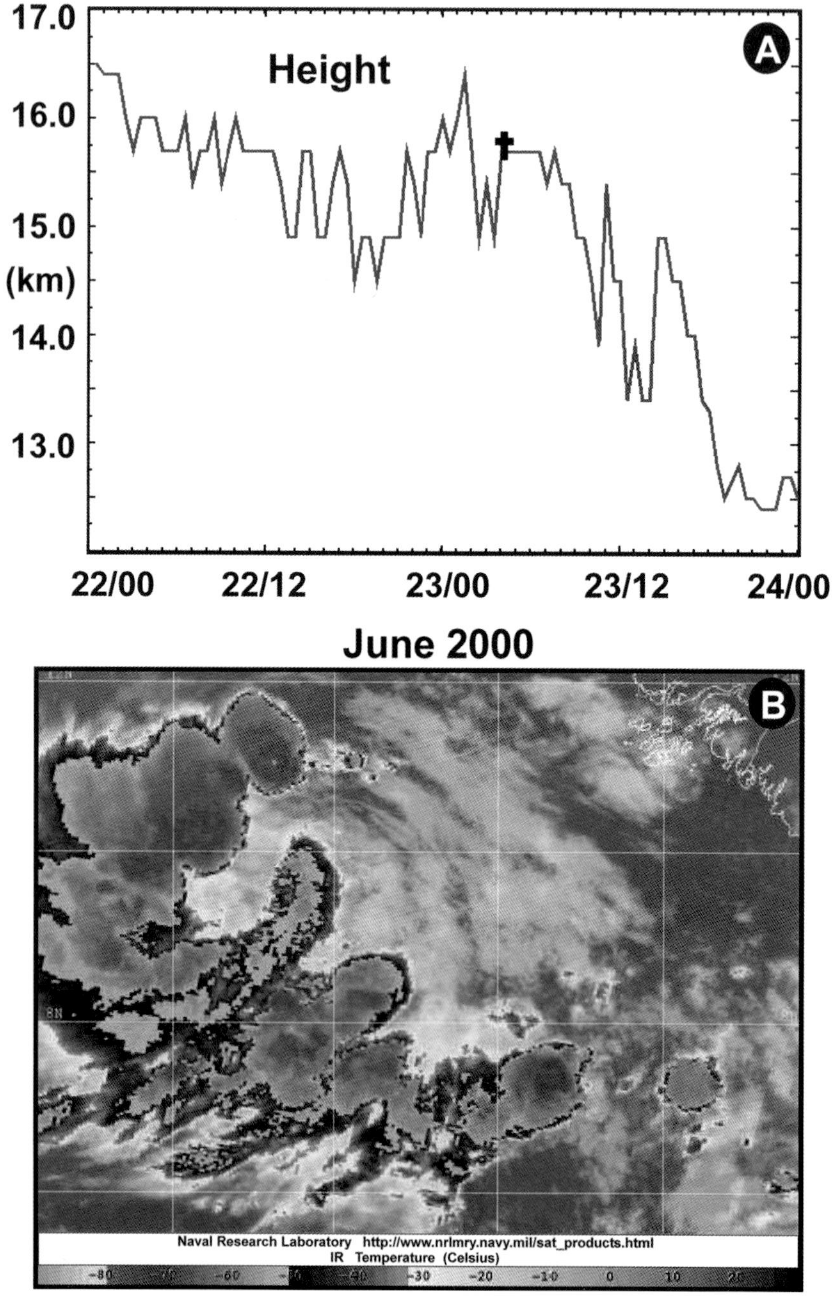

Figure 12. A) Maximum cloud height estimates near TD 2 from METEOSAT 7 for June 22 through June 23, 2000; and B) GOES IR imagery valid for 0500 UTC June 23rd.

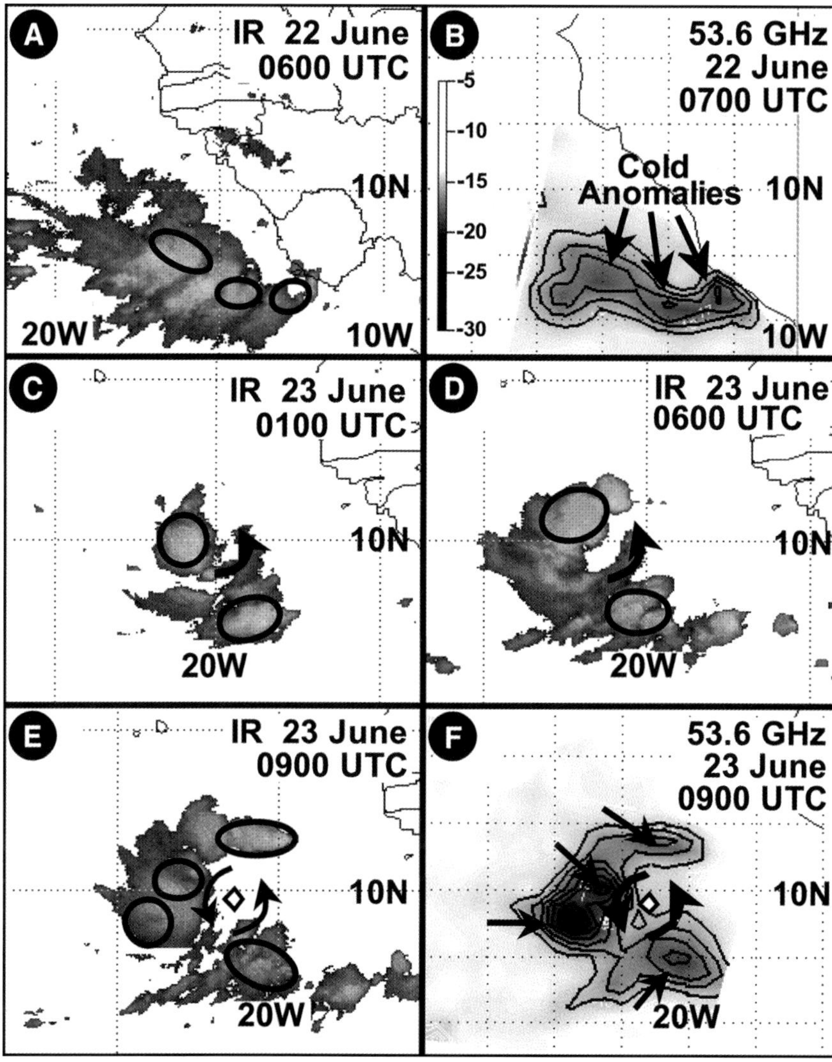

Figure 13A. A – F) Sequence of METEOSAT IR and AMSU-A 53.6 GHz images. Times and type of image are indicated on each panel. The estimated location of the surface center based on QuikSCAT data is indicated (◇). The 54 GHz images correspond to the IR image next to them. The thick black ovals indicate inferred vortices, and black arrows indicate where rotation was observed in the IR movie.

vective systems with cold IR tops and rain showers (not shown) loosely organized to the southeast of the surface center of the disturbance. Although we identified three vortices by the cold anomalies in the AMSU imagery (Figure 13B), we did not see any merger in the cloud patterns at this time. By late on June 22nd,

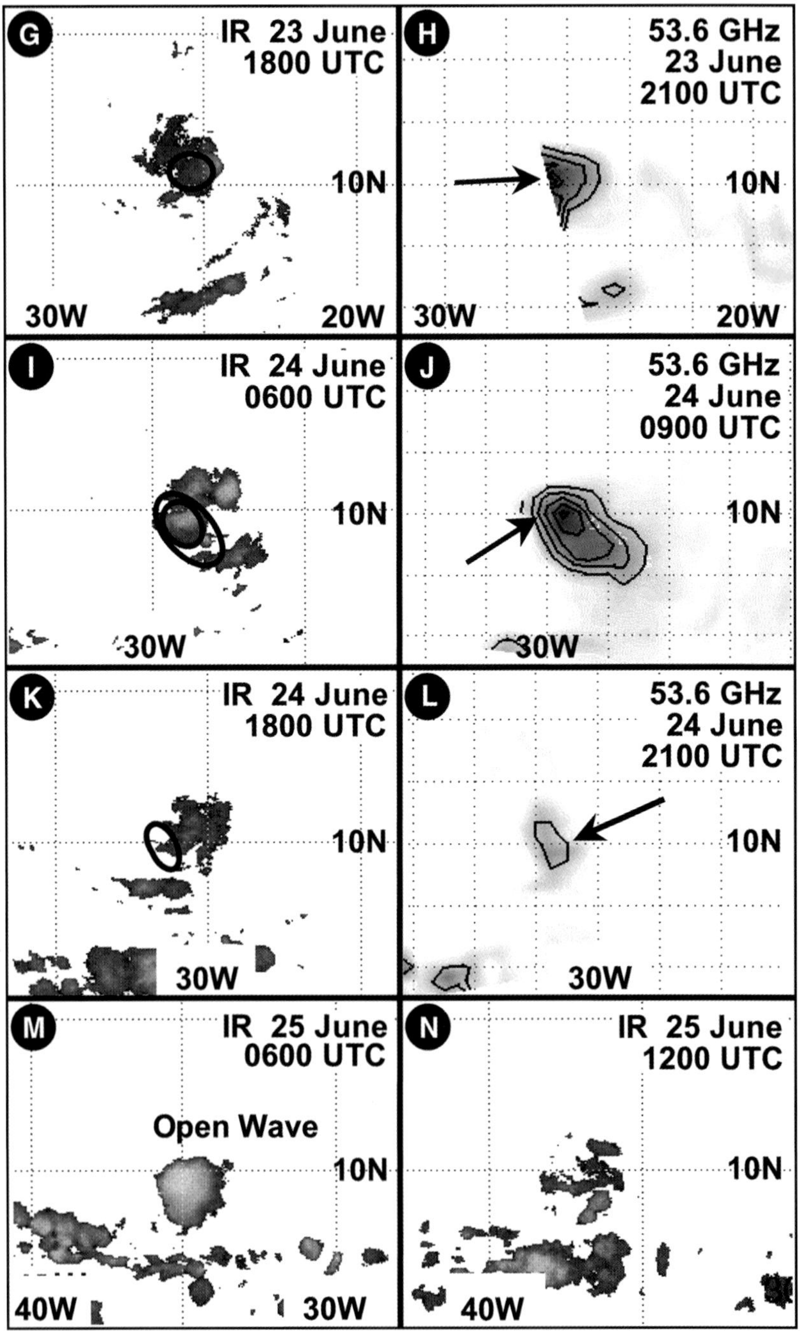

Figure 13B. G – N) As per Figures 13A, A- F.

the first of several new mesoscale convective systems developed just to the south of the surface circulation. As a result of lower θ_e in this region, there were fewer convective cells, and the associated stratiform anvils were smaller in extent (Figure 13C compared with Figure 13A). However, the comparable AMSU images show that several of the mesoscale convective systems had temperature structure associated with them consistent with the development of vortices (Figures 13C-13H). During the day of June 23rd, TD 2 reached its peak winds of approximately 20 m/s, as shown in Plate 5A. Note that some of the observations in Plate 5A are "rain-flagged" in black, meaning that they are contaminated by rain and may be an overestimate of the actual surface wind strength in that region. However, at least two key observations over 20 m/s are not rain flagged. In addition, a combination of the TRMM TMI 85 GHz channel (Plate 5B) revealed a structure that resembled a hurricane eye. The "eyewall" had several towers above 16 km northwest of the center and one directly to the east. On the 37 GHz combination (not shown) the "eyewall" was virtually closed by clouds topping at 6-7 km. The mesoscale interactions, which produced this event, are shown in Figures 13C-13F.

As the TD 2 disturbance moved west of about 25°W late on June 23rd, the sea surface temperatures decreased to less than 26° C (Figure 9), and both the low- and mid-level θ_e ahead of the disturbance had decreased (not shown). Whereas some convection continued for more than 24 h (Figures 13G-13L), the tops were shallower (~12 km indicated in Figure 12A), corresponding to the lower θ_e. A corresponding decrease in the strength of the surface vortex of more than 8 m/s was indicated in scatterometer data (Plate 6A). By 2000 UTC June 24th, the disturbance moved briefly back over marginal (> 26°C) sea surface temperatures (Figure 9), and convergence of higher low-level θ_e (> 352 K) air was indicated to the north of the surface disturbance (not shown). Another rather impressive cloud feature began to develop in this location (M in Figure 13B). However, close examination of the IR and 85 GHz temperatures revealed that only one small convective tower was associated with the development. The observed cloudiness was probably a cirrus shield that spread evenly because of the low shear over the disturbance at this time (~3 m/s in Figure 8). By 1200 UTC on June 25th, the disturbance had once again moved back over sea surface temperatures lower than 26°C. The wind shear over the system had dramatically increased to 8 m/s (Figure 8), and scatterometer data indicated the surface circulation had degraded to an open wave structure (Plate 6B).

Discussion

The outstanding feature of TD 2 was how close its wind speed came to tropical storm intensity without starting the hurricane heat engine. It emerged from Africa stronger than pre-Floyd and appeared to strengthen from mesoscale vortex interactions, which require enough convection for the vortices to develop in

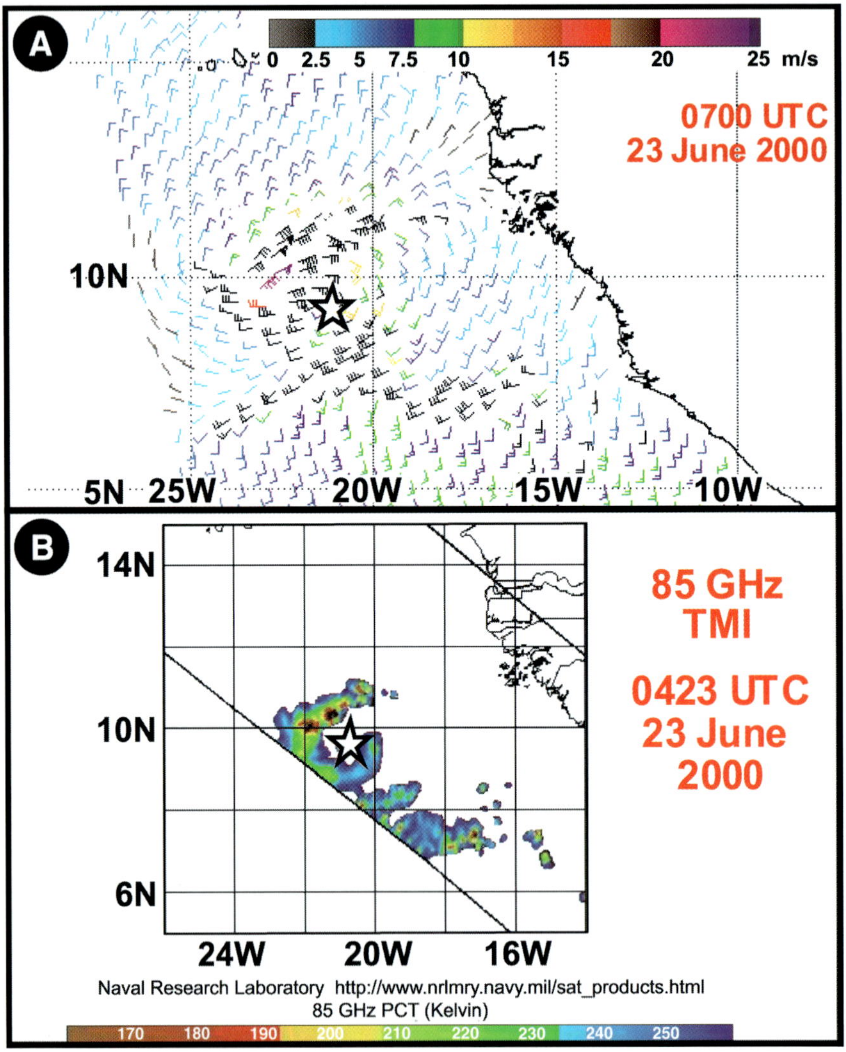

Plate 5. A) QuikSCAT scatterometer winds at 0700 UTC June 23, 2000; and B) Derived product using the polarization of the 85 GHz channels of the TRMM microwave sensor at 0423 UTC June 23, 2000 (for explanation see NRL Web Site).

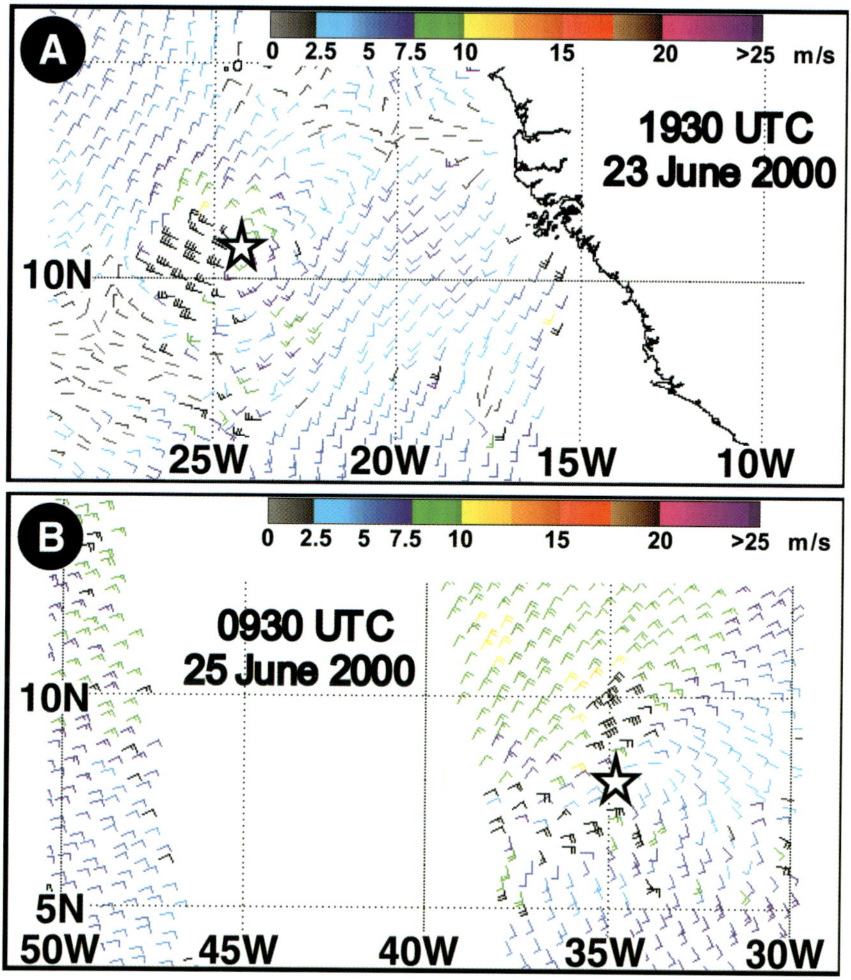

Plate 6. QuikSCAT scatterometer winds for A) 1930 UTC June 23, 2000; and B) 0930 UTC June 25, 2000 showing the degraded open wave structure. A half barb = 2.5 m/s, and full barb = 5 m/s. Black wind barbs are possibly rain contaminated.

anvils, but do not require highly penetrative hot towers in large numbers, as in a convective burst. This conclusion is made because there were no other mechanisms to strengthen the vortex apparent from the data available. In hindsight, TD 2's chances to reach named status had an extremely low probability, in view of the low ocean temperatures and lack of high-energy (high θ_e) air that is typical in the Atlantic during June. However, it is interesting to reflect, given how strong TD 2 became without a real convective burst, how rapidly it might have intensified to hurricane had there been unseasonably warm SSTs providing the energy for penetrative convection along its track.

CONCLUDING REMARKS

This chapter reports only the preliminary results of an extensive study. A main advantage of satellite technology for tropical cyclone research is that storms can be observed in all the ocean basins of the world, including those where no aircraft campaigns have yet been made. We began with Atlantic hurricanes in view of the knowledge and database and the needs of forecasters for the United States and Central America. This type of study is highly labor-intensive, and the work of many other groups is needed to examine other basins and storms. We hope that the examples and diverse use of satellite data in this chapter will inspire similar exploitation by researchers in other countries threatened by tropical cyclones. The technology used in this study will be available for at least a decade, and more will become operational in 2010.

The remainder of this section will be divided into three parts, namely: (1) conclusions which are fairly well supported; (2) hypotheses and early evidence which require many more cases and further studies to test; and (3) valuable data sources that support, and data voids that hinder, satellite studies of tropical cyclones like this one, which attempt to examine causality and the physics of scale interactions rather than statistics.

Conclusions

1. Data from a number of satellites fit well together and are able to suggest explanations or confirm or discard older postulates regarding the formation of hurricanes.
2. The consistent picture constructed using TRMM, AMSU, and QuikSCAT, in concert with recent advances in existing satellite tools, lends confidence to the observational capabilities and usefulness of all three observing systems, although each has limitations.
3. Mesoscale processes are crucial in the formation of all tropical cyclones studied here and those studied previously by the authors.
4. These tools are supplementary to and not a replacement for aircraft field programs, air reconnaissance, and surface and upper air observations,

such as pressure, temperature, humidity and winds throughout the troposphere.

5. The question in the Atlantic is really, "How do we get so many hurricanes in such a hostile environment?" Our initial conclusion is that the vorticity of the environment is enhanced by disturbances originating in Africa that reduce the Rossby radius.
6. This has two implications: (a) the small amount of midlevel warming and moistening that occurs in the weaker convective elements will be trapped within the disturbance; and (b) stable mesoscale vortices can form in the anvil regions of mesoscale convective complexes, which develop within the disturbances. Through merger, these midlevel vortices can help maintain, and even strengthen, the vorticity within the African disturbances. Thus, a few disturbances may persist far enough west to reach high-energy regions where convective bursts can develop and create a warm core and partial eyewall. Then pressure falls may ensue to accelerate the surface winds, which, in turn, increase the fluxes from a warm ocean, and the hurricane heat engine is started.

However, most seedlings fail to develop. By means of mesoscale vortex mergers, some can survive a few days as a TD, even gaining nearly strong enough wind speeds to be named, but then fail to get the hurricane heat engine going because of unfavorable thermal conditions, and so eventually dissipate in the cold waters of the mid-Atlantic. Many seedlings develop and are then killed by enemies such as wind shear (with or without African dust) or a cold underlying ocean.

Hypotheses, Early Evidence, and Suggestions for Future Work

The main theory presented here for why fledgling systems not only survive, but also even strengthen during their passage across hostile portions of the Atlantic Ocean, can only be partially confirmed by current satellite technology. We emphasize that the presence of midlevel vortices are sometimes directly observed by the cloud rotation in the IR movies. At other times, particularly when the mesoscale convective systems are younger and the cloud tops are still building and are at their peak, any underlying vortices would be masked and beyond our current ability to observe them without some other observational platform. Thus, more studies, preferably with supporting airborne observing platforms, need to be made to ascertain whether as many of the mesoscale convective systems that develop in African seedlings contain midlevel vortices as we presume here. Remote sensing technology that can directly observe particle motion through clouds would be a great advance.

In June and July, there appear to be occasional rather strong African seedlings, which travel westward a few days with strong winds but fail to find conditions that enable the development of deep penetrative convection. Because of the possibility that vortex mergers contribute to the survival of fledgling storms as they cross the hostile eastern Atlantic, additional studies need to be made of such seedlings. By August, SSTs and high-energy, low-level air raise the probability of intensification in the mid-Atlantic.

In the genesis and TD stages of a tropical cyclone, less vigorous convection with its associated stratiform anvils is all that is required for the development of midlevel vortices. Thus, in order to maintain the disturbance over the cooler SSTs, midlevel vortices can develop as long as there is enough convection to develop stratiform anvils. These can, through vortex dynamics, help to maintain or strengthen the vorticity of the disturbance so that it does not just die. However, in the intensifying and mature hurricane, the environment must be able to support the development of strong convective bursts — hot towers with nearly undiluted ascent. Although hot towers also have stratiform anvils and associated midlevel vortices, at the initiation of the heat engine and intensification to tropical storm stage, this dynamic process may no longer be as important as the thermodynamic processes triggered by convective bursts. The development of convective bursts was one major difference between TD 2 and Hurricane Floyd.

A connection between a low-level, jet-like inflow of high-energy air and the first convective burst was observed in Floyd. In order to investigate properly the physics of these surges, many more cases need to be found in QuikSCAT and other data. Mesoscale models can also be used to examine the connection between wind surges, convective bursts and warm core formation.

In Floyd we showed a connection between the first convective burst on September 9th and the appearance of a 6 K warm core in AMSU data. In a study of 12 named storms in the Atlantic during 1999, Jeff Halverson (personal communication) found that 11 had a convective burst just prior to or concurrent with first intensification. In eight of these 11 storms, eye formation occurred during the burst. The relationship between convective bursts and warm core development observed in AMSU needs to be studied both for developing storms and also for the failures, using both observations and numerical calculations.

Study Expediters and Obstacles

The data made available on a routine basis on websites, such as the CIMSS and NRL, represent a huge investment in resources and permit any scientist, without large staff or computer resources, to conduct productive tropical cyclone research. To offset the lack of upper air sounding observations of pressure, temperature, humidity and winds, NOGAPS model analyses were substituted, which make use of innovative observations such as winds derived from satellite images. Unmanned Airborne Vehicles (UAV) and high altitude balloon driftsondes are being developed that may help to fill this gap. It would be immensely valuable to fly either of these platforms from sites in Africa, Australia, and the Philippines during their respective tropical cyclone seasons.

REFERENCES

Brueske, K. B. and C. S. Velden, Satellite-based Tropical Cyclone Intensity Estimation Using the NOAA-KLM Series Advance Microwave Sounding Unit (AMSU). To appear in *Mon. Wea. Rev.,* 2001.

Chen, S. and W. M. Frank, A Numerical Study of the Genesis of Extratropical Convective Mesovortices. Part I: Evolution and Dynamics. *J. Atmos. Sci.,* 50, 2401-2426, 1993.

Frank, N. L. and P. J. Hebert, Atlantic Tropical Systems of 1973. *Mon. Wea. Rev.,* 102, 290-295, 1974.

Fritsch, J. M., J. D. Murphy and J. S. Kain, Warm Core Vortex Amplification Over Land. *J. Atmos. Sci.*, 51, 1780-1807, 1994.

Gray, W. M., The Formation of Tropical Cyclones. *Meteor. Atmos. Phys.,* 67, 37-69, 1998.

Lau, K-H. and N-C. Lau, Observed Structure and Propagation Characteristics of Tropical Summertime Synoptic Scale Disturbances. *Mon. Wea. Rev.,* 118, 1888-1913, 1990.

Liu, W. T., Progress in Scatterometer Application. *J. Oceanogr.*, 58, 121-136, 2002.

Ooyama, K. V., Conceptual Evolution of the Theory and Modeling of the Tropical Cyclone. *J. Met. Soc. Japan*, 60, 369-380, 1982.

Riehl, H. and J. S. Malkus, Some Aspects of Hurricane Daisy, 1958. *Tellus*, 13, 181-213, 1961.

Ritchie, E. A. and G. J. Holland, Scale Interactions During the Formation of Typhoon Irving. *Mon. Wea, Rev.*, 125, 1377-1396, 1997.

Schubert, W. H., J. J. Hack, P. L. Silva Dias and S. R. Fulton, Geostrophic Adjustment in an Axisymmetric Vortex. *J. Atmos. Sci.*, 37, 1464-1484, 1980.

Simpson, J., E. A. Ritchie, G. J. Holland, J. Halverson and S. Stewart, Mesoscale Interactions in Tropical Cyclone Genesis. *Mon. Wea. Rev.,* 125, 2643-2661, 1997.

Chapter 13

Continuous Monitoring of the Hurricane Core

Availability of New Long-Endurance Small Drone Aircraft

Greg Tyrrell and Greg Holland

INTRODUCTION

Previous chapters have shown that many of the most crucial processes driving the hurricane heat engine take place at the air-sea interface. The danger to manned aircraft becomes acute below about 1000 ft. At these very turbulent low levels, loss of an engine can lead to uncontrollable ditching with inevitable loss of life and millions of dollars in property damage. Another serious problem facing improvements in the understanding of variability in intensity and motion results from the very short times that piloted aircraft are able to stay in the storm, at most six to eight hours. Unoccupied drone aircraft could help provide solutions to both of these problems and perhaps contribute additional benefits as well.

DRONE AIRCRAFT

Drone aircraft include the family of Unoccupied Aerial Vehicles (UAVs). They vary in size from that of a commercial airliner to tiny experimental systems about the size of a thumb. Examples include:

1. The Black Widow from AeroVironment, which is about the size of a hand, is electric powered and used for local photographic reconnaissance.

Hurricane! Coping with Disaster

10.1029/055SP16

2. The 15 kg Aerosonde, a small propeller aircraft that was developed specifically for meteorological observations.
3. The General Atomics Predator, about the size of a private aircraft, also propeller driven, and in regular military use.
4. Northrop Grumman's Global Hawk, which is the size of a 737 aircraft, jet powered, and used in military and civilian surveillance.

The annual survey of drones in operation or under development by Shephard's [2001] provides an excellent overview of this rapidly developing area. Their results are summarized in Figure 1, which is a plot of wing span versus endurance. We see an approximately linear relationship between size and endurance for the majority of drones; the larger the aircraft the longer it flies. However, this relationship disappears at the long endurance scale, which is critical for hurricane reconnaissance. Aircraft capable of operation in excess of 30 hours vary from the small Aerosonde, with its 3-m wingspan, to the 34-m Global Hawk (Figure 2).

Control and command are accomplished in a variety of ways. Some aircraft are remotely piloted, with a person on the ground in complete control either by direct sight (e.g., model aircraft) or by telemetry of onboard video cameras. Others are

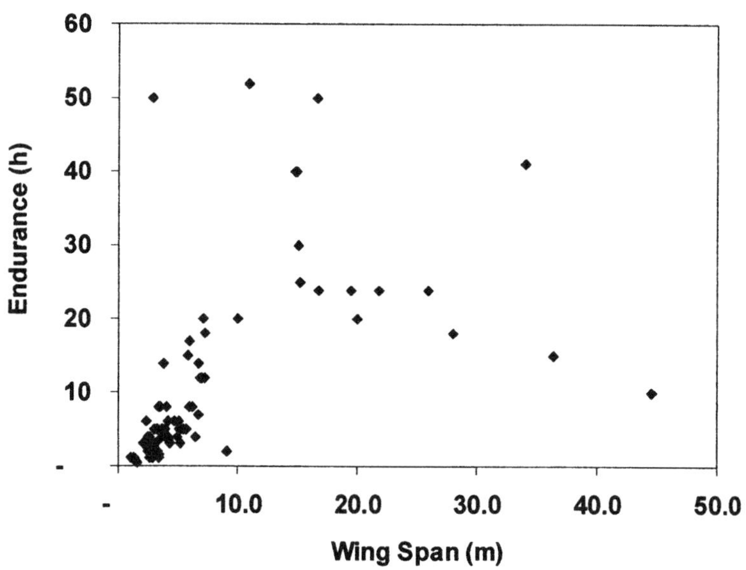

Figure 1. Scatter diagram showing the range of sizes and endurances of currently available or planned drones (after Shephard's, 2001).

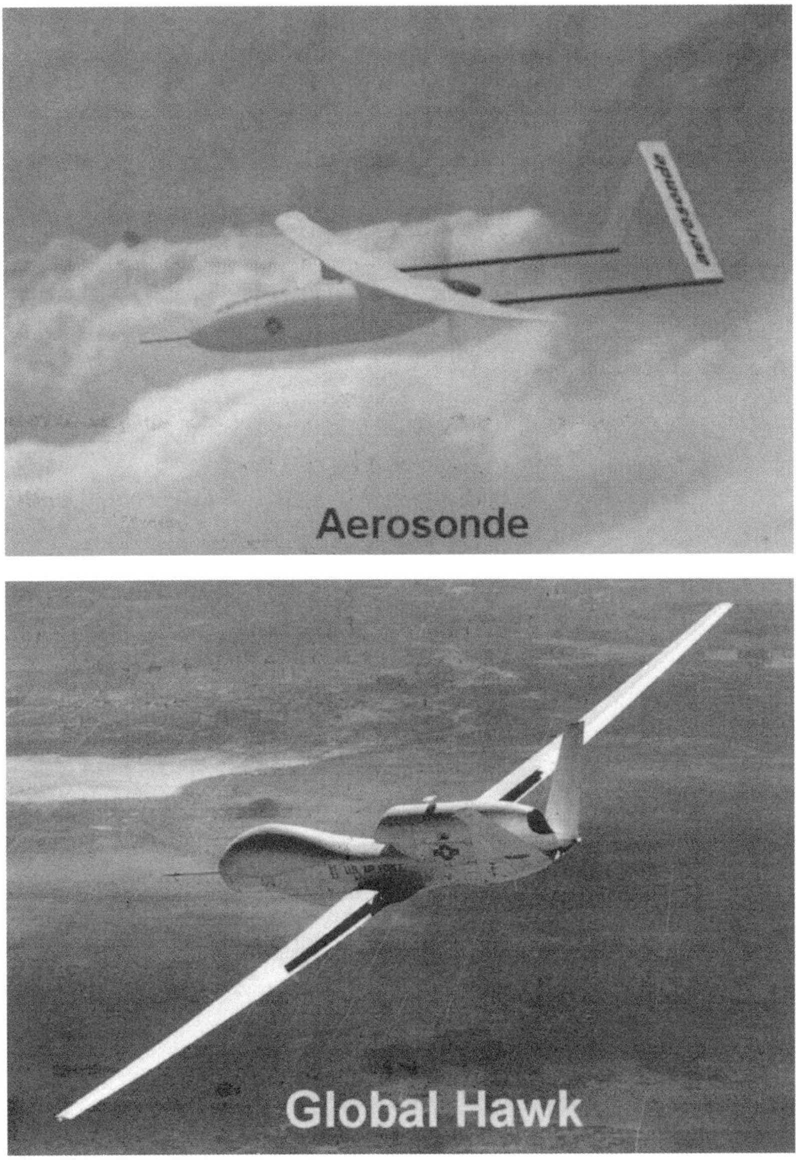

Figure 2. The Aerosonde (Aerosonde Ltd., Australia) and the Global Hawk (Northrop Grumman, USA).

fully autonomous in flight, with an onboard autopilot carrying out either predefined flight plans or new instructions that are uploaded from a ground commander. Many aircraft have the capacity for either remotely piloted or autonomous flight at the choice of the ground commander.

Not having a pilot has several advantages and a few disadvantages. The major advantage is the capacity for so-called dull, dirty, and dangerous jobs. Drones can operate for long periods undertaking repetitious tasks. They can also be used in conditions that would be difficult or considered too dangerous for a human being, such as the near surface layer of hurricanes. Disadvantages include the absence of a highly trained person in the cockpit. The need for onboard flight systems to handle the fault checking and in-flight emergencies has been a major sink of development funds.

These requirements have made a myth of the popular conception that drones are "cheaper" than manned aircraft. Generally, for the same size aircraft, a manned aircraft is cheaper than a drone to operate. However, not having a pilot allows the use of a smaller aircraft and thereby introduces economy of operation.

Until quite recently almost all drones were developed for military use. This development provided remarkable advances in the capacity of onboard systems for flight control and in military payloads. But such drones are generally too expensive and unnecessarily complex for civilian use. The Aerosonde aircraft was originally developed explicitly for meteorological operations. It combines economy of operation with sufficient endurance for use in tropical cyclones.

DRONE AIRCRAFT FOR HURRICANE RECONNAISSANCE

Any new observational system must provide one or both of two major advantages. The first is providing observations that cannot be taken, or are difficult to obtain, with the existing observing system. In the North Atlantic and eastern North Pacific, with their excellent manned aircraft reconnaissance, drones can excel in observing the near-surface boundary layer or lower stratosphere. Elsewhere in the world, drones may be able to provide a flexible and economic method of obtaining some of the observations provided by manned aircraft.

The second advantage is that drones can take some observations more efficiently than other methods currently in use and have remarkably long endurance. A small drone could remain in the cyclone core region, taking continuous observations of position and intensity, for extended periods. It could also loiter in a region of potential cyclone genesis to provide early warning of development. Alternatively, a high-flying drone, with a substantial load of small dropsondes could take soundings in the hurricane environment to support the requirements of the numerical forecast models.

There are several major questions concerning the operation of a UAV in a tropical cyclone environment:

1. What is the optimal mix of UAV and other observing systems?
2. What instrumentation suites are available?
3. Can the UAV survive and also operate safely in conjunction with other aircraft?

We examine each of these issues in the remainder of the chapter. But first we describe the basic characteristics of a current operational small drone, the Aerosonde.

THE AEROSONDE

Tad McGeer and I (Holland) conceived and developed the Aerosonde as a new system for taking meteorological observations in remote and difficult locations [Holland et al., 1992]. A particular operational focus was hurricane reconnaissance, especially for regions not observed or accessible by larger manned aircraft such as described in Chapter 3. The result is a small, robotic aircraft that has the endurance and range to reconnoiter tropical cyclones [Holland et al., 2001].

The aircraft is the size of a large hobby aircraft (Figure 2) with a wingspan of 10 ft and gross takeoff weight of 15 kg, but it has an endurance of over 30 hours in general operations. It can fly to over 20,000 ft with a low fuel load and can routinely operate up to 15,000 ft. Maximum endurance is obtained at low altitudes, but range is unaffected by altitude. Observational resolution is a few tens of meters in the horizontal and several meters in the vertical. The aircraft navigates using GPS and communicates by a combination of radio, telephone, Low-Earth-Orbit satellites and the Internet. The communications enable both transfer of data back for operational use and the upload of new mission files. The unique combination of economy of operation, flexibility, range, and endurance enables the Aerosonde to excel in remote data collection and reconnaissance missions. Its niche is in circumstances where piloted aircraft are too expensive, do not have the endurance, or are exposed to unacceptable risks, and where satellite data are not available or are inadequate. Backing up the aircraft is a sophisticated global reconnaissance facility. Aerosondes are deployed from designated launch and recovery sites, operate under a Global Command Center, and provide data to users through a PC and Internet-based Aerosonde Virtual Field Environment (AVFE). The Global Command Center is the focus of the entire operation, where:

1. All missions are coordinated.
2. A monitoring, regulatory and safety watch is maintained.
3. Multiple mission requirements are coordinated.
4. The flow of data is monitored to ensure that the user's requirements are met.

The Center can be located anywhere there are sufficient communications systems in place. The aircraft operate from launch and recovery sites that require only a large field or open area, a small shed or a vehicle for a transient command center, and a vehicle from which the aircraft is launched (Figure 3). The AVFE is a web-based data display software system, which enables users to access data and monitor the progress of missions in real time from their own personal computers or office workstations. With the AVFE, users can develop modified missions and upload these to the Global Command Center for implementation.

OPTIMAL MIX OF UAV AND OTHER OBSERVING SYSTEMS

The optimal mix of observing systems varies from region to region. In most areas, forecasters have access to a remarkable variety of remotely sensed observations from satellites and radars. As described in Chapter 11, satellites now provide a comprehensive suite of observations that cover the overall cloud, moisture, and precipitation structure of the storm; an approximation of the three-dimensional wind structure by tracking small cloud elements and moisture variations; the surface wind field for speeds below storm force; and the overall thermal structure. Such observations are usually provided by polar orbiting satellites, which leave substantial periods with no observations available. In the future, satellite-based lidars (see Chapter 14) may provide high-resolution vertical profiles of the winds and water vapor above and around the cyclone, clouds permit-

Figure 3. The Aerosonde Launch and Recovery Site used for tropical trials in Western Australia in 2001.

ting. Still there will be the need for observations from cloud-obscured regions, in particular, within the oceanic and atmospheric boundary layers. Direct observations also provide a useful ground truth for some forms of remote sensing systems, including space-based lidars.

Many of these observations are currently taken by piloted aircraft, but these are generally confined to the North Atlantic and eastern North Pacific basins. In the majority of tropical cyclone regions that do not have routine manned aircraft reconnaissance, UAVs may provide the only real potential for direct, targeted observations. A major role that small UAVs can play is to take continuous observations of the center location and intensity from within the cyclone core.

In regions without routine aircraft reconnaissance, the only means of estimating tropical cyclone intensity are the satellite pattern-recognition techniques derived by Vern Dvorak and subsequently updated and extended by many others [Velden et al., 1998]. Unfortunately, the only available algorithms based on hard data have come from aircraft observations in the western North Pacific and North Atlantic Ocean. There may, thus, be substantial errors in the analysis of intensity in other regions. Direct observations by UAVs would provide such ground truth.

Vertical soundings in the periphery of the tropical cyclone provide a valuable supplement to existing observing systems and substantial improvements in cyclone forecasting, especially track forecasting [Aberson and Franklin, 1999]. Such observations are currently being taken by dropsondes deployed in synoptic flight missions by NOAA and the U.S. 53rd Weather Reconnaissance Squadron in the North Atlantic and eastern Pacific. Dropsondes and remote-sensing equipment also can be deployed on UAVs, but are best suited to larger aircraft. Smaller UAVs, such as the Aerosonde, are used as "flying sondes" which take direct operations during horizontal transects and vertical soundings. When deployed in a swarm configuration, small UAVs can provide comprehensive soundings from the near vicinity of the storm. When one considers the additional impressive suite of equipment carried by the manned aircraft, it is obvious that a UAV does not provide a better overall service. However, use of UAVs for long endurance soundings is demonstrably more cost effective than using manned aircraft. They also free up the more sophisticated manned aircraft for other, more complex observing duties.

Deploying high-flying UAVs above the cyclone clouds and circulation could provide useful observations of the upper troposphere, lower stratosphere for intensity, and intensity change forecasting [e.g., Emanuel, 2000]. Dropsondes from these aircraft also would provide highly useful sampling of the entire depth of the storm. At this time, only large UAVs such as the Global Hawk are capable of such deployment.

A major advantage of smaller and cheaper UAVs is that their survival is not necessarily a factor in the operation. Because there is no danger to air crew, the loss of a UAV essentially becomes an economic question. They may, therefore, be used in areas, such as the boundary layer, where manned aircraft cannot safely operate and where other remote sensing systems do not provide observations. Indeed, UAVs can be deliberately used in conditions in which they are unlikely to survive if the requirement for special observations is sufficiently high.

Detailed observations of the near surface layer are of importance for several reasons. The extent of gale, storm, and hurricane force winds is of considerable importance for forecasting. While these are observed in part by satellite remote sensing, for example, QuikSCAT (see Chapter 11), aircraft observations can provide both ground truth for the satellites and additional valuable information. Recent, high-resolution dropsonde observations [Aberson and Franklin, 1999] have shown that the structure of the layer nearest the Earth's surface is highly complex, with localized jets of high speed air and conditions in which the winds may actually be supergradient [Kepert, 2001; Kepert and Wang, 2001]. Since this process is particularly evident on the "weak wind" side of the storm, substantial errors in forecasting the impact of storms may result. Small UAVs are more effective in this environment than dropsondes or satellite remote sensing as they can remain on site for long periods and selectively map out the horizontal and vertical structure of the boundary layer. We expect that this will improve both our understanding and analysis capacity of this important region. We consider that they can also be deployed to fly around the cyclone in the maximum wind region to provide a comprehensive set of observations of the structure and variations of the peak winds.

INSTRUMENTATION

Small UAVs do not have a large payload capacity and are, thus, not able to carry complex and heavy instruments and observing systems. However, there have been major advances in electronics and miniaturization of systems over the last decade, and these are expected to continue into the future. These systems are also becoming less expensive, more robust, and able to handle a variety of tasks. Thus, while advanced Doppler radars and similar observing systems will remain out of reach of small UAVs, a wide variety of instruments will be usable. Those that are under active consideration, or currently in use by the Aerosonde, are listed in Table 1. Large manned aircraft, such as the NOAA P3s and USAF Hercules, generally operate with a wide range of specialized instrumentation for tropical cyclone reconnaissance. This is a logical consequence of their excellent payload capacity and overall cost of operation. Small UAVs have a restricted capacity, but their economy of operation lends itself to individual aircraft specializing in specific observing systems. Thus, a swarm of aircraft can be deployed, some with standard meteorological sensors, others with specialized remote sensing and cloud physics instruments. The manner of such deployment is examined later.

SURVIVAL AND RISK AVOIDANCE STRATEGIES

Survival of the UAV depends on its inherent capacity for operating in severe weather environments and its mode of operation. There is no inherent reason why

Table 1. Instrumentation in use or Planned for the Aerosonde of Relevance to Tropical Cyclone Reconnaissance.

Measurement	Instrument
Aircraft navigation and operations	
Position	GPS, Differential GPS
Aircraft icing	Icing rate detector
Meteorology	
Air temperature, pressure and humidity	Vaisala RSS901 sensor
Winds	Proprietary
Radiometers	
Shortwave flux	Pyranometer UVradiation
Pyranometer(UV)	
Infrared flux	Pyrgeometer
Surface temperature	Pyrometer
Cloud and Aerosol	
Liquid water content	Hot wire probe
Cloud Physics	A.Heymsfield, NCAR
Surface Remote Sensing	
Altitude, wave height	Laser altimeter
Surface visible imaging	Camera, video camera
Infrared imaging	Infrared camera
Ocean surface winds and waves	GPS reflectance

a UAV cannot survive any conditions that can be flown by a piloted aircraft. However, UAV technology has not yet advanced to the point where it can adjust to the variety of conditions in which an experienced pilot can operate. The strong winds in tropical cyclones can cause navigation difficulties, especially for smaller and slower moving aircraft. And, finally, operating UAVs in hurricane conditions, where there are also manned aircraft, presents a potential danger of mid-air collisions.

We undertook a comprehensive risk assessment for NASA in preparation for a 2001 field experiment in the North Atlantic Ocean (Figure 4). Ten major risks were addressed and mitigation plans for each element prepared.

Aircraft collision is avoided by carefully drafted procedures developed in conjunction with regulatory authorities and local airspace users. The UAV utilizes defined corridors when traveling to and from the launch and recovery site across high-use civilian traffic areas. In the cyclone environment, when other aircraft are present, a specific slice of airspace is set aside for the UAV operation — for exam-

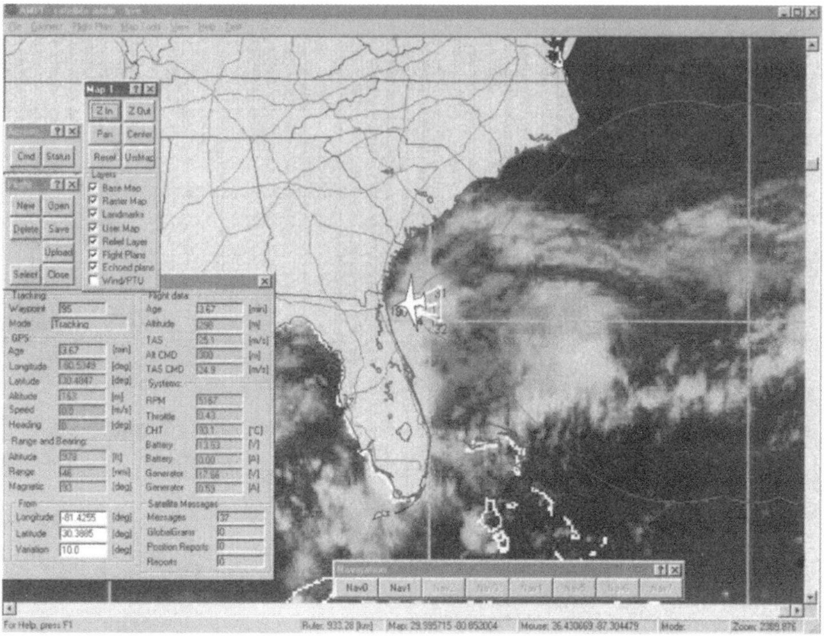

Figure 4. The display and interactive screen from the Aerosonde Virtual Field Environment used during the NASA CAMEX 4 experiment in 2001.

ple, the near-surface layer below 2,000 ft. Onboard flight code monitors the aircraft location and health and can autonomously take action if preset bounds are surpassed. For instance, if high winds take the aircraft out of its defined airspace into a high-risk zone, an engine kill and ditch into the sea may be ordered.

When operating at very low altitude, which is one of the strengths of small UAVs, the potential for collision with ships has to be taken into account. Staying above ship height mitigates this, unless there is a certainty no vessels are in the experimental zone.

Failure of a critical component on the Aerosonde, such as loss of avionics power, communications or GPS, or an engine failure, is also accommodated by the aircraft invoking special procedures. For loss of communications, the aircraft would normally return to base via a predefined route, which is cleared of traffic by the local aviation regulators, and make an autonomous landing. If an engine fails, the aircraft will attempt to glide home. Normally this is impossible, however, and it will ditch in the sea. Loss of GPS is serious, as the aircraft no longer has information on its location. In this case a slow spiral descent into the ocean is immediately ordered.

Our experience has been that atmospheric turbulence and general weather conditions are not a major problem for the Aerosonde, which has weathered very severe turbulence, flown into severe microbursts and thunderstorms, and operated in a wide range of weather systems. However, icing is a potential hazard. While this is a low probability in tropical cyclone conditions, it can occur in midlevel missions. The aircraft have icing alert strategies, which provide an assessment of potential icing conditions and, where needed, icing sensors to monitor the development of ice. The strategy is to descend when icing occurs and allow it to melt in the warmer low-level conditions.

CYCLONE DEPLOYMENT SCENARIO

Small UAVs are able to weather any turbulence found in a tropical cyclone, but their slow speed and size require care in designing missions to minimize the effects of strong winds and convective up-down drafts. For example, the Aerosonde cruises around 25 m/s and can climb or descend at a maximum of 2-3 m/s. Horizontal winds can exceed 70 m/s in severe tropical cyclones and vertical drafts of over 10 m/s are often recorded.

The best method of countering convective drafts is to avoid them by focusing on boundary layer missions. If higher level missions or specialized cloud missions are required, the aircraft is allowed to rise or descend until it exits the draft and autonomously returns to its reference altitude.

Horizontal winds are used to advantage by careful flight planning. For instance, when conducting missions in the cyclone periphery, the aircraft circulate cyclonically, using the circulation as a tailwind to extend range and provide faster response times. For missions to measure cyclone location and intensity, the aircraft enter the cyclone near the surface, where there is a maximum radial tailwind, and exit at higher altitudes. A characteristic mission into the cyclone center is illustrated by the flight shown in Figure 5. This was taken from the sophisticated Aerosonde flight simulation system using a cyclone east of Darwin, Australia. The aircraft initially flies a slightly tangential route, minimizing attempts to directly cross the wind field and taking advantage of a slight tailwind. Closer in to the cyclone core, the Aerosonde tracks on a bearing straight to the center but is swept around by the high winds. The result is a spiral path into the center that effectively utilizes the cyclone winds to best advantage. A return track out of the center follows a similar spiral, but in this case, the aircraft operator first determines the optimal path from the wide variety of possibilities. This would normally mean the aircraft starting out on a track *away* from its intended destination, allowing the cyclonic winds to sweep it around as it exits the core region.

Because the center of a cyclone is not always well known, especially in weaker systems, the aircraft can be deployed to a location in the vicinity, then tasked to find the center by itself. This is accomplished by a combination of flying across the local winds and continuously toward low surface pressure. Thus, for a Northern Hemispheric cyclone, if the winds are on the left wing and the pressure is falling, the UAV would keep its current track. If the pressure starts rising, it would turn left

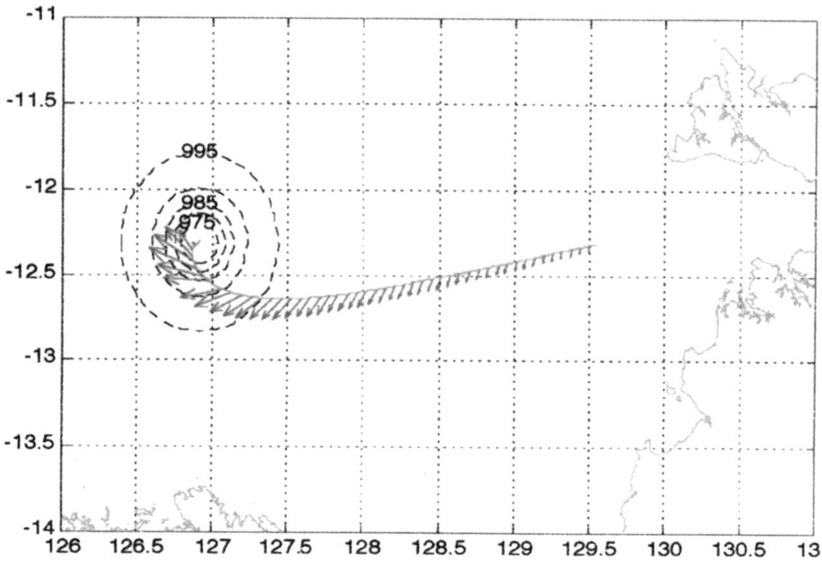

Figure 5. Simulated flight of an Aerosonde UAV into 960 hPa tropical cyclone west of Darwin, Australia

and resume tracking toward falling pressure, ensuring that the winds are still from the left side.

Small UAVs are best used in a swarm capacity. In this case, a group of aircraft, each carrying specific instrumentation systems, is deployed to jointly obtain the observations that are required. We have called this the operation of Smart Aerosonde Clusters, in which a group of Aerosondes communicate with each other and make independent decisions on how to arrange themselves to conduct a defined mission. The defining aspect is the capacity for a group of aircraft to operate together to undertake a mission defined at the macro level. Humans remain in overall command but do not dictate the details of the operation. A simple example is one aircraft proceeding to the cyclone center, navigating itself relative to the winds and derived surface pressures. A cluster of Aerosondes then commence operations to undertake defined missions, such as soundings in the near periphery or mapping the boundary layer and navigating directly from the observations passed by the first aircraft.

CONCLUSION

We have described the range of UAVs that are now coming into use for a variety of civilian and military uses. New developments with GPS and microcomputers combined with technological advances in engines, have enabled a new class of small UAV that has the endurance and capability to supplement current observing systems in providing data in critical areas of tropical cyclones. We consider that small UAVs, such as the Aerosonde, have a major role in observing tropical cyclones. In basins without manned aircraft reconnaissance, they could provide an economical and achievable means for obtaining critical *in-situ* data on cyclone location and intensity. They would also complement satellite and other observations with synoptic soundings and detailed boundary layer observations. Where manned aircraft are available, the UAVs can undertake tasks such as peripheral soundings and release the larger aircraft for more complex operations. Alternatively, the UAVs can be deployed in the boundary layer, where it is too dangerous for manned aircraft to operate.

While there is always a level of risk associated with the deployment of any airborne system, the risk analysis conducted for the NASA field program indicates that risks can be maintained at an acceptably low level when compared with other operations. Specialized onboard procedures are utilized to ensure that system failures or unforeseen events do not lead to increased risk to third parties.

The available instrumentation either in operation or being considered for the Aerosonde covers a wide range of observing potential and reflects the increasing trend toward small, low-powered electronic and remote sensing systems. Because of payload considerations, small UAVs are best deployed in a swarm, with each operating a major instrument in conjunction with its neighbors. We look forward to operations, in which a group of aircraft operates in what we have termed, "Smart Aerosonde Clusters." The aircraft in the group work together to accomplish a specified mission. This simplifies the command structure and enables optimal operations within defined parameters. Will small drones have a major effect on how we observe tropical cyclones? We believe so.

REFERENCES

Aberson, S. D. and J. L. Franklin, Impact on Hurricane Track and Intensity Forecasts of GPS Dropwindsonde Observations from the First-Season Flights of the NOAA Gulfstream-IV Jet Aircraft. *Bull. Amer. Meteor. Soc.*, 80, 421–428, 1999.

Emanuel, K., A Statistical Analysis of Tropical Cyclone Intensity. *Mon. Wea. Rev.*, 128, 139–1152, 2000.

Holland, G. J., T. McGeer and H. Youngren, Autonomous Aerosondes for Economical Atmospheric Soundings Anywhere on the Globe. *Bull. Amer. Met. Soc.,73,* 1987-1998, 1992.

Holland, G. J., P. J. Webster. J. Curry, G. Tyrrell, D. J. Gauntlett, G. Brett, J. Becker, R. Hoag and B. Vaglienti, The Aerosonde Robotic Aircraft: A New Paradigm for Environmental Observations. *Bull. Amer. Met. Soc.*, 82, 889–902, 2001.

Kepert, J., The Dynamics of Boundary Layer Jets Within the Tropical Cyclone Core. Part I: Linear Theory. *J. Atmos. Sci.,* 58, 2469–2484, 2001.

Kepert, J. and Y. Wang, The Dynamics of Boundary Layer Jets Within the Tropical Cyclone Core. Part II: Nonlinear Enhancement. *J. Atmos. Sci.,* 58, 2485–2501, 2001.

Shephard's, *Unmanned Vehicles Handbook 2001*. The Shephard Press, ISSN 1351-3478, 140pp, 2001 (also available at www.uvonline.com).

Velden, C. S., T. L. Olander and R. M. Zehr, Development of an Objective Scheme to Estimate Tropical Cyclone Intensity From Digital Geostationary Satellite Infrared Imagery. *Wea. Forecast.*, 13, 172–186, 1998.

Chapter 14

Hurricane Upper Level Wind Measurements

Applications of Doppler Lidar

G. David Emmitt

INTRODUCTION

As is the case for many aspects of weather prediction, the requirements of models for input data outpace the capabilities of existing observing systems. Increases in model resolution, enabled by the recent rapid increase in computational capability and resources, justify the investment in observations with higher resolution and accuracy. In this chapter we focus on one technology that promises advances in resolution and accuracy: the space-based Doppler wind lidar (DWL). The DWL can provide significant advances in observing wind fields at upper levels of a hurricane and in its immediate surroundings.

OBSERVATIONAL GAPS

As mentioned in Chapters 3 and 4, observing wind fields within a hurricane and its immediate environment is critical to improved storm track forecasting. In Chapters 3, 10, and 11, the primary current wind observing systems were discussed. They include dropsondes from piloted aircraft, scatterometers in low-Earth orbit (LEO), cloud motion vectors (CMV), and water vapor motion vectors (WVMV) from passive instruments in geosynchronous orbit (GEO). The observing systems and their data products are summarized in Table 1.

Hurricane! Coping with Disaster

10.1029/055SP17

TABLE 1. Current Sources for Wind Data Over the Open Ocean

Source	Platform	Spatial Coverage	Spatial Resolution	Temporal Resolution	Accuracy
Dropsondes	Aircraft	100-1000 km	10-100 km	1-8 h per aircraft between soundings	~1 m s^{-1}
Scatterometer	Spacecraft	1000 km swath	12-25 km	12-24 h	~2 m s^{-1}
Cloud Tracking	GOES	2 - 10 km global	5-20 km	30 min	3-7 m s-1
Water Vapor Tracking	GOES	4-10 km global	15-30 km	30 min	4-7 m s^{-1}

From Table 1 we can conclude that the only direct vertical soundings of the wind currently available over the open ocean waters are obtained from dropsondes. Near-surface winds are inferred from the scatterometer signals and that inference becomes problematic at high wind speeds. The CMV and WVMV inferred winds are limited to areas where there are clouds and water vapor structures viewable from a space perspective. These "motion vector" winds have the common problem of inaccurate height assignments of the wind estimates.

In earlier chapters, examples of future wind observing systems were also described. Table 2 provides a listing of several potential observing systems along with their expected data products.

PRINCIPLES OF DOPPLER LIDAR WIND MEASUREMENT

Lidar stands for LIght Detection And Ranging. In many ways, lidar is like radar (RAdio Detection And Ranging). However, instead of radio waves, a lidar transmits a beam of photons and detects the return of those photons that have been reflected off the target. The target may be "hard," such as a building or the Earth's surface, or the target may be "distributed," such as dust particles, cloud particles, or even molecules. Besides the scattering medium, one major difference between lidars and radars is the size of the illumination cross-section or footprint. At ranges of several 100s of km, the very small (approximately 10-100 microrad) lidar beam divergence produces illumination spots only a few 10s of meters in diameter. For equal ranges, radar would require a very large antenna to achieve similar resolution.

TABLE 2. Potential Future Sources of Wind Data

Source	Platform	Spatial Coverage	Spatial Resolution	Temporal Resolution	Accuracy
UAV instruments	UAVs	100-1000 km range	10-1000 m	varies	~1 m s^{-1}
Dropsondes	Stratospheric balloons	0-15 km	200 km	2/day	~1 m s^{-1}
Motion Vectors	GIFTS in GEO	2-10 km	4 km	30 min	3-5 m s^{-1}
Doppler Wind Lidar	LEO platform	0-25 km targeted or global	25 -100 km	1/day	<1 m s^{-1}

Doppler lidar is engineered to detect the frequency shift of the returned signal caused by the motion of the target relative to the lidar. As is also the case for radar, only the relative motion along the line-of-sight (LOS) perspective is detected. There are several lidar techniques for detecting the Doppler shift. The three primary techniques are direct detection of returns from molecules in the atmosphere, direct detection of returns from aerosols and clouds, and coherent detection of aerosol and cloud returns. Briefly, direct detection derives a LOS Doppler shift by observing the intensity of the backscattered photons as they pass through "band pass filters" or their equivalent. If molecules are the primary scatterers, thousands of photons are required to obtain a reasonably good accuracy (errors less than 2-3 m/s). For direct detection, the velocity measurement accuracy improves with the square root of the number of returning photons detected.

Coherent detection works in the frequency domain and thus directly measures the Doppler shift. Coherent DWLs are more efficient in that they only need a few tens of photons to make a very accurate (errors usually less than 0.5 m/s) wind measurement. While assured of an accurate measurement once the threshold number of coherent photons is obtained, the challenge for coherent DWLs is to increase the detection threshold to lower concentrations of aerosols. Although the differences between direct and coherent detection of the wind are several and significant, both approaches have similar technology issues. One is the need for multiple perspectives, and another is the issue of representativeness. Just as critical is the accounting for clouds in the field of view.

To resolve the total wind vector, several LOS perspectives are required. From the ground or from an aircraft, the lidar can be scanned conically (continuously or step-stare) to obtain many different viewing angles, producing a sine wave of Doppler frequency shift that enables the computation of u, v, and w. From space, however, the diameter of the cone near the Earth's surface would be ~ 100s of kilometers and, thus, yield very coarse resolution of the wind field. However, if small (less than a few m/s) and locally random vertical motions are assumed, only two orthogonal (or nearly orthogonal) perspectives are needed to resolve the horizontal wind components. Obtaining several perspectives on the wind field is a major consideration in the design and operation of a space-based DWL. The vertical wind assumption may not be appropriate for all regions near the hurricane.

Given the small beam divergence, the sample volumes of DWLs are small compared to the target sample volumes (TSV) over which they are trying to estimate the winds. Furthermore, a DWL LOS sample is only sampling one component of the wind projection. In practice, the need to average the returns from several shots and the need for multiple perspectives usually result in spreading the sampling over more of the TSV than would be the case for a full vector observation with rawinsonde. Consequently, observations from practical DWLs will tend to be more representative than rawinsondes but still fall short of being truly representative. In a "targeting mode" of sampling, a space-based DWL could be scanned to provide TSVs with horizontal dimensions of approximately 25 km.

At all the laser frequencies being considered for DWLs, the backscatter effects of clouds are only weakly dependent upon wavelength. Thus, the primary cloud issue is one of obscuration (attenuation). Given that the global cloud coverage is approximately 80% (including subvisual cirrus), clouds play a major role in determining where DWLs will make wind observations. The backscatter intensities experienced during the shuttle-based Lidar In-space Technology Experiment (LITE) and the airborne Lidar Atmospheric Sensing Experiment (LASE) suggest that DWLs can obtain wind observations below cirrus and broken/scattered cloud scenes. In the case of direct detection, the presence of clouds reduces the number of photons in the return signal and, thus, decreases the accuracy below that which would have been obtained in a cloud-free scene. For coherent detection, the presence of clouds does not decrease the accuracy of "below cloud" observations but does decrease the threshold sensitivity to aerosols. What this also means, in the coherent case, is that when a signal is obtained from below clouds, its accuracy is undiminished. For hurricanes, the issue of clouds is clearly in play. Experience has shown, however, that lidars can obtain useful observations even in the presence of clouds. In the next section, some observations taken near hurricanes demonstrate the "porosity" of clouds for the narrow lidar beam.

PAST APPLICATIONS OF DWL

Doppler lidars have been used to probe the atmosphere for several decades. Ground-based DWLs have made observations of winds well into the stratosphere [Hall et al., 1984; Chanin et al., 1989; Hawley et al., 1993]. Airborne DWLs have been used to study thunderstorm outflows [Emmitt, 1985; McCaul et al., 1987], convective storms and hurricanes [Rothermel et al., 2000]. In Figure 1 some DWL vectors are overlaid on the outline of the satellite cloud images associated with a line of storms. NASA's Airborne Doppler Lidar (ADL) was flown within the subcloud layer and its beam transmitted horizontally into the storm outflow region. These data were taken during an experiment to visualize the collisions of downbursts and the subsequent response of the convection.

A more recent version of the NASA ADL known as MACAWS (Multi-center Airborne Coherent Atmospheric Wind Sensor) was flown on the NASA DC-8 aircraft during the 1998 CAMEX III (Convection And Moisture Experiment) field campaign. As part of the experiment, the DC-8 and MACAWS flew into several hurricanes, including Hurricane Bonnie. The winds near the eyewall of Hurricane Bonnie were observed by MACAWS and are presented in Figure 2. While DWLs are not known for their ability to operate fully within cloudy environments, the amount of cloud penetration experienced has usually been more than expected prior to these field studies.

An example of the vertical coverage by a laser-based instrument operating over a hurricane is shown in Plate 1. In this case, the lidar was the LITE system flown on the space shuttle. The ability to detect multiple cloud layers is clearly

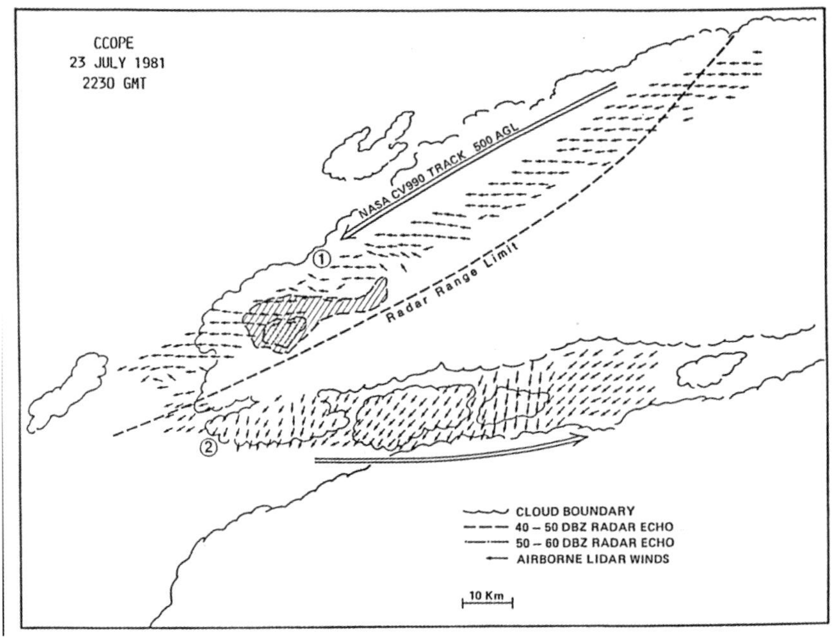

Figure 1. 330 m AGL horizontal wind vectors as determined by NASA's airborne Doppler wind lidar during the 1981 CCOPE. Wind vectors overlain on GOES E satellite imagery cloud boundaries and the radar echo map (> 40 dBZ) [Emmitt, 1985].

demonstrated. This mission data suggests that a space-based DWL may be able to observe the hurricane wind field even below what is considered totally cloudy scenes.

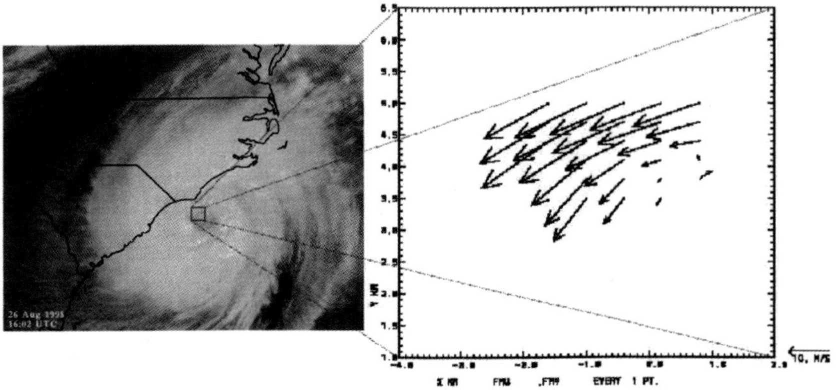

Figure 2. Eye wall wind vectors in Hurricane Bonnie as measured by MACAWS during the 1998 CAMEX III [Rothermel et al., 2000].

Plate 1. Lidar aerosol observation over a hurricane during the 1994 LITE experiment.

DWL IN SPACE

The idea of a space-based lidar (WINDSAT) was studied in the late 1970s by NOAA and the DoD [Huffaker, 1978; Huffaker et al., 1980; Lockheed, 1981]. In the late 1980s, the Laser Atmospheric Wind Sounder (LAWS) was being designed to be a component of NASA's Earth Observing System (EOS) until budget realities led to the de-selection of the wind sounder [Baker et al., 1995]. In the mid 1990s, NASA's New Millennium Program (NMP) selected a DWL for one of its missions, the Space Readiness Coherent Lidar Experiment (SPARCLE). Again, budget considerations led to a termination of that mission as well. More recently, the European Space Agency (ESA) selected a DWL for one of its Atmospheric Dynamics Missions (ADM).

It is noteworthy that the Integrated Program Office of NPOESS lists the direct observation of winds from space as its number one data need that is not currently being met. A very long-range look at the potential of DWLs in space was conducted by NASA in its feasibility study of placing lidars on geosynchronous spacecraft [Emmitt, 2001]. The obvious advantage of a GEO perspective on the hurricane's wind field is the temporal resolution. However, the technology requirements clearly place such a mission at the end of a multiple decade roadmap.

Since the power, mass, and size of a DWL scales to R^2 (R is the range to the target), the most likely insertion of DWL in space will occur in LEO (Low Earth Orbit) onboard the space station or as a free flier. As indicated in Table 2, the potential accuracy of LOS observations is better than 1 m/s. The height assignment errors would be less than 100 m. The horizontal location errors would also be less than 100 m. While these performance numbers are the trademarks of DWL observations, it must be remembered that a single perspective provides incomplete information on the local wind vector. As mentioned earlier, the challenge is to obtain two or more perspectives within a local area. For a hurricane, local may be defined as a few 10s of km. For a space-based DWL the spacecraft's motion is used to provide the opportunities to sample the TSV from different perspectives. Depending upon the coverage of the hurricane desired, trades can be made between broad area (~1000 km wide) coverage and more targeted (~100 km) observations. Figure 3 illustrates the broad-coverage sampling mode over a hurricane.

DWLs do not have any fundamental limitations on the magnitude of the wind speed to be sensed. The accuracy of observing a 200 kt wind is no different (at least for coherent DWLs with ample aerosols) than measuring a 1 kt wind. There is, of course, the potential for some random errors due to any failure to account fully for the satellite's motion and the Earth's rotation. In Table 3, the sequence of obtaining a wind observation from a space platform is outlined. DWL concepts proposed for space (LAWS, ADM, SPARCLE, and Zephyr) have included means to keep the platform and the Earth's motion errors to values less than 1 m/s.

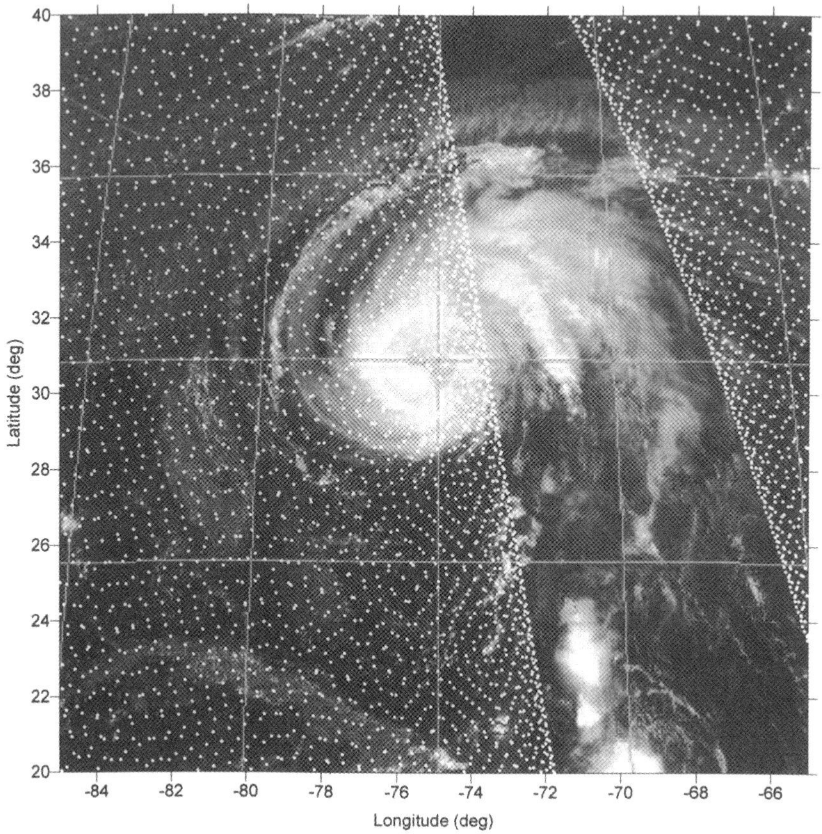

Figure 3. Simulated 24-h space-based DWL coverage from a 833 km orbit over Hurricane Bonnie (August 28, 1998) [Greco et al., 2000].

While DWLs hold great promise for both high accuracy and spatial resolution wind profiles globally (Table 4), their shortcomings must also be acknowledged. The two major shortcomings of lasers in LEO, as far as hurricane observations are concerned, are the 12-14 h temporal sampling from a single satellite and the partial global coverage resulting from ground swaths of less than 1500 km. Both shortcomings can, of course, be remedied with multiple satellites.

Another issue that is frequently raised about the performance of a DWL from space is the matter of clouds. For a hurricane environment, this is a major issue with clouds obscuring significant portions of the storm's wind field. On this point, however, it should be noted that clouds in the field of view offer opportunities for observations as well as impediments. Cloud returns for a DWL can provide very accurate velocity signals that are not compromised by cloud propagation effects or radiance contamination (as is the case for passive sounders). There

TABLE 3. The Space Based Doppler Wind Lidar Measurement Sequence

1. Transmit a pulse of light towards the earth's surface at an off nadir angle (e.g., 40 degrees).
2. Determine the LOS magnitude of the satellite motion relative to the earth's surface. This relative motion includes both the orbital ground speed and the earth's rotation speed at the latitude of the lidar shot.
3. Adjust the DWL receiver to account for this relative motion, leaving only the sensed motion of the air (or cloud) particles for data recording.
4. Record the returned signal in a fashion that allows for different amounts of signal accumulation along the LOS (determines the range resolution)
5. Repeat the process for a desired number of LOS observations to improve the signal to noise ratio. In some DWL concepts, 100's of LOS samples are added together before a final estimate of the LOS wind speed is obtained.

is, however, the matter of vertical motions within the diverging outflow in the tops of hurricanes. Given the proper distribution of many lidar samples, the vertical velocity component should be removable, leaving only information on the horizontal flow. If the DWL data are assimilated into models using four-dimensional variational (4DVAR) techniques, it is likely that the individual LOS observations can be quite useful even if they contain information on the vertical, as well as the horizontal motions. This leads us to explore the potential impact of DWL observations on hurricane forecast models prior to there being a real DWL in space. Efforts to simulate space-based DWL observations of hurricanes are discussed below.

SIMULATIONS OF DWL OBSERVATIONS OF HURRICANES

Both NASA and NOAA have been actively engaged in evaluating the potential for space-based DWL [Atlas and Emmitt, 1995; Lord et al., 2002]. Both agencies have recognized the critical importance of direct wind observations for science investigations and improved weather forecasting. One tool used to evaluate the merits of such observations is the Observing System Simulation Experiment (OSSE). The OSSE quantifies the impact of a new, yet to be launched, instrument on current model analyses and forecasted parameters [Atlas, 1997]. To conduct these model experiments, the proposed instrument's

TABLE 4. Strengths and Limitations of Space-based DWLs

Strengths	Limitations
1. Potential for global high accuracy observations of wind profiles	1. Optically thick clouds can preclude penetration to lower levels
2. Precise horizontal and vertical location information	2. Need two local perspectives to resolve horizontal wind vector
3. High horizontal resolution	3. Regions of correlated vertical motions can compromise the "average zero vertical motion" assumption needed for 2. above
4. Measures motion of particles, not propagation of atmospheric features such as clouds	

performance must be simulated. Key to this process is providing realistic errors due to both instrument as well as atmospheric causes. The atmospheric factors that usually dominate the simulation of space-based DWLs are the clouds, subgrid scale wind variability, distribution of aerosols, and absorbing gases.

Impact studies for future DWLs have been conducted using global models with spatial resolution ranging from 500 km down to 60 km. In those experiments, hurricanes were not resolved sufficiently to study the mesoscale impacts of the simulated DWL observations. At present, NASA is conducting OSSEs using the FVCCM (Finite Volume Community Climate Model) that contains realistic features including several Atlantic hurricanes. NOAA is also preparing for OSSEs using models (e.g., the Penn State/NCAR mesoscale model MM5) that contain realistic mesoscale features, including a hurricane. An example of a simulated 24 h DWL coverage of a real hurricane (Bonnie in 1998) is shown in Figure 3. DWL simulations using mesoscale model fields such as those from the NCEP meso-Eta model have illustrated the expected penetration of the lidar sampling well down into the mid and lower portions of the hurricane. While these simulations are encouraging, such coverage has yet to be demonstrated, either from aircraft or space.

SUMMARY

Accurate, high resolution wind observations of much of the upper tropospheric regions of hurricanes and portions of the mid and lower areas can be expected of space-based Doppler wind lidars, clouds permitting. Complete wind profiles between the surface and the tropopause would be routine in the cloud free and partly cloudy regions surrounding the hurricane. Such profiles would complement the dropsonde wind observations near the center of the hurricane and extend the coverage far out into the regions of the environment that influence the track and evolution of the storm. While airborne DWLs have demonstrated the ability to observe winds within and near hurricanes, a space-based DWL has yet to be deployed. However, recognition of the critical importance of wind profiles by both the research and operational communities is behind the current efforts of not only U.S. agencies, such as NOAA, NASA and the DoD, but also other entities, such as ESA (Europe) and NASDA (Japan), to move forward on launching a DWL within the next decade.

REFERENCES

Atlas, R., Atmospheric Observations and Experiments to Assess Their Usefulness in Data Assimilation. *J. Meteor. Soc. Japan*, 75, 111-130, 1997.

Atlas, R. and G. D. Emmitt, Simulation Studies of the Impact of Space-based Wind Profiles on Global Climate Studies. *Proc. AMS Sixth Symp. Global Change Studies*, Dallas, TX, January, 1995.

Baker, W. E., G. E. Emmitt, F. Robertson, R. M. Atlas, J. E. Molinari, D. A. Bowdle, J. Paegle, R. M. Hardesty, R. T. Menzies, T. N. Krishnamurti, R. A. Brown, M. J. Post, J. R. Anderson, A. C. Lorenc, and J. McElroy, Lidar Measured Winds from Space: A Key Component for Weather and Climate Prediction, *Bull. Amer. Meteor. Soc.*, 869-888, 1995.

Chanin, M. L., A. Garnier, A. Hauchecorne and J. Porteneuve, A Doppler Lidar for Measuring Winds in the Middle Atmosphere, *Geophys. Res. Lett.*, 16, 1273-1276, 1989.

Emmitt, G. D., *Convective Storm Downdraft Outflows Detected by NASA/MSFC's Airborne 10.6 μm Pulsed Doppler Lidar System*, NASA Contractor Report 3898, 1985.

Emmitt, G. D., A Geosynchronous Lidar System for Atmospheric Winds and Moisture Measurements, presented at *IGARSS 2001*, Sydney, Australia, 2001.

Greco, S., S. A. Wood, G. D. Emmitt, M. Nicholls and R. Pielke, Lidar Simulations over Hurricane Bonnie Using CAMEX-3 Data, a Lidar Simulation Model and Numerical Model Analysis, *Preprints 24th Conf. on Hurr. and Trop. Meteor.*, 354-355, Ft. Lauderdale, FL. 2000.

Hall, F., Jr., R. M. Huffaker, R. M. Hardesty, M. E. Jackson, T. R. Lawrence, M. J. Post, R. A. Richter and B. F. Weber, Wind Measurement Accuracy of the NOAA Pulsed Infrared Doppler Lidar, *Appl. Opt.*, 2505-2510, 1984.

Hawley, J. G., R. Tag, S. W. Henderson, C. P. Hale, M. J. Kavaya and D. Moerder, Coherent Launch-site Atmospheric Wind Sounder: Theory and Experiment. *Appl. Opt.*, 4315-4322, 1993.

Huffaker, R. M., *Feasibility Study of Satellite-borne Lidar Global Wind Monitoring πSystem*. NOAA Tech. Memo, ERL WPL-37, 276 pp., 1978. [Available from NOAA/Environmental Technology Laboratory, Boulder, CO 80303.]

Huffaker, R. M., R. T. Lawrence, R. J. Keeler, M. J. Post, J. T. Priestly, and J. A. Korrel, *Feasibility Study of Satellite-borne Lidar Global Wind Monitoring System, Part II.* NOAA Tech. Memo, ERL WPL-37, 276 pp. 1980. [Available from NOAA/Environmental Technology Laboratory, Boulder, CO 80303.]

Lockheed Missiles & Space Company (LMSC) for NOAA/WPL, *Global Wind Measuring Satellite System: Final Report* LMSC-D767868, Contract NA 79 RAC00127, 1981.

Lord, S. J., M. Masutani, J. C. Woollen, J. C. Derber, G. D. Emmitt, S. A. Wood, S. Greco, R. Atlas, J. Terry and T. J. Kleespies, Impact Assessment of a Doppler Wind Lidar for NPOESS/OSSE, *AMS Sixth Symposium on Integrated Observing Systems*, Orlando, FL, January, 8 pp., 2002.

McCaul, E. W., H. B. Bluestein and R. J. Doviak, Airborne Doppler Lidar Observations of Convective Phenomena in Oklahoma, *J. Atmos. Ocean. Tech.*, 479-497, 1987.

Rothermel, J., D. R. Cutten, J. N. Howell, L. S. Darby, R. M. Hardesty, D. M. Tratt and R. T. Menzies, Hurricane Wind Field Measurements With Scanning Airborne Doppler Lidar During CAMEX-3, *Preprints of the AMS 24th Conf. Hurr. and Trop. Met*eor, 354-355, Ft. Lauderdale, FL, 2000.

Chapter 15

A Century After Galveston, 1900

A Personal Perspective and Outlook

Greg Holland

INTRODUCTION

Disasters have a way of sneaking up on communities. Towards the end of the 19th century, the combination of a brash pioneering spirit, a burgeoning economy, and generally benign weather produced a rapid growth in cities along the tropical coast of the United States. Unfortunately, the building, communications, observing and forecasting technology, as well as the community capacity to rapidly respond to a threat, had not kept up. Thus, when the infamous hurricane arrived at Galveston in September 1900, widespread death and destruction was inevitable. As the sea rose, buildings were shattered in the onslaught of winds and waves, and approximately 8,000 lives were lost. Poignantly, many died on a train that could easily have turned back from its approach had there been adequate communications.

The Galveston tragedy has been repeated many times in history. A year earlier, Tropical Cyclone Mahine arrived unannounced and annihilated a pearling fleet in far northeastern Australia, with a resulting 300 deaths. The Labor Day Hurricane devastated the Florida Keys in 1935, and in an ironic twist to Galveston, many died in a train trying to escape the onslaught. Hurricane Tracy visited Darwin on Christmas Day 1974 and showed that a modern town could be

Hurricane! Coping with Disaster

10.1029/055SP18

destroyed by winds alone. Hurricane Andrew proved this again at Miami in 1992.

The death toll in some storms has been incredible. Up to one million people died in tropical cyclones in Bangladesh in the 1960s and 1970s. Bob Southern reported that floods drowned five million people in China in the mid-19th Century as a typhoon moved inland and transformed into a rain system. As recently as 1998, Hurricane Mitch killed 10,000 people in Central America.

Such danger and its associated tragedies require action. As a result, all aspects of understanding, observing, predicting, and reacting to hurricanes has improved substantially over the past 100 years. This closing chapter provides a summary of the advances that have occurred during the 20th Century and concludes with a perspective on future developments.

The century that has passed since the infamous hurricane tore the heart and soul out of Galveston has seen remarkable progress in all aspects of observing, analyzing, understanding, predicting, and warning of tropical cyclones.

OBSERVING SYSTEMS

Surface Observations

In the early 19th Century, FitzRoy and Redfield established the first surface networks and their important communications systems. However, expansion of these networks was slow, and at the dawn of the 20th Century, forecasters had little more than a few ship and island observations to aid them in assessing the presence and structure of a hurricane. Even these limited observations were often not communicated in sufficient time for the forecast cycle. Careful observations of the sky and ocean at the forecast site often provided the best information of the impending threat. Hurricanes were poorly analyzed or even missed altogether, and forecasting was accomplished largely by dead reckoning and persistence.

In 1900, Isaac Cline, Galveston's astute meteorologist and student of Viñes' methods for tracking hurricanes (discussed in Chapter 2) was deeply troubled by changes he observed in cloud and sea conditions as the great hurricane approached. But he lacked authority to warn of the impending disaster, an action that had to originate in the Washington office. A year earlier, Clement Wragge in Australia could only issue a general alert of potential cyclonic conditions before the most severe cyclone ever observed in the Coral Sea came ashore and destroyed all in its path.

The surface-observing network has expanded substantially since Cline and Wragge's time. Specialized instrumentation on offshore structures and reefs, ocean buoys, a coordinated program of ship observations, and excellent communications mean that modern forecasters have access to hundreds of observations in the vicinity of a tropical cyclone. Nevertheless, these surface observations are generally insufficient, as the small intense core of a tropical cyclone is rarely directly observed by surface platforms, and even the extent of gale and storm force winds are difficult to gauge[1]. As a result, serious errors of analysis and interpretation would still occur from use of surface observations alone.

Upper-air Observations

Aside from local observations of the motion of clouds, hurricane forecasters in 1900 had no information whatsoever on the structure of the wind and thermal fields aloft. Thus, they could not apply even the most basic techniques of assessing cyclonic movement from steering by the deep environmental flow. Indeed, tropical cyclones were viewed as shallow phenomena of only a few kilometers deep, overlain by the normal, undisturbed atmospheric motions. In Chapter 2, Ed Rappaport and Bob Simpson noted that a comprehensive upper-air-observing network was established during and following WWII. Included were land and island stations and dedicated weather ships strategically located to provide data in critical places. This network provided substantial improvements in the forecasters' ability to analyze the environment in which the cyclone was embedded. However, it provided very few direct observations of the cyclone core. Even when an observing station found itself in the cyclone core, soundings were often prevented by the difficulties of launching a balloon in high winds. There have been only half a dozen radiosonde soundings ever recorded in the eye and maximum wind region.

Ground-based Radar Observations

The first comprehensive observations of the structure of the hurricane core came with the development of radar. Originally designed to detect military aircraft and ships, radar also proved to be very effective at observing the water and ice in clouds by the magnitude of the reflected signal. Perhaps the earliest recorded image of the characteristic spiral clouds and circular eyewall of a tropical cyclone was that of Typhoon Cobra, which inflicted substantial damage to Admiral Halsey's fleet in the western North Pacific Ocean during WWII. The remarkable potential of radar for observing weather systems was soon recognized, and a comprehensive radar network was rapidly established in developed countries in the 1950s and 1960s and then steadily expanded.

Radar provided unprecedented information on the location of tropical cyclones (Chapter 2), together with general characteristics such as the radius of the eye and the location of bands with their accompanying high winds and rainfall. Only minimal information was provided on the cyclone intensity until the 1990s, when Doppler radars were deployed. These radars measure the change in frequency of

[1] Paradoxically, the improvement of cyclone forecasts has caused fewer direct surface observations in the cyclone core, as ships now steer well clear!

the reflected microwave energy and, thus, can estimate the speed of the reflector by the well-known Doppler effect. When two radars have overlapping areas of coverage or when a radar is moving in a circular motion (e.g., on an aircraft), a full three-dimensional map of the wind fields can be obtained with high resolution.

The main limitation of weather watch radars is their range, which is limited by two major factors. The wavelength, power, and the antenna size define the degree of absorption of the emitted signal. A 3-cm wavelength radar with a small antenna will barely see beyond the nearest convective band, whereas a 10-cm radar with a large antenna can observe through multiple bands. Regardless of the degree of absorption, the curvature of the Earth provides the ultimate limitation. This means that coastal radars can observe a cyclone, at most, around 400 km out to sea and, at this distance, only at the upper levels. Doppler radars have a much shorter range of only 100 km or so. Thus, a cyclone directly approaching shore will typically be observed for less than 24 h before it makes landfall.

Aircraft Reconnaissance

Full information on the intensity, location, and inner structure of tropical cyclones only became routinely available with the advent of aircraft reconnaissance. In Chapter 3, Bob Sheets describes in detail the routine reconnaissance of tropical cyclones that was first undertaken by the U.S. military in the North Atlantic and western North Pacific, driven by the considerable damage done to the Pacific fleet by two typhoons during WWII. These were followed by civilian flights sponsored by NOAA and pioneered by Bob Simpson and his colleagues.

Aircraft initially provided flight-level observations, together with estimates of the surface winds, based on pattern recognition of the sea state. Driven by the establishment of the U.S. National Hurricane Research Project (now the NOAA/AOML Hurricane Research Division), aircraft instrumentation has improved substantially. The efforts to develop a hurricane modification program provided substantial funding for the two NOAA P3 aircraft, which are now among the most sophisticated observing systems on Earth (Chapter 3). The U.S.A.F. 53rd Weather Reconnaissance Squadron operates C130s fitted with both *in-situ* and dropsonde capability, and a recent addition to the NOAA fleet, the Gulfstream IV, provides an extended-range dropsonde survey of the cyclone environment. For special programs, other aircraft, such as the NASA DC-8 and ER-2, have been deployed with sophisticated remote sensing instrumentation.

Satellite Remote Sensing

The explosion of observations available from remote sensing satellites has been the defining aspect of observing tropical cyclones in the latter third of the 20th Century. Since the first grainy image of a tropical cyclone taken over the South Pacific on April 1, 1961, satellite observations have advanced to become the dominant provider of information on tropical cyclones around the globe. In Chapter 11, Chris Velden and colleagues describe an array of polar orbiting and geostationary satellites that now provide multi-spectral information on clouds and water vapor fields, direct downward-looking radar images, microwave observations of the ocean surface and derived surface winds, cloud-drift winds, vertical temperature profiles, and rainfall. Perhaps the greatest contributions from satellites are their ability to provide data globally and an easily accessible means of detecting cyclones and estimating their intensity. Because of the enlightened global policy instituted by the World Meteorological Organization of distributing weather observations free of charge, satellite data are available to all forecast offices. This is not the case for other observing systems. For example, aircraft reconnaissance is now only routinely available in the North Atlantic and eastern Pacific Ocean.

COMMUNICATING THE DATA

Meteorological data grow old and quickly become useless for real-time forecasting. Therefore, rapid electronic communication and processing of data are critical to the forecast process. For the Galveston hurricane, telegraphic transfer of data and warnings was available but not helpfully used (see Chapter 2). But such systems were prone to fail in severe weather conditions. Even with the advent of radio around 1910, it was very difficult to get observations from important locations, such as ships at sea.

Increasing use and expansion of radio and telegraphic systems resulted in steady, but unspectacular improvements in the first half of the 20th Century. After WWII, the International Meteorological Organization, and its successor the World Meteorological Organization, established the Global Telecommunications System. This is a massive and complex system operating through several major collection points and gateways that collects and distributes meteorological observations around the globe. National networks supplement this global system. Together, these systems generally ensure that local data are available within one to two hours, and global data are distributed within several hours of the observation time.

Most recently, the establishment of the Internet has led to a revolution at least the equal of the Global Telecommunications System. A variety of data are routinely available from specialized websites and the use of ftp transfer. This communications medium will probably grow to become the dominant mode of data and forecast communications in the future.

UNDERSTANDING TROPICAL CYCLONES

In 1900, knowledge of tropical cyclones was rudimentary. That tropical cyclones were large atmospheric vortices was well known. Informed meteorologists like Isaac Cline also knew of the strong winds at the core, the heavy rainfall that could cause flash flooding, and the potential for huge waves and rising seas at landfall. However, he and his contemporaries incorrectly considered that cyclones were shallow systems of only several thousand feet deep and had virtually no knowledge of the mechanisms that lead to formation, intensification, or motion.

As we move into the 21st Century, tropical cyclones have been observed and analyzed in great detail, and our understanding of both their structure and internal mechanisms is quite profound. The advances in the science were largely led by improvements in observing systems. As new observations became available, the theoretical and knowledge base adjusted accordingly. This has changed in the last decade or so as the use of numerical models and powerful analytic tools has provided substantial improvements in knowledge and understanding beyond those of observations.

Kerry Emanuel, Hugh Willoughby, and Liz Ritchie and her co-authors (Chapters 9, 10, and 12) have described these advances in detail in earlier chapters. They show that tropical cyclone understanding has moved through several phases.

Perhaps the first comprehensive analysis of the surface wind structure came from the detailed observations taken by the Japanese Imperial fleet of a typhoon in the western Pacific in 1935. Subsequent analysis clearly showed the high-wind core, the extent of damaging winds, and the asymmetric structure, which could be related to the cyclone motion. Following WWII, the research and operational community rapidly developed the capacity to observe and analyze in detail the basic structure of the core and nearby environment. They analyzed the low-level wind structure and found it to be similar to a well-known type of fluid vortex, called a Rankine or potential vortex. Radar and aircraft observations enabled detailed analysis of the rainband and eyewall structure, the balanced nature of the vortex (with its central warm core and winds decreasing with height), and the "in-up-and-out" radial circulation. More and more sophisticated observing systems have enabled detailed analyses of the dynamics and thermodynamics of the cyclonic core. For example, we have documented the family of rainbands, how they develop and evolve, and their considerable similarity to general tropical squall lines. We understand the role of Rossby waves propagating through the main cyclonic vortex in the rainband generation and evolution.

Surprisingly, given the excellent aircraft observational base, the presence and importance of small vortices in the development and decay of tropical cyclones only became widely studied in the last decade. A solid body of theoretical and observational research has shown that complex vortex interactions across a wide range of scales are critical to the entire lifecycle of tropical cyclones. The essentially stochastic nature of these interactions provides a limit on their predictability.

That the tropical cyclone gathers its energy from the ocean is beyond question. However, the precise manner in which this occurs remains uncertain. Effects such as waves, spray, and other changes that occur with high winds have yet to be fully quantified. Cyclones also are affected by their immediate environment. Passing weather systems, upper-level troughs and ridges, nearby cyclones, and even quite subtle systems in the cyclone vicinity have been shown to be capable of marked effects on intensity and motion. Here again the precise mechanisms are still only vaguely understood, with a complex soup of wave dynamics, vortex interaction, and energy exchanges all contributing to the final result.

Cyclone motion was considered for many years to be essentially that of a "cork in the stream." The cyclone was steered by the larger-scale atmospheric flow averaged over some depth. We now understand that wave dynamics, interactions with nearby land, asymmetries in the cyclone circulation, and even the vertical tilt of the eye region have a substantial impact on motion.

The cyclone effect on the ocean is well understood and relatively easy to reproduce, provided the detailed ocean bottom topography is known. For example, storm surges, like the one that devastated Galveston can be predicted to a high degree of certainty with coupled ocean models. However, small errors in the track or the surface wind field lead to large errors in the surge assessment.

Ultimately, perhaps the most surprising aspect of a tropical cyclone is not its complexity but the large degree of inherent consistency. The surface wind field can be derived to a first order by quite simple radial equations. These form the basis of much of the development of cyclone-resistant building codes in affected regions. The energetics of the cyclone can be modeled as a simple heat engine, and very simple thermodynamics have proven capable of estimating the maximum intensity that can be reached. Vern Dvorak developed a remarkable cloud pattern recognition scheme that provided the first step in analyzing intensity for tropical cyclones around the world. And to a first order, a tropical cyclone does move with the integrated flow of its environment.

Unfortunately, the small scale of the severe part of a tropical cyclone means that the details become very important. A 100-km error in a track forecast may be an excellent first order effort. But in many cyclones, 100 km is the difference between devastation and a minor inconvenience for those in its path. It is, therefore, the outliers, those cyclones that suddenly intensify or change direction or develop strong asymmetries, which cause the major problems, and the reasons for this behavior lie in the details.

FORECASTS AND WARNINGS

Early in the 20th Century, forecasts were largely empirical and based on forecaster experience. This produced practitioners who are legendary for their ability to consistently produce good forecasts, remarkably so when one considers the paucity of information and the lack of detailed understanding of a hurricane's inner workings. Outstanding examples include Father Benito Viñes in Cuba, who analyzed the infamous Galveston hurricane as it passed into the Gulf of Mexico;

the Australian Clement Wragge, a crusty individual who instituted the naming of tropical cyclones (and was not averse to naming severe storms after politicians who had raised his ire); Father Depperman in the Philippines, a genuine forecaster-scientist who contributed much to our early knowledge of cyclone characteristics; and Grady Norton at the U.S. Hurricane Center in Miami, who steered the center through several major hurricane strikes (and eventually died on the job).

These early forecasters have one common failing — even the best eventually had a major forecast bust. Capricious storms, combined with a lack of quantitative information and relatively unsophisticated forecast techniques, eventually led to a poor warning of a destructive storm and, thus, a major career disaster. Grady Norton provided an eloquent summary of the fate of forecasters from this era in his unpublished soliloquy, where he provided the following advice to junior forecasters:

> *All you can hope for, when you finally retire a broken and dejected man, is that the flags were always raised*[2] *and you have no congressional enquiries to your credit.*

The major improvements in forecasting up to WWII relate more to our ability to detect and inform, in the most general ways, rather than explicit improvements in our ability to forecast formation, track, and intensity. Following the war, however, several initiatives combined to provide a major upturn in forecast skill.

The invention of the electronic computer has been by far the greatest advance. It has complemented the observing system advances described above, enabled the major upgrades to communications, and provided the capacity for new analytical techniques used in improving our overall understanding of tropical cyclones. Most importantly, modern computing systems have enabled the development of advanced forecast techniques, such as statistical methods and sophisticated numerical models of the atmosphere. In addition to these advances, we must include the establishment of specialized cyclone forecast centers and the use of targeted research programs aimed at specific forecast requirements.

Many countries initiated specialized forecast centers during the 1950s and 1960s. In addition to the obvious benefits arising from a concentration of expertise on tropical cyclones, these centers commenced careful reanalysis of tropical cyclones at the end of each season, in many cases also going back to earlier times, to provide a good database for use in research and technique development. They provided an enhanced political and scientific push toward development of new techniques and systems. Specialist forecasters also provided a "public face" that vulnerable communities could identify with in interpreting and understanding warnings.

[2] Early warnings were communicated by raising the storm warning flags.

Targeted research programs have helped the community substantially improve its understanding and development of new techniques. The oldest and most successful of these programs has been the hurricane project established by NOAA, originally known as the National Hurricane Research Project and now as the Hurricane Research Division in the Atlantic Oceanic and Meteorological Laboratories. This group has contributed widely to the science, provided the basis for the instrumented P3 aircraft that are used in both research and forecasting activities in the North Atlantic and elsewhere, and developed a range of advanced forecast and analysis techniques. A second program of note has been the international effort to increase our understanding of cyclone track and drive improvements in forecasting sponsored by the U.S. Office of Naval Research. This program has substantially increased our understanding of cyclonic movement. It also produced a number of forecast aids including a systematic approach for forecasters to use in interpreting numerical model output, and it initiated the development of the Aerosonde UAV described in Chapter 13.

The first moves toward quantitative forecast techniques came with the development of statistical forecasting aids (Chapter 4). Charlie Neumann at the National Hurricane Center pioneered much of the work in this area. He introduced the basic CLIPER technique in which climatology (CLI) of past storms and persistence (PER) of the current track are used to forecast the future track. This technique remains in use today as the benchmark against which all other techniques are measured. There followed various types of statistical regression techniques including those that use the output of numerical models.

Lewis Fry Richardson conceived the idea of numerical models of the atmosphere in the 1920s, but it was not until after the ENIAC electronic computing system was developed that Jule Charney and John von Neumann were able to demonstrate the feasibility of undertaking the required intensive calculations. Since that time, we have seen a remarkably creative period of development in everything from numerical integration techniques to methods of representing physical processes (such as release of latent heat in clouds) as well as methods of objectively analyzing data from widely disparate sources. These models have led to major improvements in the accuracy of forecasting cyclone tracks in the past decade. Specialized ocean wave and storm surge models also provide solid quantitative estimates of these important parameters.

FORECAST SKILL

Statistics on forecast skill have been kept for most of the globe since the 1960s and in some regions earlier than that. Forecast track and intensity are routinely monitored and, in some cases, formation as well. Other critical factors such as extent of gale or storm force winds, rainfall, and storm surge are rarely monitored, largely because of the difficulty in providing effective data for comparison.

Formation

The skill of predicting tropical cyclone formation is difficult to estimate. The main difficulty lies in the somewhat arbitrary definition of formation. We could use the onset of tropical cyclone winds or the initiation of a closed (circular) wind pattern, but either definition is arbitrary. There also are no coherent long-period statistics that can be drawn upon, and indeed it is only the few occasions where cyclones develop close to major population centers that there is a significant forecast problem. If we define formation as being the capacity to forecast an initial closed circulation where none currently exists, then a subjective estimate of skill is that genesis is forecast 24 h ahead of time approximately 50% of the time. However, there are a significant number of false alarms, and the skill rapidly drops off with time. Compared to persistence and climatology, genesis forecasts show modest but real skill.

A major component of genesis forecasting skill is the remarkable capacity of global numerical models to catch the time and place of hurricane formation. While there is a high false alarm rate, there are many examples of these models forecasting development up to 48 h ahead. The implication is clear: most good numerical models implicitly contain the major mechanisms for cyclone development.

Intensity

Cyclone intensity is forecast largely by persistence and climatology combined with some statistical and empirical schemes. Although some numerical models now have the resolution and physical processes to enable at least rudimentary forecasts of intensity, none show real skill in this area yet. Perhaps as a result, the official forecasts are also little better than climatology and persistence, if at all.

Track

After a period of steady improvement at around 1% per year, track-forecasting improvements have been impressive in the last decade. This has especially been the case for longer time periods. As shown by Mark DeMaria and Jim Gross in Chapter 4, North Atlantic forecast errors improved to the level that the 72-h forecasts are now more accurate than the 48-h forecasts were ten years ago. This improvement is due to three major factors: observations, which have been targeted to the major forecast issue; numerical forecast guidance; and understanding of cyclone dynamics, as well as the methods derived by Russ Elsberry and his collaborators to provide a logical forecast framework for applying this understanding.

Ocean and Rainfall

Waves and storm surge can be accurately forecast if the track and surface wind structure are known in sufficient detail, which unfortunately is not the case. Rainfall is poorly forecast in all but a general sense. In particular, extremely high rainfall amounts associated with local topography are not forecast well at all. As with oceanic components, rainfall forecasts are highly sensitive to errors in track and details of the cyclone structure, such as the development of rain bands.

COMMUNICATING THE MESSAGE

Even the best forecast becomes effective only when it has been adequately communicated to the vulnerable communities that lie in the path of the storm. In Isaac Cline's day, it was difficult to even communicate the message, as was evidenced by the train that continued towards Galveston as the cyclone approached, leading to great loss of life. Phenomenal advances in communications occurred during the 20th Century, and most regions threatened by tropical cyclones now receive saturation coverage, perhaps even to the point of information overload. Effectively communicating the relevant information in a manner that evokes the best community response appears to be the biggest remaining challenge.

Sophisticated warning systems have evolved to alert communities of the danger. The alerting systems are similar around the world and generally have four phases:

1. A general outlook, which can consist of alerting specialized users to the potential for an early or more active hurricane season, or reminding communities to take prudent pre-season actions;
2. An early warning or watch phase, aimed at alerting communities of the possible threat, especially those that need longer lead times to prepare for a cyclone;
3. A direct warning of the likelihood of tropical cyclone conditions being experienced in the near future;
4. A running update on the storm movement, development, and effects as the event unfolds.

Despite the sophistication of alerting systems, communications are not always fully effective. Cultural differences and individual perceptions mean that even carefully crafted messages may not always evoke the required responses. Sometimes this is due to the poor judgment of certain members of the public. All cyclone events produce plenty of examples of "hurricane parties," groups coming to the beach to take advantage of the high surf, enjoy the excitement, or those who just simply refuse to take notice. A commonly-told story is the effective message that was relayed by a local policeman, who carried an indelible pen with him and asked those who refused to evacuate in the teeth of the storm if they would mind writing their names on their wrists to aid in later identification of their bodies!

However, there are many more examples of community groups that are unable to fully comprehend the risk or are incapable or too frightened to take effective action. The complex mix of economic, cultural, language, and age groups in most coastal communities makes it very difficult for a general message to have the optimal impact.

The societal aspects of reducing impacts and effectively communicating the threat and degree of vulnerability to communities, as well as evoking adequate responses, have seen significant advances in the last couple of decades. But this societal aspect of the cyclone warning process is barely out of its infancy.

THE NEXT 100 YEARS

Some may claim that the surprise arrival of a Galveston hurricane could not occur in the future. Or could it? There remains the potential for rapid intensification on the coast. The disastrous impact of rain and flooding from Hurricane Mitch is a recent memory. And many countries do not have the degree of sophistication of the United States and other developed countries in observing and forecasting systems.

We now attempt to address how observing, forecasting, and societal systems may improve and also estimate the ultimate limitations of forecast accuracy. While there are a range of issues that could, indeed should, be considered, such as inland rainfall and tropical transition to extratropical cyclones, we will concentrate on the specific issue of coastal impacts by a tropical cyclone making landfall or moving close to vulnerable communities. As shown in Plate 1, forecasting coastal impacts involves all aspects of the cyclone — from track and intensity to direct impacts of high winds, rainfall, and oceanic inundation.

Any attempt at foreseeing the future is fraught with difficulty. It is also somewhat akin to cyclone forecasting: persistence of current trend is usually fine for the short term, but longer-term forecasts must take into account new developments, some of which may be beyond our predictive capacity or imagination. To illustrate the problem, let us place ourselves in Isaac Cline's time and try to envision the state of forecasting at the end of the millennium. We would probably have made good estimates of the expansion of the surface-observing network. But would we have foreseen the invention of the computer? What of satellite remote sensing? Would we have predicted aircraft reconnaissance?

Seeking to peer into the future, therefore, requires a canny combination of extrapolation, identification of outstanding requirements, and a feel for potential changes. For example, in 1855 Henry Piddington called for reconnaissance of tropical cyclones using sailing ships. He could see the requirement for detailed observations of the cyclone core, but the means were inadequate. One hundred years later aircraft reconnaissance turned Piddington's dream into reality.

Needs, Potential, and Driving Forces

So where are the major needs and potentials? We expect the major driving forces to include:

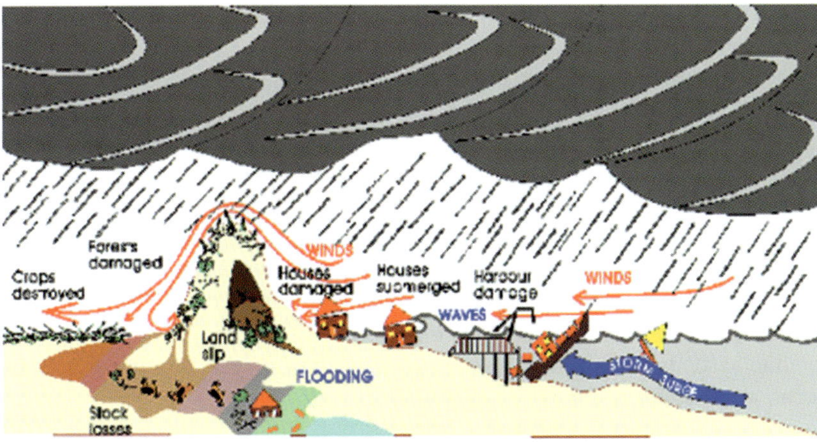

Plate 1. Some coastal impacts of tropical cyclones (Rex Falls, personal communication, 1997)

1. Economic pressures,
2. Societal changes,
3. Commercialization of services,
4. Technological advances
5. Improvements in scientific understanding
6. Improved forecast techniques

Economic Pressures. Governments around the world are all seeking ways of reducing costs. As such, the cost of maintaining even basic observational systems is becoming increasingly difficult to sustain. While some high-profile or advanced technology systems have expanded in the last thirty years, several observing systems have shrunk. Examples include the upper-air radiosonde network and aircraft reconnaissance of tropical cyclones. Even the basic surface-observing network is under threat in many regions. Unless there is a major disaster that causes policy makers to reverse these reductions, the cost of observing systems will become increasingly a driving factor.

Currently, satellite systems provide a buffer against shrinkage of observing systems in areas other than the United States. However, a substantial component of the expansion of the satellite system is not directly related to weather issues. National prestige and the manner in which space developments drive other technological advances are significant factors. A few advanced countries now supply all the satellites used around the world. Other countries are able to access the data free under international agreements on exchange of weather data. There are many similarities to the rapid expansion of the upper-air observing network in the 1950s and 1960s. It is reasonable to expect that the cost of maintaining major satellite systems will eventually be questioned, and a reduction in service will occur, as has happened with the upper-air network.

Societal Changes. Over half the world's population is subject to tropical cyclone threats, and some of the largest population increases are occurring in threatened regions. People like living in coastal regions. This desirability means that coastal property values are increasing. It is somewhat paradoxical that countries are cutting back on meteorological observations and services, when the value of property and the number of lives at risk are increasing. We may see a reversal, but one in which funds are much more directly aimed at solving specific problems. For example, the cost of hurricane evacuation averages many hundreds of thousands to a million dollars per mile of coastline. Substantial support, therefore, may become available for reducing the warning zones.

There is, at present, a wide range in educational level and degree of sophistication of coastal communities. It will be important to increase the general level of education, awareness, and the ability of these communities to interpret the complex information underlying hurricane watches and warnings. This is necessary so that people will respond logically when evacuations are ordered and not over-respond or panic in the face of a tropical cyclone threat.

Another looming problem is the withdrawal of insurance coverage for cyclone-prone regions. One or two major tropical cyclone catastrophes could bring down the global insurance industry, and they may well withdraw before this happens.

Commercialization of Services. Some meteorological services have already been commercialized, and given the attitude of economic rationalization prevalent in governments around the world, this trend may well continue. In addition, major television and radio networks already provide many direct services to the public. Although forecasters with these networks refrain from issuing their own hurricane warnings, they do provide a significant degree of interpretation. The rise of global television networks, such as CNN, has produced world forecasts from the one center. One wonders how long it will be before a few major forecast services evolve to cover the entire globe.

Technological Advances. People born near the middle of the century have lived through amazing technological advances in computing systems and communications. These have led to the numerical modeling advances described earlier and the establishment of personal communication systems through the web and cellular phones.

Computing advances will continue at a great rate. Should proven systems such as quantum computers become technologically feasible and be combined with room temperature superconductors, global weather analyses and forecasts will become available at a horizontal resolution of meters, instead of the tens of kilometers of today's models.

What other technological advances will we see? Will computers become intelligent? Will humans merge with computers so that surgically implanted systems will enable true personal access to data and forecasts? Suffice it to say that, if we achieve nothing more than what we can currently perceive, the world will be a very different place. For example, although many people in hurricane-prone areas do not yet use computers or advanced communications, they are benefiting indirectly in improved forecasts and warnings. We can expect that the entire world will be "wired" into the global computing and communications systems within the next century.

Alternative forecasts and analyses for specific tropical cyclones can already be found by checking relevant web pages. While this has the potential to cause a degree of confusion if the information is spread throughout the general community, it also provides the potential for substantial improvements in forecasts. Small forecast offices, that even a decade ago would have struggled to receive basic observing information, can now access analyses and forecasts from some of the most sophisticated systems on earth. Already we are seeing discussions by a wide range of "experts" on dedicated bulletin boards. Forecasts and warnings are dissected and argued over in real time. One hopes that this chaos and creativity will lead to more informed and accurate forecasts and warnings rather than increased public confusion.

The 20th century was the time of computers and communications; the 21st Century will be the era of automation. We are seeing the beginnings of this now, with the advent of robotic aircraft (Chapter 13), increasingly automated surface observing systems, oceanic buoys, and satellites. It is likely that robotic systems will provide a degree of specificity and capacity for targeted observations that we can barely dream of at present.

Improvements in Scientific Understanding. While we now understand much of the large-scale characteristics of the tropical cyclone, we have not yet adequately addressed the finer scales. Improvements in forecasting will increasingly require a better understanding of the detailed structure of tropical cyclones. One example is the interaction of small-scale vortices and perturbations addressed in Chapter 12. The interactions between hurricanes and the ocean, the surrounding environment, and the interior physical processes in clouds are of increasing importance. In addition, the economics of building standards and insurance will place a greater emphasis on the manner in which high winds interact with urban boundary layers to cause destruction.

Given the expected advances in observing capacity, computing systems, and the powerful analytic tools that are now becoming available, there are no inherent reasons why we cannot understand all relevant aspects of the hurricane. The only real limitation will be the availability of resources, especially since many further advances may require sophisticated and expensive observing systems.

One issue that we have barely addressed is the ultimate limit of forecast skill. Lance Leslie and his collaborators have shown that the limit for short-range track forecasting may be around half of the current forecast errors. But there are no similar estimates for intensity, structure, or genesis; and there has been no real assessment of the limits of forecast accuracy beyond 72 h. Such estimates are increasingly important as we try to develop better observational and modeling systems in an effort to extend the range of accurate forecasts and warnings sufficiently to allow all in harm's way to find safe shelter when a hurricane approaches.

Improved Forecast Techniques. Much of the current trend toward more accurate forecasts has resulted from the maturation of numerical modeling systems. This has included improved numerical methods, ways of handling physical and sub-grid-scale processes, and increasingly sophisticated techniques for using data from diverse sources. Such trends will continue for the near term, with the modeling community benefiting from both the advancements in computing technology and the continuing strong research programs of model development. Global models will soon have the capacity to resolve the core of tropical cyclones. Improved data assimilation techniques and, in particular, the use of targeted observations will revolutionize the initial analysis effort. And ensemble or multiple forecasts, using slightly different initial conditions, will continue to both improve the most likely forecast and provide an assessment of the potential error or uncertainty in this forecast.

Forecasts, Warnings, and Community Responses

The accuracy of forecasts is likely to continue improving over the next several decades. Given this scenario, we can expect to approach the fundamental limits imposed by the nonlinear nature of the atmosphere. Certainly this is a broad assessment since inherent predicability limits have yet to be resolved, possibly because we perceive ourselves as far from such limits. Nonetheless, track and intensity forecast errors may decrease by half, and formation may be entirely predictable, but at some cost of false alarms. Substantial improvements of intensity forecasts will be difficult to obtain. However, forecasting the bounds or limitations of intensification has substantial promise. The state of the ocean (high seas, storm surge, etc.) is highly predictable now, provided the cyclone track and structure are known. These ocean conditions will continue to be limited by the forecast uncertainties associated with these parameters.

The time period of useful forecasts will increase, but this period will be limited by the ability to predict cyclone development and decay.

Numerical models will be the backbone of forecasts, using ensembles and combined statistics. Forecasts will increasingly include uncertainty as well as the best estimate. This uncertainty will be used in the preparation of warnings and assessing the action that should be taken by vulnerable communities. That is, when the uncertainty is low, the warning areas will be reduced and vice versa. While the use of statistical methods has gone out of fashion at present, these will be reintroduced to provide a basis for the warning preparation. Included will be a combination of local cost-loss assessments within the warning process.

Continuing emphasis on reducing cost-benefit ratios will see a rationalization of the forecast and observing systems. Observations will be increasingly tailored to fit the current requirement, with direct interactions between numerical models and the observing systems as well as increasing the use of automated or robotic systems to complement remote sensing systems. It is even likely that the next decade will see the number of observations peak and then decline. A move may be expected to occur away from numbers to quality, as we become smarter at using the observations with targeted approaches.

Exciting though these scientific and technological advances are, they will be dwarfed by new ways of using the forecasts. In addition to the traditional parameters of track, intensity, and structure, forecasts will provide estimates of impacts in which local vulnerability and capacity to react is directly coupled to cyclone parameters. Inclusion of uncertainty, via ensemble or other approaches with local cost-loss estimates, will enable substantial improvements in the capacity of communities and industry to react in an optimal way. Internet communication provides the capacity for this to occur down to the individual household or even personal level. Indeed, there are no inherent reasons why a personal profile could not be stored in a supercomputer and used to provide a direct assessment to a cellular phone or other personal communications system. The next century is likely to be even more surprising and exciting than the century inaugurated by the Galveston hurricane.

APPENDICES
RECOMMENDED FURTHER READING

Anthes, R. A., *Tropical Cyclones: Their Evolution, Structure and Effects.* Meteorological Monographs 19, Amer. Met. Soc., Boston, 208 pp., 1982 (ISBN 0-933876-54-8)

Cline, I. M., *Storms, Floods and Sunshine, An Autobiography.* Pelican Publishing Co. Gretna, LA, 352pp, 2000.

Dunn, G. E. and B. I. Miller, *Atlantic Hurricanes.* Baton Rouge, Louisiana State University Press, 326 pp., 1960.

Dunn, G. E., A Brief History of the United States Hurricane Warning Service. *Muse News,* 3, 140_143, 1971. Available From the Museum of Science, 3280 S. Miami Avenue, Miami, FL 33133.

Elsner, J. B. and A. Birol Kara, *Hurricanes of the North Atlantic.* Oxford Univ. Press, 488 pp., 1999.

Fernandez-Partagas, J. J., *Father Benito Viñes, S. J. and his Role in Hurricane Knowledge.* Unpublished manuscript, 19 pp., 1989. Available from the National Hurricane Center, 11691 S.W. 17th Street, Miami, FL 33165-2149.

Fitzpatrick, P., *Natural Disasters.* Hurricanes, A Reference Handbook. ABC-CLIO Press, 287 pp, 1999.

Gray, W.M., J. D Shaeffer and C. W. Landsea, Climate Trends Associated with Mulitdecadal Variability of Atlantic Hurricane Activity. Chapter 2 in *Hurricanes: Climate and Socioeconomic Aspects: A Current Perspective.* H. S. Diaz and R. S. Pulwarty, Eds., Springer-Verlag, 292 pp., 1997.

Grice, G. K., *National Weather Service Snapshots — Portraits of a Rich Heritage.* The Service: Silver Spring, MD, 93 pp., 1991.

Larsen, E., *Isaac's Storm.* Crown Publishers, N.Y., 323 pp, 1999.

Ludlum, D. M., *Early American Hurricanes, 1492-1870.* Amer. Met. Soc., Boston, 160 pp. 1963.

Sheets, R. C. and J. Williams, 2001: *Hurricane Watch: Forecasting the Deadliest Storms on Earth.* Vintage Books, 331 pp.

Simpson, R. H. and H. Riehl, *The Hurricane and its Impact.* Louisiana State Univ. Press, Baton Rouge, (ISBN 0-8071-0688-7) 398 pp., 1981.

Tannehill, I. R., *The Hurricane Hunters.* Dodd, Mead & Company, N.Y., 271 pp., 1955.

Tannehill, I. R., *Hurricanes.* Princeton University Press, Princeton, N. J., 308 pp., 1956.

Whitnah, D. R., *A History of the United States Weather Bureau.* The University of Illinois Press. Urbana IL. 267 pp.,1961.

ACRONYMS AND ABBREVIATIONS

4DVAR	4-D Variational Analysis
ADL	Airborne Doppler Lidar
ADM	Atmospheric Dynamics Mission
AMSU	advanced microwave sounding unit
ATS	Applications Technology Satellite
AVFE	aerosonde virtual field environment
CAMEX	Convection and Moisture Experiment
CIMSS	Cooperative Institute for Meteorological Satellite Studies
COSMIC	Constellation Observing System for Meteorology, Ionosphere and Climate
CMV	cloud motion vectors
CISK	conditional instability of the second kind
DMSP	Defense Meteorological Space Program
DoD	Department of Defense
DWL	Doppler Wind Lidar
ECMWF	European Centre for Medium-range Weather Forecasts
EIR	enhanced infrared
ERS	European Remote (Sensing) Satellite
ESA	European Space Agency
FEMA	Federal Emergency Management Administration
FVCCM	Finite Volume Community Climate Model
GFDL	Geophysical Fluid Dynamics Laboratory
GHz	gigahertz
GMT	Greenwich Meridian Time
GOES	Geostationary Operational Environmental Satellite *Atlantic*: GOES-EAST, *Pacific:* GOES-WEST
GPS/MET	Global Positioning System/Meteorology
GPS	global positioning system
hPa	hectopascal
HRD	Hurricane Research Division, NOAA
INS	inertial navigation system
IR	infrared
JPL	Jet Propulsion Laboratory
JTWC	Joint Typhoon Warning Center
LASE	Lidar Atmospheric Sensing Experiment
LEO	low earth orbit
LFM	Limited-area Fine-mesh Model (NMC/NCEP model)
Lidar	light detection and ranging
LITE	Lidar In-space Technology Experiment
LLCC	low level circulation center
LOS	line-of-sight

MACAWS	Multi-center Airborne Coherent Atmospheric Wind Sounder
McIDAS	Man-computer Interactive Data Access System
METEOSAT	European Geostationary Meteorological Satellite
MSLP	Mean sea level pressure
MM5	Fifth Generation Pennsylvania State/NCAR Mesoscale model
NASA	National Aeronautics and Space Administration
NCEP	National Centers for Environmental Prediction
NHC	National Hurricane Center
NHRL	National Hurricane Research Laboratory
NHRP	National Hurricane Research Project
NIMBUS	Sun-synchronous Meteorological Satellite
NMC	National Meteorological Center
NMP	New Millennium Program
NOAA	National Oceanographic and Atmospheric Administration
NOGAPS	Navy Operational Global Atmospheric Prediction System
NPOESS	NOAA Polar Orbiting Environmental Satellite System
NRL	Naval Research Lab
ODT	Objective Dvorak Technique
ODW	Omega dropwindsonde
OSSE	Observing System Simulation Experiment
PR	Precipitation Radar
QuikSCAT	Conical scan scatterometer that estimates sea surface winds
Radar	radio detection and ranging
RO	radio occultation
SLP	sea level pressure
SPARCLE	Space Readiness Coherent Lidar Experiment
SSMI	Special Sensor Microwave Imager
SST	sea surface temperature
STORMFURY	hurricane seeding experiment
TC	tropical cyclone
TCM	Tropical Convective Mission
TD	tropical depression
TIROS	Television and Infra-Red Observation Satellite
TMI	TRMM Microwave Imager
TOGA-COARE	Tropical Ocean-Global Atmosphere-Coupled Ocean-Atmosphere Response Experiment
TRMM	Tropical Rainfall Measuring Mission
u, v, w	x,y,z components of wind, in a Cartesian coordinate system
UAV	unmanned aerial vehicle
USAF	United States Air Force
USWRP	U.S. Weather Research Program
UTC	coordinated universal time
UTWA	upper tropospheric warm anomalies

VIRS	visible infrared radiometer
VIS	visible wavelengths
WMO	World Meteorological Organization
WRF	Weather Research and Forecast Model
WVMV	water vapor motion vectors

BRIEF METEOROLOGICAL GLOSSARY

The following definitions of technical terms are working, practical explanations for non-scientists, not complete scientific definitions of all the meanings and uses of the words.

adiabatic — no heat added to or subtracted from a system

aerosol — small (less than one micrometer) solid or liquid particle. Haze is made up of aerosols.

angular momentum — the product of the tangential velocity about a center of rotation and the distance from the center of rotation

anticyclone — an atmospheric circulation associated with a high pressure center that rotates clockwise in the northern hemisphere and counterclockwise in the southern hemisphere

anvil — top part of a thunderstorm or cumulonimbus cloud that resembles an anvil

assimilation — see *data assimilation*

asynoptic — not occurring at a *synoptic* time; see *synoptic*

attenuation — a general decrease in signal intensity or strength as a signal is transmitted from one point to another

backscatter — dispersal of radiant energy into a hemisphere; for radar it is radiation returned from a target at 180° to the radar beam

band pass filter — a blocking of all signals except those in a selected frequency range

barb — also called a feather. A means of representing wind speed at points on a weather map.

baroclinic — pertaining to or characterized by horizontal temperature differences

baroclinic instability — a large-scale instability in the atmosphere that occurs in the presence of horizontal temperature differences, for example, as associated with fronts. Extratropical cyclones grow from small disturbances in regions of horizontal temperature differences through the mechanism of baroclinic instability.

barotropic — pertaining to the absence of horizontal temperature differences, the case of zero *baroclinity*

bathymetry — study of water depth profiles

beta effect — the effect of latitudinal variations in the Earth's rotation about the vertical as given by the Coriolis parameter; contributes to a westward drift of atmospheric vortices in the tropics

boundary conditions — see *lateral boundary conditions*

boundary layer	the layer of atmosphere near the Earth's surface in which frictional forces are important; typically the lowest 1 km of the atmosphere
buoyancy	the property in fluids associated with density differences that causes a fluid volume (parcel) to rise or sink relative to its surrounding environment
C-band	radar frequency band lying between 4 and 8 GHz with a wavelength of 2.5 to 3.75 cm
central surface pressure	the pressure at the center of a cyclone or hurricane, the lowest surface pressure associated with the cyclone
centroid relative frame of reference	frame of reference relative to the center of the cyclone
coherent detection	waves with a continuous relationship among phases
compressibility	the condition that a fluid volume will decrease (contract) as external pressure increases or increase (expand) as external pressure decreases
conditional instability	a state of a layer of the atmosphere that is unstable with respect to small vertical displacements of a saturated (100% relative humidity) parcel of air, but is stable with respect to small displacements of unsaturated parcels.
confluent/confluence	the property that describes the coming together or approaching of horizontal air streams without *convergence*. Confluence of two air streams is not associated with compression or compensating vertical motion.
convection	small-scale motion in the atmosphere that is primarily vertical, as in updrafts and downdrafts, of cumulus clouds
convective instability	instability due to buoyancy forces in the atmosphere
convective scale	length and time scales typical of cumulus cloud motions, e.g., 100 to 10,000 m or 10 to 100 min
convergence	the property that describes the compression or flowing together of horizontal air streams. Horizontal convergence is associated with vertical motions. Convergence is the opposite of *divergence*.
Coriolis force	also called deflecting force; the apparent force on a rotating platform such as the Earth in which moving objects, including air, are deflected to the right of the direction of motion in the northern hemisphere and to the left in the southern hemisphere. The Coriolis force is proportional to the *Coriolis parameter* times the speed of motion.
Coriolis parameter	a parameter equal to twice the Earth's angular velocity times the sine of the latitude. The Coriolis parameter is maximum at the Poles and zero at the Equator.
cut-off low	a region of low pressure isolated from the main flow in the atmosphere

data assimilation	a powerful analysis process in which diverse data collected at possibly different times are incorporated into a numerical model of the atmosphere or ocean, thereby causing the model to adjust toward the observations
divergence	the property that describes the expansion or flowing apart of horizontal air streams. Horizontal divergence is associated with vertical motions. Divergence is the opposite of *convergence*.
Doppler effect or Doppler shift	usually used to mean frequency shifts in electromagnetic radiation due to motion of the air relative to the observer
Doppler radar	detects and interprets the radial velocity of the target (such as clouds and precipitation)
downburst	an area of strong winds produced by a convective downdraft over a small area (several kilometers in diameter), usually associated with a thunderstorm
drag	frictional force on air passing over a surface such as land or water, or an airplane wing
drop size spectra or distribution	the frequency distribution of the size (diameter) of cloud or rain drops
dropwindsondes	an instrument dropped from aircraft which measures and reports wind direction and speed as well as temperature and pressure
dry adiabat	a line of constant potential temperature on a thermodynamic diagram. In the absence of addition or removal of heat energy, the temperature of a parcel of air rising or sinking follows a dry adiabat on the chart.
dynamics	related to the motion of a fluid and considering the forces that cause the motion
dynamical model	a model that is based on the laws of fluid dynamics
easterly wave	a large-scale westward-migrating wave-like disturbance in a deep-layer of barotropic easterlies (or tradewinds)
Ekman boundary layer	a hypothetical model of the boundary layer in which the motion of the air is a result of a balance among the pressure gradient, Coriolis and frictional forces
enthalpy	a property of a gas that is proportional to its temperature
entropy	a property of a gas that is a function of the temperature and pressure of the gas. In the absence of heating or cooling, entropy is conserved.
equivalent potential temperature	a property of air that is conserved during a reversible moist-adiabatic process.
Exhaust cloud	outflow of cloud material, usually at the top of deep convective clouds
eye	the central region of the hurricane inside the region of max-

	imum winds where the pressure is lowest, the winds are relatively light and the skies are often clear
eyewall	the ring of convective clouds, maximum winds and torrential rainfall surrounding the eye of the hurricane
finite difference	the difference between the values of a function or variable at two discrete points on a grid, used in numerical models
forward model	a model or a function that computes an observed variable, such as a satellite-observed radiance, from the variable predicted by a numerical model. Forward models are used in data assimilation schemes.
freezing level	the lowest level of the 0 degree (Celsius) isotherm in the atmosphere
geopotential height	the height of a given point in the atmosphere in units proportional to the potential energy of a unit mass at this level (geopotential) relative to sea level. For most practical purposes, the geopotential height is approximately equal to the geometric height above sea level.
geostationary/ geosynchronous	a satellite in east to west equatorial orbit at an altitude of 35,786 km (21,580 mi) and remaining stationary relative to a point on the earth's surface
geostrophic	referring to a balance between the Coriolis and horizontal pressure gradient forces in straight (uncurved) flow
gradient	difference, as in pressure gradient
gradient flow	referring to an imbalance between the Coriolis and horizontal pressure gradient force in curved flow, where the difference in forces is the centripetal force, or the force required to accelerate the air in a curved path
gravity waves	waves in the atmosphere or water in which the restoring force is gravity. Ordinary ripples or waves on the surface of lakes are gravity waves
grid	a set of points on which specified variables are analyzed or predicted
ground speed	speed of motion relative to the ground
heat engine	a system such as the atmosphere which can produce work (or motion) as a result of a temperature difference
hectopascal	a unit of pressure, abbreviated hPa. A hPa is equal to one millibar (mb). The mean surface pressure over the Earth is about 1013 hPa or 1013 mb.
horizontal wind shear	*(see wind shear)*
hot towers	towering cumulus or cumulonimbus clouds in which the temperature inside the cloud is higher than outside the cloud
hydrometeor	any ice or liquid water particle in the atmosphere
hydrostatic	describing a balance between the vertical pressure gradient

	force and gravity. Large-scale atmospheric flows are very nearly hydrostatic.
hydrostatic equation	the equation describing the relationship of pressure, height and density in hydrostatic balance
inertial navigation system (INS)	guidance system that uses gyroscopic and accelerometer data processes by a computer to maintain a predetermined course
inertial stability	a state of a flow that is stable to small perturbations, one that does not transfer kinetic energy from the mean flow to disturbances
initial conditions	the values of all the predicted variables in a numerical model at the start of the forecast, obtained from an analysis of the available observations
inversion	a condition in an atmospheric layer in which the temperature increases with height rather than decreases with height, which is the normal situation. Inversions are stable because warm air overlies colder air.
isotach	a line on a map connecting points with equal wind speed
isothermal	of equal or constant temperature, either in space or time
Kalman filter	a mathematical method of four-dimensional data assimilation that considers the errors of the observations and the errors of the model state
kinematic	related to the motion of a fluid without regard to the forces producing the motion
kinetic energy	energy of a body associated with its motion, proportional to the square of the velocity
Ku-band	radar frequency between 12-18 GHz and a wavelength between 1.67 and 2.5 cm designed to detect clouds
lapse rate	rate of decrease of temperature with height
laser altimeter	a ranging instrument using the reflection of a light beam to determine altitude
latent heat	the heat energy associated with a change of phase of water
lateral boundary conditions	the specification of atmospheric or oceanic variables on the edge or boundary of a limited-area domain of a forecast model. There are no lateral boundaries for global atmospheric models because the global atmosphere is cyclical (continuous) in space
lidar	light detection and ranging. An instrument analogous to radar that propagates light rather than radio waves and measures the return of the reflected signal
longwave radiation	radiation in the infrared wavelengths, longer than about 4 micrometers, and generally associated with radiation from the Earth and its atmosphere
low level jet	a high velocity core of winds between a few hundred meters

	to 3000 m elevation in the atmosphere, often occurring at night
maximum sustained wind	in Atlantic hurricanes, fastest wind speed sustained for one minute
mechanical mixing	vertical and horizontal mixing in a fluid derived from the kinetic energy (as opposed to buoyancy) of the fluid
mesoscale	pertaining to atmospheric flows with horizontal space and time scales ranging from 10 to 100 km and 100 to 1000 min, including aggregations of thunderstorms, squall lines, fronts, precipitation bands, mountain waves and land-sea breezes
mesovortices	atmospheric circulations on the mesoscale
microwave	electromagnetic radiation having wavelengths between approximately 1 mm and 1 m
mixed layer (ocean)	the upper layer of the ocean in which turbulent mixing takes place and temperature and salinity are approximately constant; typically 50 m
moist adiabat (also called saturation adiabat)	a line on a thermodynamic chart that expresses the rate of change of temperature with height of a saturated parcel of air.
moist adiabatic	A reversible process in which water is condensed as a parcel of air is cooled or evaporated into the parcel as it is warmed, so that the parcel remains saturated at 100% relative humidity throughout the process
momentum	the mass times the velocity of a body
monsoon/monsoon trough	derived from Arabic *mausem*, a season, it is the large scale seasonal reversal of winds such as occurring over the Arabian Sea
omega dropwindsondes	*(see dropwindsondes) using the Omega navigation network*
overwarn, overwarning	the practice of issuing warnings for a larger area than actually experiences the hazard being warned against
pendulum day	the time required for a suspended pendulum to complete a rotation about the vertical, approximately equal to one day divided by the sine of the latitude and thus varying from one day at the poles and approaching infinity at the Equator
pitot static pressure	*(see static pressure)*
polar orbit	a satellite orbit that traverses polar latitudes on every orbit
potential temperature	the temperature a parcel of air would have if brought to a pressure of 1000 mb (the pressure close to the pressure at sea level) without the addition or subtraction of heat.
potential vorticity	a property of the atmosphere proportional to the absolute vorticity times the vertical gradient of potential temperature. In the absence of friction and heating, the potential vorticity of a parcel of air is conserved

precipitable water	the total amount of water vapor in a column of air expressed as the depth of water that would result of all the water vapor condensed and fell out as precipitation
pressure gradient	the difference in pressure in the vertical or horizontal
primitive equations	the basic unfiltered set of equations describing the motion of a fluid (dynamics) and the temperature and density (thermodynamics)
pseudo-adiabatic	see *moist adiabatic,* "wet" and "pseudo" being nearly synonymous
pyranometer	a radiation instrument measuring total direct and diffuse solar radiation
quasi-geostrophic	referring to large-scale flows that evolve slowly compared to the rotation rate of the Earth and are in approximate geostrophic balance at all times
radar reflectivity	see *radar*
radiance	rate of radiant energy transfer across a unit area of surface. Satellites measure radiances from the atmosphere in various wavelengths, and hence obtain information about the temperature of the part of the atmosphere emitting radiation in these wavelengths.
radio occultation	a method of determining vertical profiles of density in the atmosphere. Atmospheric soundings are obtained as satellites in low-Earth orbit rise and set relative to the Global Positioning System (GPS) satellites and measure the change in frequency of the GPS dual-frequency radio waves. The changes in frequency can be related to the atmospheric density, and hence the temperature and water vapor profiles
radiometer	an instrument that measures radiation (radiated electromagnetic power)
rain band	an elongated system or band of rain/convective clouds often used to describe those spiraling in towards the eyewall of a hurricane
refractive index	the ratio of the speed of light in a vacuum to the speed of light in a fluid, a function of the density of a fluid
refractivity	a measure of the refractive index of the atmosphere. Refractivity in the troposphere is a function of temperature, pressure and water vapor
relative vorticity	a measure of the rotation about the vertical axis of an atmospheric circulation
rime ice	white or milky deposit of ice formed by the rapid freezing of supercooled water (usually -10°C or colder)
Rossby radius of deformation	a parameter that describes how far a cold pool of air will spread out and under warmer air in the presence of the

	Coriolis force until a geostrophic balance is obtained and the spreading stops. It is generally large in the tropics and small near the poles
Saffir/Simpson scale	a classification scheme for hurricanes based upon maximum surface wind speed and the type and extent of the damage potential of a storm ranging from 1 (least intense) to 5 (most intense)
Saharan(dust) layer	layer of dust/dry air usually below the trade wind inversion at altitudes of between 1000 and 4000 meters stretching at times across the Atlantic
saturated adiabat	see *moist adiabat*
scatterometer	a radar system that infers wind speed over the ocean surface by measuring the backscattered radar signal
seeding agent	a material such as silver iodide used for cloud seeding
seedling	a weak tropical disturbance with wind and thermal structure leading to hurricane formation, given the appropriate environment
shallow water equations	the equations describing the behavior of a shallow incompressible fluid; these are useful in studying simplified models of the atmosphere and ocean
shear	see *wind shear*
skill (of forecasts)	a measure of the accuracy of a forecast, often compared against a simple standard such as persistence or climatology. A forecast is said to have skill if it improves upon the accuracy of these simple methods
spectral	referring to waves, or the wavelength and frequency of waves
spectral model	a numerical model in which the variables are expressed as sums of a finite number of wavelike functions such as sines and cosines rather than values at discrete points on a grid
spectrum	(1)radiant energy per unit wavelength, (2)quantitative description of the energy or power in different wavelengths present in a function
static pressure	the pressure of a barometer moving with the fluid (air)
storm canopy	the horizontally extensive layer of cloud in the high outflow region of a hurricane
storm surge	a sudden rise in sea level responding to the drag (friction) of the wind at the sea surface and the bottom topography (bathymetry) of the coastline. The storm surge is the single most deadly component of a hurricane.
streamline analysis	a map with lines showing the direction of the flow, but not the speed
supercooled water	liquid water (in the atmosphere) whose temperature is colder than 0°C

surface layer a layer of air between 10 and 100 m thick above the earth's surface in which mechanical turbulence (generated by wind shear) dominates over buoyant mixing (generated by vertical temperature differences)

synoptic originally pertaining to fixed times (0000 GMT and 1200 GMT) when observations were taken simultaneously world wide. More generally, pertaining to large scales of atmospheric phenomena such as cyclones and anticyclones.

synoptic scale pertaining to time and space scales in the atmosphere of days to 10's of days and 100's to 1000 km such as hurricanes and mid-latitude cyclones

synoptic time fixed times when many international observations are made (0000 GMT and 1200 GMT)

thermodynamic having to do with or describing the temperature, pressure and density of a fluid

thermodynamic inversion *(see inversion)*

thunderstorm outflow *(see downburst)* cold (dry) air flowing out (both forward and in the wake) of a thunderstorm

transform grid the grid in spectral models that is used to compute certain terms in the model equations, such as those involving physical processes (e.g. radiation, latent heat release, friction). After computation of these terms on the grid, the variables are transformed back into spectral space where they can be applied in the spectral equation

tropical wave a large scale wave-like trough in the low and middle layer tropical easterlies frequently overlain by baroclinic flow

trough an elongated area of relatively low pressure

true airspeed in air navigation, the indicated air speed (IAS) measured from a plane's pitot-static system) corrected for changes in air density as a function of pressure altimeter indicated height above mean sea level

vertical cross section in weather analysis a vertical plane depicting some atmospheric variable (such as wind speed, water vapor or temperature) along a given line on the Earth's surface and extending up to some given altitude

vorticity a measure of the rotation rate of a fluid about an axis of rotation. The axis may be horizontal or vertical. *Absolute vorticity* includes the Earth's rotation as well as the rotation of the fluid relative to the Earth. *Relative vorticity* describes only the rotation relative to the Earth.

vorticity filament a relatively thin, sinuous ribbon of high vorticity (spin)

wake (see *thunderstorm outflow*) in the atmosphere usually referring to the region of outflow of cold dry air reaching the surface after a thunderstorm has moved on

wet adiabatic	See *moist adiabatic*
wet bulb temperature	the temperature a parcel of air would have if it were cooled at constant pressure to saturation by evaporation of water into it; the lowest temperature attainable by a thermometer wrapped in a wet gauze and cooled by evaporation
wind shear	the variation of wind in the horizontal or vertical
wind surge	significant increase in wind speeds, usually in or near large storm systems, owing to changes in circulation structure not reflected in regional scale weather analyses

GLOSSARY OF UNITS OF MEASURE

Systems of Units and of Measure

Quantity	*S.I. Unit*	*Early Metric*	*Maritime*	*English*
Length	meter (m)	centimeter (cm)	foot (ft)	foot (ft)
Distance	meter (m)	kilometer (km)	nautical mile (nmi)	mile (mi)
Depth	meter (m)	meter (m)	fathom (fa)	foot (ft)
Mass	kilogram (kg)	gram (g)	pound (lb)	pound (lb)
Time	second (s)	second (s)	second (s)	second (s)
Speed	m /s	cm/s or km/h	knot (kt)	miles/hour
Temperature				
Sensible	degree Celsius (C)	degree Celsius (C)		degree Fahrenheit (F)
Potential	degree Celsius (C)	degree Kelvin (K)		degree Kelvin (K)
Force	Newton (N) kg m/s^2	dyne (dy) g cm/s^2	poundal (pl)	poundal (pl)
Pressure	Pascal (Pa) N/m^2	millibar (mb)	inches of Hg (in)	inches of Hg (in)
Energy				
Motion	joule (J) kg m^2/s^2	joule (J)	footpoundal (fp)	footpoundal(fp)
Heat	joule (J)	calorie (cal)	cal (4.187J)	BTU (252 cal)
Power	watt (W) J/s	cal/s	horsepower (hp)	hp

GLOSSARY OF UNIT CONVERSIONS

Unit	*Conversion*	*Unit*	*Conversion*
	Length		*Mass*
1 inch	2.54 cm	1 kg	2.205 lb
1 ft	0.3048 m		*Force*
1 m	3.281 ft	1 Ñ	10^5 dy
	Distance		*Pressure*
1 nmi	1.151 mi	1 hPa	1 mb
	1.853 km		0.0295 in. (hg)
1 mi	1.609 km	1 lb / sq ft	4.88 kg/ sq m
1 deg lat	60 nmi		*Speed*
	69 mi	1 m/s	1.94 kt
	111 km		3.6 km/h
	Depth	1 deg lat/6h	10 kt
1 fa	6 ft	10 kt	11.51 mi/h
	1.83 m		

Index

D

E

F

G

H

S

T

U

V

W-X-Y-Z

About the Editor

Robert Simpson, senior editor of this book, had his first (and near fatal) encounter with hurricanes in Corpus Christi, Texas, in September 1919, before his seventh birthday. His lifelong curiosity and concern for severe weather and its impact led to a professional career of more than 62 years studying severe storms, exploring, planning and conducting research, and forecasting storm behavior. Along the way he earned his doctorate in Geophysical Sciences from the University of Chicago, served the National Weather Service as Deputy Director of the Research Division (Severe Storms), Director of the Office of Operations, Director of the National Hurricane Center, and founding director of the National Hurricane Research Project. During these assignments, he traveled widely, exploring severe weather the world over—from the arctic to Antarctica—but particularly in tropical regions. In 1945 he planned and conducted one of the first complete research flights through a hurricane, and in 1947 the first research overflight of a major hurricane.

In collaboration with Herbert Saffir, Simpson helped design and implement the Saffir/Simpson damage potential scale for classifying expected damage from hurricanes, now widely used in both Atlantic and Pacific areas.

For his contributions to severe storm science, Simpson received the Cleveland Abbe Award of the American Meteorological Society, of which he is a Fellow, the U.S. Department of Commerce Gold Medal Award, and an honorary DSc from Southwestern University (Texas). He is a Fellow of the Explorers Club of New York, and is or has been associated with the American Geophysical Union, the Royal Meteorological Society, the New York Academy of Sciences, the Pacific Science Association, the Hawaii Academy of Sciences, and the Washington Academy of Sciences.